The Sea Fisheries of Scotland

To my parents

The Sea Fisheries of Scotland

Scotland

A Historical Geography

JAMES R. COULL

JOHN DONALD PUBLISHERS LTD
EDINBURGH

ISBN 0 85976 410 9

A catalogue record for this book is available from the British Library.

Typeset by Pioneer Associates Perthshire
Printed and bound in Great Britain by
The Cromwell Press Ltd, Melksham, Wiltshire

Contents

Preface   vii
Acknowledgements   xii
List of Figures   xiv
List of Plates   xvi

1   Introduction   1
2   The Resource Base of Scottish Fisheries   12
3   Fishing in Prehistory and Early History   22
4   The Development of Fishing Settlements   33
5   The Early Herring Fisheries   54
6   The Early White Fisheries   79
7   The Herring Fishery in the Nineteenth Century: the Rise to Pre-eminence   104
8   The Herring Fishery at its Peak 1893–1914   126
9   The Advent of Trawling: White Fishing for Industrial Markets   139
10   The Herring Fishery Between the Wars: Crisis and Readjustment   153
11   The White Fisheries in the Inter-war Period   174
12   From World War II to the European Community: Modernisation and Diversification   185
13   The Development of Mobility   200
14   The Shell Fisheries   223
15   Fishing Boats   235
16   Fishing Piers and Harbours   251
17   The Development of the Administrative Framework: Fisheries Districts   275
18   Epilogue: Conservation and Extended Regulation   291

Select Bibliography of Main Sources   298
Index   301

Preface

I have been conducting research for over thirty years, and the main topic of that research has been Scottish fisheries. This has allowed the linking of library research with information and views from various family and other contacts: the latter has been particularly stimulating in 'bringing the topic to life'. My view is that there is much that is distinctive in fishing, and in particular in Scottish fishing; and I accordingly welcome the opportunity to put the results of decades of academic work before a wider public.

There is ample evidence that fisheries have been important in Scotland during the whole history of human occupation of the country. As well as being clear from archaeological evidence from the earliest times, there is sufficient documentary evidence to indicate their importance in the historic period. Moreover the importance of fisheries is mentioned by a series of travellers and commentators at different periods.

While there is in print a considerable amount of material on Scottish sea fishing, the great part of it relates to local situations, and relatively little of it has included detailed use of primary sources. To date also a comprehensive treatment of the topic through the whole time spectrum from prehistoric to modern has not been made. This is not to say that there have not been some penetrating analysis within the field, and here Malcolm Gray's economic history of the transition to the modern situation 'The Fishing Industries of Scotland 1790–1914' is outstanding. The object of the present work is different, in focusing on the location and geographical distribution of the action, and in covering the whole time spectrum.

In any academic work, there is inevitably a measure of debate about content and balance. With the importance of environmental considerations in any sector of human geography, it is obviously necessary to discuss the biological basis of the resources on which Scottish fisheries have depended. However marine biology is a specialised field, and the amount of published material relating

vii

even to the waters around Scotland is immense. It has been deemed essential here to aim at giving an overview of essentials only, and among other things, in the names of fish the everyday ones have been employed and zoological names have been avoided. The interested reader will not find it difficult to find fuller accounts in the literature should these be desired.

In a historical geography of fishing, there is the issue of how far the treatment should be by different periods (including, for example the Prehistoric and Medieval periods) and how far it should be thematic, treating separately, for example, the herring and white fisheries. It has been judged here best to use a fusion of these approaches. In the prehistoric period, the limited evidence can reasonably be treated as a whole; but the development of a major national fishery, such as that of the herring, is best for the great part treated as a separate topic on its own; and in the case of the herring and white fisheries such is the change in scale and distribution of activity over time that they can best be treated by division into different periods. Shell fisheries too are a distinctive group, although the scale of importance here is lesser, and it has been judged best to treat the whole time span in a single chapter. While this has been the main organisational division of the work, it is realised that no such division can be perfect, and at the level of the individual fisherman, fishing family and fishing community there have often been activity in different fisheries at the same time. Also, with the diversification and rapid change that there has been since World War II, it has been judged best to treat the phase up to the UK joining the European Community as an integrated chapter: during this phase as well as the ending of the long-term leading importance of the herring fisheries, there have been far-reaching changes in emphasis and relative importance in the other fisheries. One of the complications of tracing trends which extend into the 20th century is the big increase in inflation, especially since World War II, which renders the quotation of money values a problem. For that reason, trends have been mainly shown by other values, using such data as numbers of boats and tonnages of landings.

It has been judged that a separate chapter on the development of mobility is required, as this over time came to be a major factor in the geography of the fishing. This is closely related to the herring fishery in particular, although it features in other fisheries as well.

There are, however a number of relevant topics related to the changing geography of the fisheries for which a separate thematic treatment is desirable. These include the resource base of the fisheries; the development of specialised fishing settlements; the development of boats and harbours; and the spatial organisation of the administrative framework for the industry.

All historical work is inevitably conditioned by the amount and character of available evidence. It is recognised too that while there are different emphases and view-points in academic fields interested in the past, there are no sharp divisions between historical geography, economic history and social history. How representative or well-balanced is the evidence is never an easy question, and inevitably leaves much to the individual judgment. In approaching this work, in addition to access to available records, I at least have had the privilege of contact with many people who have been personally involved in the fisheries; and typically their knowledge derives not only from their own experience but also from those of their families and forebears. Even so it has to be admitted that right until the late 18th century, the picture is inevitably clouded by scarcity of comprehensive or systematically gathered information.

In the 19th century fishery records in Scotland become abundant and detailed to the point that they have very few parallels anywhere. Good although the information base is, the effort devoted to data collection has decreased considerably in the past half century as the relative importance of the fisheries has become less. In addition, various parts of the data base change from time to time; and while in general this reflects important changes in the fisheries, there is a tendency for the earlier stages of changes to be more poorly re-corded. As a consequence, in the tracing of trends, various arbitrary decisions have to be made in the use of available statistics.

Published material can of course be supplemented by oral evidence and folk memory. In my own time there has been consid-erable material preserved in memory and tradition from especially the late 19th century. For the 20th century the documentary record can be more fully supplemented by recollections, and for the inter-war period this is especially vivid: this time of crisis was inevitably imprinted in the memory of those concerned with the fishing, as well as being the subject of several special reports.

Another relevant issue is where historical geography ends and contemporary geography begins. I have felt it apposite to deal in some detail with the period after World War II, which could be cat-egorised as the ultimate period of the old and traditional freedom of fishing on the high seas, and which now from this standpoint in time can be seen as a separate historical phase. In that same period however the build up of pressures in the fisheries has led to the much more regulated modern regime, which I have covered briefly as a concluding chapter. I have taken this phase after World War II as ending in the 1970s: during that decade there was the general extension of national fishery limits to 200 miles, which brought to an end most international freedom of fishing on the high seas; and it was also the decade in which Britain joined the European Common Market (now the European Union). This also means in essence that fisheries policy in Britain is now part of, and subordinate to, that of the European Union. It also happens that after 1974, most national statistics were given in metric rather than the old Imperial units; and as a general rule 1974 has been taken as an end point in the tracing of trends.

There is also the matter of how far the historical legacy is preserved in the present, and indeed the question of what effort needs to be made to sustain that legacy in the day of consciousness of heritage in a mass-production and mass-consumption society in which many traditional ways and practices have lapsed in the recent past. Even in our fast-changing modern world, there is a big legacy from the past in many of the patterns in fishing: this applies to such basic matters as settlement patterns and to fishing grounds exploited. The link with the past is now being systematically preserved and sustained: there has been something of a modern proliferation of museums and heritage centres, and at various places on the coast there are now fisheries museums, or museums with an important fishing content.

On content of the work, it is clear that the exploitation of the main groups of fish stocks in the sea – demersal, pelagic, and shell – should be included; there is however the issue of whether the salmon fisheries are appropriate here. It is recognised that in the historical record of fishing in Scotland, salmon has a prominent place. However, salmon was unusual and apart, in that the fisheries were and are a Crown prerogative, and that fishing for them in the open sea was for centuries specifically forbidden by law. They had to be fished where the sea ebbs and flows, in the river estuaries as they entered them on their spawning runs. Although they did also come to be fished outside the river mouths in the 19th century, the organisation of them has always been separate. In addition, to do justice to material in the historical record, any treatment of them within a general work on Scottish fishing would either be disproportionately long, or over-simple. While there is recognition here of the importance of salmon in the fisheries, they have not been given detailed systematic treatment.

In dealing with fishing boats there is another specialised field in which it would be possible to go to considerable length on such matters as the details of hull construction, types of sail rig, and in the later period details of engines and equipment. I have felt that these are matters in which the interested reader is better to consult specialised literature in the field; and my objective has been to give for the general reader an account of a topic that is obviously germane to this book, which does have important areal differences within the country, and in which the marine environment has obviously affected the boats employed.

Fishing has always been a distinctive activity, and it is fortunate that some of the boats, people and of action associated with fishing can be illustrated over a period of more than a century by photographs. There is a rich store of relevant photographs in Scotland, and within the confines of one book it has been necessary to be highly selective because of pressures of space. Something of the same might be said of maps and diagrams: again with the amount of material on record in Scotland, it would also have been possible to multiply these illustrations several times.

Aberdeen, 1996   J.R.C.

Acknowledgements

In the preparation of this book, my thanks are due to a big number of people and institutions.

I should wish to acknowledge in the first place a number of library sources. Over many years, the staff of the Queen Mother Library in the University of Aberdeen, and of the library of the Torry Marine Laboratory in Aberdeen have helped find many documentary sources. My thanks are due to the staff of the Scottish National Library, and of the Scottish Record Office; in the latter I have to thank particularly the staff of West Register House for access to the extensive material of the Fishery Board. A number of other libraries and archives have been consulted at different times, and I should especially here wish to recognise the Shetland Archives.

In the preparation of drawings, I should wish to thank Susan Powell, Alison Sandison and Jennifer Johnston, the cartographers in the Geography Department of the University of Aberdeen.

I have to thank the Earl of Dalhousie to include in ch. 4 material from the Panmure Papers.

I have to thank Ms. Caroline Wickham-Jones for permission to include fig. 2.1 and Dr. Hance D. Smith to include the material of fig. 6.2.

The material shown in a number of the figures is relevant to more than one chapter, and the general practice has been to insert each figure at the point where it is first referred to.

It is fortunate that many of the methods and scenes of the fisheries which are now history did not predate the age of the camera, and I have been happy to include a selected number of photographs: these often show more eloquently than text ever can what fishing involved. The origins of the photographic illustrations are detailed with the captions. I wish to thank the following for allowing photographs to be included:

The Scottish Fisheries Museum, Anstruther
Shetland Museums Service
The North East of Scotland Library Service
Aberdeen Journals
City of Aberdeen, Art Gallery and Museums Collection
Shetland Development Department
Economic Development and Planning Department, Grampian
Regional Council
Wick Society
Illustrated London News
Yarmouth Mercury
Mr. J. Livingston, Geography Department, University of
Aberdeen
Mr. John H. Goodlad of Burra Isle, Shetland
Mr. I McGeachy, Campbeltown
Dr. I.B.M. Ralston, Department of Archaeology, University of
Edinburgh
The George Washington Wilson Archive, Aberdeen University
Library.

My special thanks are due too to the Carnegie Trust for the
Scottish Universities for financial aid in underwriting this publication, and to the University of Aberdeen for its generous contribution
towards the cost of publication.

Finally I should wish to thank Mr Russell Walker, Commission-
ing Editor of Messrs. John Donald for his encouragement and for
attending to the publication.

List of Figures

3.1   Distribution of main Mesolithic Sites in Scotland (*After Wickham-Jones*).

4.1   Development of Fishing Settlements in North-East Scotland, 14th–19th Centuries.

5.1   Main methods of Herring Fishing.

6.1   Main methods of White (or demersal) Fishing.

6.2   'Haaf' Stations and Cod Bases in Shetland, 18th and 19th Centuries (*After H. D. Smith*).

6.3   Production and Export of Cured Cod, Ling and Hake by Five-year Means 1822–1886.

7.1   Production of Cured Herring by Five-year Means 1809–1938.

7.2   Numbers of Boats at Herring Fishery in Peterhead District 1830–1914.

7.3   Daily Landings in Wick District for 1860 Season.

8.1   Herring Landings in Fisheries Districts, along with Boats and Personnel at Peak Season, 1910.

9.1   Numbers and Tonnage of Steam Trawlers Landing in Scotland 1892–1914.

9.2   Scottish Demersal Landings, including Trawl Landings 1889–1914.

9.3   Distribution of Trawl Catches landed at Aberdeen 1912.

9.4   Closure of Scottish Inshore Waters to Trawling 1885–1892.

10.1   Herring Landings and Disposal 1907–1938.

10.2   Numbers and Values of Scottish Steam Drifters and Liners 1913–1938.

10.3   Area and Value of Herring Netting in Scotland 1913–1938.

11.1   White Fish Landings by Method of Capture 1919–1938

12.1   Numbers of Fishermen, 1887–1974.

12.2   White Fish Landings by Method of Capture 1938–1974.

12.3   Pelagic Landings 1938–1974.

13.1   Herring Curing Stations in Shetland 1884.

13.2 Numbers of Boats at the Shetland Herring Fishery, 1875–1914.

13.3 Herring Curing Stations in Shetland 1913.

13.4 Numbers of Scottish and English Boats at East Anglian Herring Fishery 1884–1968

13.5 Scottish Migrant Shore Labour at East Anglian Herring Fishery 1899–1957.

14.1 Lobster Landings, 1892–1954 (after H. J. Thomas).

14.2 Landings of Shell Fish 1958–1974.

15.1 Fleet Composition by Main Fishing Method 1963–1974.

15.2 Development of the Section of the Fleet over 40 Feet 1955 – 1974, with Numbers Registered in the North-East 1965–1974.

16.1 Pier and Harbour Works undertaken by the Fishery Board 1828–1913.

16.2 Grant and Loan Expenditure by Fishery Board on Piers and Harbours 1910–1938.

17.1 Fisheries Districts In Scotland 1850.

17.2 Fisheries Districts in Scotland 1893–1953.

List of Plates

1   Cairnbulg, Aberdeenshire
2   Crovie, Banffshire
3   Sandend, Banffshire
4   Whinnyfold, Aberdeenshire
5   Bullers of Buchan village, Aberdeenshire
6   Baiting lines with mussels, Cruden Bay, Aberdeenshire
7   A Shetland sixern being rowed
8   The steps at Whaligoe, Caithness
9   At sea aboard a line boat from Gourdon, Kincardineshire
10 'Haaf' station at Stenness, Northmavine, Shetland
11 Fisherrow fishwives
12 The harbour at Keiss, Caithness
13 Burghead, Morayshire
14 Peterhead harbour and town from the air
15 A good herring catch, Wick
16 Fifies being rowed out to catch the wind, Wick
17 The sailing herring fleet with dipping lug sails
18 Boats pulled up on the beach, Burnhaven, Aberdeenshire
19 Loch Fyne skiff under sail, Campbeltown Loch
20 Steam drifter *Golden Rod*, Peterhead
21 Steam drifters entering harbour, Great Yarmouth
22 Hauling drift herring nets at Shetland
23 'Redding up' herring nets in port on a steam drifter
24 'Barking' herring nets, Peterhead
25 Women mending nets, St Monance
26 Curing scene at Wick in the early 1870s
27 A big fleet of herring sail-boats at Castlebay, Barra
28 Concentration of curing yards at the north end of Lerwick
29 Cooper putting hoops on a barrel, Lerwick, 1950s
30 Aberdeen steam trawler *Lord Learney*
31 Steam trawlers in the fish market basin, Aberdeen
32 Buckie seine-netter *Carinthia* in locks at Fort Augustus
33 Peterhead seine-netter *Fidelia*
34 Shetland purse-seiner *Charisma*

1

INTRODUCTION

The Importance of Fishing in Scotland's Past

Fishing has always been important in contributing to the food supply in Scotland, and from Medieval times onwards fish have also featured prominently in Scottish commerce. The long standing importance is due to a combination of causes. In the greater part of the country – especially the hill areas – the land is of limited productivity. There was also the convenience of fish from the rivers as a main source of protein food in the days of a simpler economy, in which the needs of subsistence were paramount; but the importance of fishing is in addition related to the main concentrations of population in Scotland for many centuries being relatively close to the coast, which rendered sea fish relatively easily available to the majority of the people. In much of the country fish was one of the main sources of food protein to supplement the main food staples of oats and barley. There are also numerous records of the importance of fish in the diet of the poorest classes, and in the not infrequent cases of poor harvests, it could be vital for survival: shell fish around the coasts, easily gathered, was particularly important in this context. The contrast in productivity of land and sea in Scotland was still recognised by officialdom, in rather atypical poetic language, in the mid-19th century: 'The fisheries of Scotland present a remarkable contrast to the soil. They seem almost destined by nature to compensate for the natural infertility and insuperable difficulty for cultivation of large tracts of land' (AFBR. 1849: 3).

Evidence for something as widespread and permanent as the fisheries comes from a wide range of sources, although not till relatively modern times is the record other than fragmentary. Even so, there are ample indications that fish have been important through

1

the whole of history. The earliest evidence is of course archaeological, and there is evidence for the exploitation of fish in Scotland from the earliest days of human settlement, as shown by material from the Mesolithic as early as 7000 BC. In the documentary record, the surviving evidence is selective in form: the earlier records that indicate the existence of fisheries concentrate on which lords and corporate bodies had the rights to them; they also show a good deal of the entry of their products into trade. However information on types of boats, gear and fishing methods tends to be scant and incidental. Fishermen and fishing communities were also generally at the low end of the social scale, and very little is known about them until late in history.

In all, there are therefore basic problems in any quantitative estimate of the importance of fishing till a late date in history because of the lack of comprehensive records. There is some partial record of commercial fishing, the product of which entered was also of considerable political importance as well as entering into trade; but the part of catches which went to local subsistence must needs be a matter of conjecture.

Despite the difficulties in the way of a full assessment of the development of fisheries in Scotland, it can be stated confidently that they were of significant importance on the European scale. A major modern review of the historical fisheries of Europe has in fact recognised this (Michell 1977: 147–148; 183). In much of its history, Scotland in the European context was something of a commodity producing outpost, and along with wool and hides (the products of pastoral farming), fish was an important item that was sent to continental markets, as well as being consumed and traded internally within the country.

The early legal position appears to have been that all rights to sea fishing were vested in the Crown (Stewart 1869: 19–21). The rights to salmon fishing have continued to be an exclusive Crown right to modern times. Although both herring and white fisheries were officially Crown rights in early legal documents, they were from their nature more difficult to appropriate. It is recognised that there has never been any restriction on white fishing in Scotland, despite there being cases of grants of white fishing being given by the Crown along with coastal lands (Stewart 1869: 19–21). A useful summary of the development of the legal position has been made by Fryer, in which it is divided into three main phases (Fryer 1884: 8, 9). The first stage is categorised by legislation which was restrictive or protective; this in broad terms applies to the Medieval period, and the main effect of legislation was to prescribe which individuals or institutions had the right to the fisheries. In the second stage legislation was promotive, and was designed to foster the expansion of fisheries with a view to employing them to increase national wealth by securing from an international common property resource a greater share for the country; this was the situation in the Mercantilist period, although it also extended into the early 19th century. The third stage is categorised as administrative or regulative, in which the function of law is to reconcile conflicting interests and foster good order; here the fisheries in principle are unaided by government and left to the free play of economic forces except where these are constrained by regulation. While this three-phase division is valid and useful, it is necessarily over-simplified. Some of the Medieval legislation on matters such as closed times and net mesh sizes could be seen as administrative, while in the last of the three periods government intervention has been exercised in various matters to promote fishery development.

The sustained importance of salmon fisheries, especially in the river estuaries, stands out in early sources. Also well recorded are the herring fisheries, which were seasonally abundant on some parts of the coast, and in the Firths of Forth and Clyde were from an early date capitalised on as a commercial item. There was also a centuries-long standing ambition to mount a bigger scale herring fishery in the open sea, after the manner of the Dutch, for whom the North Sea herring fisheries were a main source of national wealth; but the organisation and the technique were insufficient until the end of the 18th century, after which the Scottish herring fishery experienced a century of expansion which carried it to the position of the world's leading fishery. More problematical in the record are the white fisheries: although their products entered to an extent into trade, including Scottish foreign commerce, for most of history they are less prominent in the known record. Yet, when the picture of Scottish fishing becomes clearer from the fuller record 17th and 18th centuries, line fishing for white fish was emphatically the main activity of hundreds of fishing villages which were dispersed along the coasts of the southern part of the country, and along the East Coast as far north as Easter Ross; and this was also true of many crofting townships in the Highlands and Islands.

Shell fish have also had a sustained direct and indirect importance. As well as featuring prominently in the diet of the earliest settlers around the coasts, they occur in the written record from Medieval times onwards. Among these the most important in the earlier written record appears to have been oyster fisheries, and it is known that those of the Firth of Forth especially were the subject of numerous disputes from at least the 17th century (Black 1951: 60–77), and for oysters there were attempts to divide up the firth between proprietors. Although long an important source of food and wealth, a sessile species like oysters is particularly prone to overfishing, and by the late 19th century they were unimportant. Lobster and crab are recorded in the food resources by Bishop Leslie and others among early chroniclers. Early fisheries for them appear to have been with nets and lines, or even by gathering them among the rocks at low tide. However from the late 18th century lobster has been important for the city markets in London and else-where, and from that time the main catching method has been that of setting baited creels. Various descriptions in the Old Statistical Account and elsewhere suggest that there was already a brisk trade which was due to English firms coming in and stimulating the fishery by creating an expanded market; and already there were cases reported of catches falling through over-exploitation.

The location of fishing is obviously governed or limited by where the fish stocks are. With the development of commercial fisheries, however there is also the issue of the location of the main seats of enterprise in maritime and other affairs. These influence access to capital, and the availability of infrastructural provision like harbours. Part of the pattern was to become by the 16th century the fitting out of boats at main seaboard towns, especially in the southern part of the country, to fish off other parts of the coast. Most prominent here appear to be the burghs on the Forth, especially in Fife, and these were to be joined in some strength by those on the Clyde.

FISHING IN THE CONTEXT OF NATIONAL SCOTTISH OBJECTIVES

In Scotland the fisheries were long seen as a main avenue for economic expansion and the increase of national wealth. From at least the 15th century, there was a general increase in these efforts at expansion; and although there were many fluctuations and set-backs, over time there was a greater measure of success. As well as providing for the food supply of the people at home, the products of the fisheries entered prominently into Scotland's international trade: here the country was a commodity producer for continental markets, and fish with wool and hides were leading exports. The scale of success was inevitably affected by the general progress of the economy, and from the 17th century onwards also reflected in important degree decisions taken in Britain as opposed to Scotland.

The union with England was to have other effects. There were the important issues of crewing the ships of the Merchant Marine and of the Navy, for which men in coastal settlements with sea-going experience were regularly wanted, and there was an interaction of both of these with fishing. While recruitment of crews to merchant ships was in general voluntary, and the rewards more dependable than those from fishing, recruitment to the navy was more problematical and there was empressment, although this was very generally resisted. In the 18th century and rising to a peak in the Napoleonic Wars, the press gang were active in pressing men, and from Shetland alone it was estimated that by the end of the Napoleonic Wars there were 3000 men in the Navy (Edmondston 1809, 2:20–23). It was also active on other parts of the coast, and fishermen at sea were especially vulnerable if they encountered the brig of the press gang. It was claimed that many an empty fishing boat was found on the Forth at this time with the fate of the crew unexplained (McGowran 1985:34).

None the less, the coming of the industrial age was attended by a greater measure of success in fishing, as is evident from the generally steepening upward trend from the latter 18th century, despite the main complication of the Napoleonic Wars from 1793 to 1815. When in the 19th century Britain attained the position of world economic leadership, the fisheries of Britain also became world front runners. Here Scotland played a very full part: if Scottish white fish trawling, though important, was overshadowed by that of England, in the herring fisheries in which the volume of production was greater, it was Scotland that was dominant.

However, in the longer term with the fuller development of the modern economy, the relative importance of fisheries became less, and their proportionate contribution to national wealth and employment declined. Scotland lost both relatively and absolutely as other countries arose as rivals in fishing and other economic sectors. The fisheries were in prolonged depression during the difficult inter-war period this century. The period since World War II has been a more prosperous time for the fishermen, and has been a phase of diversification and accelerating change. The fortunes of the industry have become part not only of those of the fisheries of Britain but also of Europe; and the greatly enhanced catching power of the modern period has produced a permanent conservation problem in which fisheries are in perpetual danger of destroying the resource on which they depend.

EVIDENCE FOR AND RECORDS OF FISHING

From the archaeological material an outline picture of the earliest phases has emerged over more than a century from organised 'digs'. In addition the contribution of more modern techniques like radio-carbon dating and pollen analysis have enhanced knowledge and aided the formulation of an accurate chronological framework; and they have also aided interpretation and understanding in various other ways. As shown by the archaeologists, fishing was particularly important to Mesolithic communities in Scotland before the advent of farming. While there was a continuing importance thereafter as an adjunct to subsistence farming, the fact that items like fish bones, and fishing lines and nets more rarely survive than animal bones and some other types of equipment, is in danger of giving the erroneous impression of a too low level importance of fishing activity. In addition, archaeological information from the Dark Age onwards is still very sparse in Scotland, and a good deal of the evidence in much of history is in fact circumstantial. There is the frequent presence of settlement in the coastal zone throughout the whole period of human occupation, and the known importance of fishing when documentary evidence becomes clearer in later times suggests that early coastal settlements must often have been involved in fishing. The situation becomes clearer in early modern times with the records of traditional practices which go back an unknown distance into history; but in view of the general tendency for such practices to be perpetuated by custom and necessity in subsistence conditions, such practices could well be hundreds – or even thousands – of years old.

While from Medieval times onwards it is national sources which give the best available overview, it is not until the 18th century that even those have much in the way of systematic reports and attempts at comprehensive coverage. The earlier records reflect the rights and priorities in a feudal society: they include as main sources the acts passed by the national Parliament, and such collections as the Exchequer Rolls and the Register of the Great Seal certainly attest the importance of fish and fishing; and this can be enhanced and corroborated to an extent by the records of ecclesiastical sources, and of those of individual estates and burghs.

The burghs of course played the leading role in commercial fisheries, and matters related to especially herring and white fish come up fairly frequently in the records of the Convention of Royal Burghs. The main early concerns was to control trading practice, and the selling of fish is in fact covered by law 27 of the Acta Quattuor Burgorum of the early Medieval period: this stipulates the tolls payable on fish at the market and also the penalties for infringement of the rules. However from the 16th century there were also efforts to stipulate that for curing of herring barrels should be branded by the burgh where they were cured, and it was also enacted that they should be of a standard size of nine gallons, measured by the Stirling pint, and that a standard reference measure should be held in Edinburgh (APS General Index 1875: 638). It is clear that for centuries development of sea fisheries was seen as a main way to increase the national wealth of the country, and there were repeated attempts to promote them, as well as various measures to regulate them. Also Scotland and its fishermen came into repeated contention with other countries and their fishermen, who also exploited the relatively abundant fish stocks which were off the coasts of what was of one of the more marginally located and backward countries of Europe.

Although inevitably impressionistic, the records of various chroniclers and travellers from the 14th century onwards are suggestive of a high order of importance for the fisheries; they are rarely omitted in descriptions of the country or of parts of it. They are emphasised, for example by native Scots like John of Fordun in the 14th century, by John Major and Hector Boece and Bishop Leslie in the 16th (Hume Brown 1893: 11; 45–46; 69–71), and by a traveller like Pedro De Ayala in the 15th century (Hume Brown 1891: 44). Bishop Leslie, in particular, is noteworthy for recording the range of fresh and salt water fish which were exploited, and says something of the types of traps which were used for fishing in fresh water and at the edge of the sea. Among the sea fish he detailed pelagic, demersal and shell species, and claimed that their abundance was such that strangers found the prices unusually low (Leslie 1893: 138–141).

The value of the salmon fisheries, which were in principle owned by the Crown, and which had fairly frequent contentious issues which attached to them, results in them being the most prominent fisheries featuring in national records before the modern period. They are also fairly prominent in the records of main East Coast river mouth towns like Aberdeen, Banff, Perth and Inverness. This resource in earlier times was little challenged by other countries, as they were more easily fished as they entered the rivers to spawn than in the open sea.

Records of fishing on coastal estates are very generally poor. This no doubt reflects for estates the much greater general concern with agriculture; and it almost certainly reflects the fact, known in Scotland and many other countries, of many of the fisherfolk originating as a depressed population sector near the bottom of the social scale, who might be virtually landless and who might be squatters. It is a striking fact that in an inquiry by the Fishery Board for Scotland at the end of the 19th century, after more than a century of general improvement, it was found that the majority of fishing families had defective legal title to their houses, and many were still technically squatters (ARFBS. 1884: xlvii–l). While there are occasional references to such matters as the rights to sea fishing and to such items as bait in estate charters from at least the 16th century, systematic records of fishing activity are prominently absent from such an estate as that of Seafield in Banffshire, which from at least the 17th century has been one of the most prominent in the country as a home and base for fishing communities. The most prominent exception here is the Shetland Islands, where commercial fisheries were a main activity on estates from the early 18th century, and where systematic records on fishing are in several cases extant.

On the verge of modern times in the 17th and 18th centuries, the general descriptions of MacFarlane's Geographical Collections, despite their partial and uneven coverage, are actually quite valuable in gauging the importance of fisheries in various parts of the country. From then on various local sources, accounts and compilations, including the Statistical Accounts and a range of works in local history, are available to show the development and importance of the fisheries in different settlements and parts of the country.

One of the most useful sources for the later phases are government reports which become of importance from the early 18th century, and these reflect a general and increasing concern at the national level for the development of the fisheries. Of great importance was the setting up of the Fishery Board in 1809: although a national

British body, Scotland was always the main theatre of its operations, and from 1868 its work was exclusively in Scotland. Its basic task was initially to promote the development of the herring fishery, and after 1820 it was also given the task of promoting the white fisheries (for cod and ling) as well. This entailed the formation of an organisation that had government officers at intervals all around the coast, and it is due to it that from the early 19th century Scotland has for the fisheries detailed records on a substantially uniform and co-ordinated basis that has virtually no rivals anywhere. The published annual reports are relatively brief, but there is a great volume of manuscript material that gives many extra details of the fisheries and of the working of the Board. The 19th century was a period of general success and expansion, and this can be traced in considerable detail.

When British fisheries had attained the position of world leadership in the late 19th century, in 1882 the Fishery Board was reconstituted and given expanded responsibilities for Scottish fisheries: as well as publishing much expanded reports on them, it was also much involved in the new field of scientific marine research. The published records of fisheries in the period from 1882 to World War I are especially voluminous, although the amount of manuscript material retained becomes much less. For the first time recording of all landings became comprehensive, and new levels of detail on fleets and equipment were gathered. There was also considerable detail at the level of individual fisheries districts and ports, and this allows the regional and local variations of the time to be seen. Although in some things later records become more detailed, the general subsequent tendency was for records to become more summary and condensed. A consequence is that some trends can be followed only for restricted periods.

A DEMANDING AND DANGEROUS CALLING

For the great part of its history, fishing in Scotland was dominated by operation in open boats in an environment in which weather uncertainties were not infrequent; it always involved hardship and discomfort. In fishing communities there were always men who never completely conquered sea-sickness, even in a life time of fishing; and in Scottish conditions being at sea was nearly always cold, even in summer. It was the usual practice for fishermen to wear several layers of clothing, and garments were formerly of oiled wool to be more waterproof. It has also been claimed that it was formerly possible to know from which part of the coast a Scottish fisherman came from by the style in which his wife or mother had knitted his jersey. While from the 19th century there were improvements in special seamen's clothing, with factory-made oilskins, jerseys and sea-boots, fishing continued to be cold and uncomfortable work.

However more serious than discomfort was danger. The main cause of this was adverse weather, particularly from wind and gales; and sudden gales were and are a great hazard. There are various records of small fishing settlements, at dates from the 16th to the 18th centuries, which did not survive till later times, and one of the contributing causes was loss of manpower through loss of life at sea.

From all around the coast there is a record of disasters with boats sinking, and of serious loss of life. Even more poignantly, the worst places for boats sinking was where they approached the shore: coming through breakers to reach an open beach, or to enter a harbour, was often the most hazardous part of a voyage. If less prominent as a news item, there are also many instances of individual fishermen going overboard from their boats. Few traditional fishing families have in fact to go back more that two generations to know of fatalities. The casualty rate was due to necessity rather than foolhardiness: in the days before the welfare state the choice between work and starvation could be stark, and boats might put to sea when conditions were less than ideal; but the most frequent source of tragedy was the sudden gale which got up when boats were at sea, and did not have time to regain the land. Although on modern fishing craft there is more safety provision and the work is less hazardous, the scale of life insurance premiums paid by fishermen shows that it is still one of the most dangerous occupations – considerably more so than mining for example.

While there are few systematic records of losses in earlier times, it is known that in the Buckie district of Banffshire, for example, there were at least 20 boats that sank in the 18th century (Hutcheson 1887:18). After 1800 the situation is generally clearer: it is known that in a big storm in 1848 the loss of fishermen on the East Coast was 100, and that in the 1870s the one community of Cellardyke lost 30 men (Young 1883:69–70). The two worst disasters known in Scottish fishing were those of 1832 at Shetland, and 1881 at Eyemouth. In the former case 17 boats and 105 men were lost (O'Dell 1939:116), and in the latter the total loss of life in south-east

Scotland was 191, 129 of them from the village of Eyemouth alone (Young 1883:70). Tragically too these are only the most heart-rending episodes in a long list.

REFERENCES

Acts of the Parliament of Scotland (APS), General Index, HM. Register House, Edinburgh.

Annual Report of the Fishery Board (ARFB) 1849.

Bell, R. (1812) *A Treatise on the Election Laws*, A. Constable and Co., Edinburgh.

Brown, P. Hume (1891) *Early Travellers in Scotland*, David Douglas, Edinburgh.

Brown, P. Hume (1893) *Scotland before 1700 from Contemporary Documents*, David Douglas, Edinburgh.

Edmondston, A. (1809) *A View of the Ancient and Present State of the Zetland Islands*, John Ballantyne and Co., Edinburgh.

Fryer, C.E. (1884) *The Relations of the State with Fishermen and Fisheries*, William Clowes and Sons, London.

Gray, M. (1978) *The Fishing Industries of Scotland 1790–1914*. A Study in Regional Adaptation, Oxford U.P., Oxford.

Bishop Leslie (1578) 'History of Scotland', in Brown, P. H. (1893) (ibid.), 113–183.

McGowran, T. (1985) *Newhaven-on-Forth. Port of Grace*, John Donald, Edinburgh.

Michell, A. (1977) 'The European Fisheries in Early Modern History', *Cambridge Economic History of Europe*, V, Cambridge UP. Cambridge, 134–184.

O'Dell, A.C. (1939) *The Historical Geography of the Shetland Islands*, Shetland News Office, Lerwick.

Stewart, C. (1869) *A Treatise on the Law of Scotland Relating to the Rights of Fishing*, Edinburgh.

Young, A. 'Harbour Accommodation for Fishing Boats on the East and North Coasts for Fishermen' in Herbert, D. *A Selection of the Prize Essays of the International Fisheries Exhibition*, Edinburgh, Edinburgh and London.

2

The Resource Base of Scottish Fisheries

The position of Scotland, along with the rest of Britain on the continental shelf of North-West Europe entails that it is surrounded by some of the most productive seas of the North-East Atlantic. On the North Sea side of the country the shelf extends for a full 400 miles and is the widest extent of shelf anywhere in Europe; and there are also fish stocks on the west side of Scotland, although the shelf here extends less than 100 miles. The waters around Scotland are classed as 'boreal' or cool temperate, and contain an assemblage of fish stocks which have important common elements with those of other parts of the North Atlantic, and this is related to there being a broadly similar eco-system across the area. Also, although the rivers and lakes of Scotland are not on the scale of those on the bigger land mass of continental Europe, they do contain an assemblage of fish stocks that have long been important for the Scottish population.

Marine scientists in Scotland have been active in studying the available resource base from the latter 19th century, and their work was some of the earliest in marine biology in Europe. Indeed it was various issues and problems in fisheries at that time that were the effective trigger that led to organised effort and resources being committed to what was largely a new branch of science (Coull 1993: 258–266). Marine science in Europe is notable for having been one of the more fruitful fields for international co-operation. This has also been fostered by important common characteristics of different waters, and also by the fact that the sea is essentially indivisible.

PRIMARY PRODUCTIVITY
IN THE SEAS AROUND SCOTLAND

The main index of marine productivity recognised by marine scientists is the rate of primary production, which is essentially a measure of the rate at which food is produced in the sea by the plankton. The vegetable plankton (or 'phytoplankton') synthesise food from dissolved minerals in the presence of sunlight by the process of photosynthesis: this occurs in the euphotic zone, the upper levels of the sea where sunlight can penetrate. The level of primary production in practice is estimated by the rate of carbon fixation during photosynthesis, and in the seas around Scotland this rate averages between 100 and 200 mg./sq.m./day: while there are considerably more productive marine areas, particularly in the deep upwelling areas in the tropical oceans, over much of the great oceans productivity is much lower and is as low as 10 mg./sq.m. per day. Productivity levels are much determined by the availablity of key nutrients, and in the sea these are mainly phosphates and nitrates. The primary production in Scottish waters represents a typical level for temperate continental shelves.

The phytoplankton are the foundation of life in the sea and form the base of the food chains and food pyramids in which the fish are. However, commercial fish are from three to five links along the food chains from the phytoplankton, and the efficiency of 'conversion' of about 10% at each stage does entail that the amount of primary production ever available as fish is only a small fraction of the primary production.

In the waters around Scotland, and indeed in most seas outside the tropics, there is a strong seasonality in primary production (Hardy 1956: 53–60). Although seasonal temperature fluctuations are less in the sea than on land, the winter temperatures in the vicinity of 5° to 6° C are sufficiently low to reduce greatly the relative abundance of phyto-plankton and to reduce photosynthesis and primary production to low levels. However there is a marked increase in primary production in the spring and early summer in response to increasing daylight and rising temperatures which eventually reach levels of around 13° C. This triggers an increase in feeding by the zooplankton, and indeed this increase in available food and in feeding is transmitted along the food chain. However in the sea in the vicinity of the British Isles, and especially in the northern North Sea, primary production does not reach a simple peak in the height of summer. The characteristic sequence is for primary productivity to reach its peak in early summer, followed by a decline. This is due to a thermocline becoming established in the water column through the seasonal warming of the upper layers, which limits the vertical exchange between the less dense warmed surface layers and the denser lower and colder water. It is as a result of this that the surface layers become short of the key nutrients of nitrates and phosphates, and this causes the fall in primary productivity in high summer. The incidence of gales in autumn usually causes more stirring of the water, destroys the seasonal thermocline, and a secondary autumn peak in primary production occurs before seasonal cooling of the water reduces it to winter levels.

The seasonal rhythm of primary production has a series of ecological effects, of which the most important is that the summer is for nearly all marine life forms outside the tropics the great season of feeding and growth. This is neatly shown in the use of the growth rings in fish scales and otoliths (ear bones) as the scientists' main indicators of the age of individual fish. However the food produced in the short-lived phytoplankton is consumed by the longer-living zooplankton; and zooplankton and other life forms near the lower end of the food chain constitute food that is available for fish in winter, when feeding rates are generally much lower; and overall the result is a relatively constant biomass.

There are also important biological rhythms in feeding and spawning migrations, and these are of basic importance for the practical fishermen; and indeed a store of knowledge based on experience attaches to such fish behaviour, and this has led to collaboration between scientists and fishermen in the understanding of fish stocks and of the marine eco-system. Such migrations render the fish more active, and often lead to shoaling behaviour, especially in the case of pelagic species like herring and mackerel. This in turn leads to fishing being more profitable, and has led to the establishment of more or less regular seasonal fisheries for various species.

TYPES OF FISH STOCKS

The main fish stocks which have been important in Scotland include an assemblage which are present through most of the North Atlantic in varying abundance and proportions. They are generally divided into three main groups of species. They include demersal (or 'white') fish such as haddock and sole which are bottom dwellers, and pelagic species which spent part of their life in the upper levels of the sea: these include especially the herring and the mackerel. Some importance also relates to 'crustaceans' (or shell fish), which include mobile species like lobster and crab and sessile species like oysters and mussels. In addition to these wholly marine species there are also fresh water fish, among which pride of place goes to the anadromous salmon, which actually spends most of the adult stage of its life cycle in the sea, but returns to the rivers to spawn. In addition there are a variety of other fresh water fish, especially the trout.

DEMERSAL FISH

There is ample evidence for the sustained importance of demersal fish throughout history. There is a series of groups of them, of which the leading group is the gadoids, which contain the cod, haddock, saithe and whiting. These have always been the most abundant fish in Scottish waters; and for long the most important of them were the haddock and saithe, which are particularly plentiful in inshore waters, although their distribution extends all over the shelf around Scotland. The haddock, the most important of all the demersal fish for Scottish fishermen, inhabits and spawns in a large area of the north North Sea, and there is a smaller spawning area to the north-west of Scotland (Hardy 1959: 217). It has been shown that the growth rate of the haddock is at its maximum off the East Coast of Scotland and on the Great Fisher Bank of the North Sea; and that its growth rates fall off in the deeper areas of the central north North Sea, due to poorer feeding opportunities (Jones 1962: 18). The cod is also of great importance, although not of the over-riding importance it has been in such locations as North Norway and Eastern Canada. The cod and haddock stocks inhabit the same waters and this is possible because their food is largely different: the haddock feeds almost entirely on small invertebrates, while the adult cod is a top predator and feeds mainly on fish, especially sand eels, nephrops, herring and mackerel. The cod has been shown to feed all year and to have two feeding peaks: the main one is in mid-summer, but it also feeds intensively in mid-winter prior to spawning in spring. It also shoals to an unusual extent for a demersal fish, and this can result in big variations in catch in the same area (Rae 1967: 49–52: 55–58).

The bigger demersal fish like the cod, ling, skate and halibut, the main distributions of which are in deeper water, have also been known and exploited from beyond recorded history. While demersal fish have all been fished to an extent the year round, there have been seasonal emphases in fishing effort which are due in the main to the greater amount of good weather and longer daylight in summer, but can also be related to seasonal abundance at other times. This is most notable in the cod, where winter spawning in especially the Moray Firth has resulted in a seasonal fishery, and the greater feeding activity of the species from the late spring (and greater readiness to take a hook) has also resulted in an offshore fishery then at various parts of the coast. Among the bigger demersal fish the most important commercially in history have been the cod and ling: such big fish give a relatively big weight of product in relation to the effort of splitting, salting and drying them. In the Shetland 'haaf' fishery the main effort was directed at the ling of which the main concentrations are on around the continental edge, which is within 20 to 40 miles from the west and north coasts of the islands.

There are also the flat fish, which are adapted to life on the sea floor and which are most abundant on sandy and muddy bottom: these include especially the plaice and lemon sole. Although in overall distribution the main concentrations of plaice are in the south-eastern North Sea, it is the most common of the flat fish in Scottish waters and is dominantly a species of shallow sandy sea bottom: it is found most in the Moray Firth and the bays of the East Coast; it is also found to a lesser extent on West Coast grounds, especially between Orkney and the Isle of Lewis (Rae 1970: 6, 7). The plaice has long been a main component in inshore line catches, but with the intensified operation of the last century and more, it was one of the first species in the North Sea to show clearly the effects of fishing pressure. In the 20th century in Scotland there has been a marked decline in the average size of plaice in the catches (Lamont 1964: 20). The lemon sole is found in rather deeper water than the plaice, but the main concentrations of the species are off the East Coast of Scotland, and it is also common off the north and west coasts: it favours rocky bottom alternating with rough gravel (Rae 1970:7,8). Of the other flat fish species landed in Scotland, most frequent are the witch, megrim and halibut: these are all species of deeper water, especially the megrim and halibut, and the main catches come from grounds towards or on the edge of the continental shelf (Rae 1970: 9–13). Such species have been caught mainly by great line and trawl.

Although less valuable than cod or haddock, the whiting is one of the most abundant white fish species in the waters around Scotland, and in modern times has come to be one of the main

landed species. Another demersal species which is widely distributed between the Scottish coasts and the edge of the continental
shelf is the Norway pout: it is found mostly in deeper water
between 50 and 100 fathoms deep (Raitt and Mason 1968: 3, 4),
and although it has never been of importance as an edible fish, it
had in modern times become important as a main species fished for
reduction to meal and oil. The blue whiting occurs in big concentrations in the deeper water to the west of Scotland, and although
it has not to date had much importance in the edible market, it has
also to an extent become the object of fisheries for reduction.

PELAGIC SPECIES

Important as are the demersal species, they are overshadowed by
the pelagic herring, which can be said to have preoccupied Scottish
fishing effort and organisation for centuries. In detail there are
actually several herring stocks, with different migration patterns
and times of spawning. Because of this herring can actually be
caught in some quantity at all times of year. However there is a
pronounced annual cycle in herring behaviour, and they are in
general most active and in best condition for catching in summer,
when they are fat from feeding. They feed directly on zooplankton,
which means that they are relatively efficient converters of primary
production into food in an edible form; and it had been computed
that in the natural equilibrium ecology in the North Sea, the herring alone constituted as much as 30% of the fishable biomass. The
herring found in the waters around Scotland for the great part
belong to the southern herring stocks of the North-East Atlantic,
which mature at ages of three to four years, and can live to ages of
ten or eleven years. These contrast with the other main group, the
Atlanto-Scandinavian which are found in the waters between
Norway and Iceland, which are slower maturing and longer-living:
they mature at six to seven years and can live to 20 years.

In the different herring stocks recognised in Scottish and British
waters, and there are variations in life cycles that include spring,
summer and autumn spawners. The outstanding group available
in the waters around Scotland are the north North Sea summer
spawning group, also known as the bank herring. These winter on
the edges of the Norwegian Deep, but with the spring warming of
the water and the accompanying multiplication of plankton, they become active and move off westwards: the outburst of photosynthesis

in the water at the western edge of the colder and less saline Baltic outflow acts as a trigger mechanism at this season (Steele 1961: 3–7). As increased photosynthesis spreads westwards with the advancing year, the herring move towards the Scottish coasts, feeding as they go. This means that in the central part of the north North Sea they become available for fishing from April, although they do not reach best condition until ready to spawn in July; and spawning continues until early September. The spawn of herring is unusual in being laid on the sea bed, usually on beds of gravel, and is unlike the spawn of most fish which is released into the water and floats among the plankton; and herring spawn is an important source of food for various other fish species. After spawning the herring return to their wintering grounds and become less active. It is because of this pattern of feeding and migration that what came to be known as the Scottish great summer fishery was conducted off the East Coast from July to September. This had also been effectively anticipated earlier by the Dutch in their great 'buss' fishery. It was after herring had become fat from feeding that they were in best condition to 'take the salt' which was necessary for curing – for centuries the main market sector.

While the bank herring were always fished to a minor extent on the Forth, it was not really until the 19th century that it became the main resource on which Scottish herring fishermen depended. Other herring stocks were available, and the most important of these was the Clyde herring which are available through the second half of the year. These are associated with the famous Loch Fyne herring, Loch Fyne being a main feeding ground, and also with the fishery on the spawning grounds of the Ballantrae Bank and the south coast of Arran. To the north-west and west of Scotland, in areas open to direct oceanic influence from the Atlantic, are stocks which spawn in spring from February to April. These are distributed in an area which extends from the Shetland Islands, the waters west of the Hebrides and in the Minch, to the waters north of Ireland (Hodgson 1957:16–17). The Clyde herring have been shown to consist of a mixture of autumn spawners with the main spring spawning stock: they start being captured at ages of 1.5 to 2 years, and seldom reach ages above 5 or 6 years. After spawning in the Clyde the adults move to grounds between Barra Head and the north of Ireland (Wood 1960:23,24). There is also a local stock which spawns in the spring at the entrance to the Forth; and in addition both the entrance to the Forth and the inner Moray Firth have been the location of fisheries for immature herring in the early months of the year.

A significant, if subsidiary pelagic species is the sprats (or 'garvies'), which has long been the subject of a fishery in the estuaries, especially in the Forth, and has been widely reported in other inshore waters. One of the pioneer ventures in marine biology which began as early as 1836 was the establishment of the distinction between sprat and young herring in the Forth. This was the work of the anatomist Robert Knox, undertaken at the invitation of the Fishery Board to help resolve a dispute (ARFBS Sc.I. 1883: xiii). In later times the sprat has been fished more widely in inshore waters along much of the East Coast, but has never made other than a limited impact on the edible market.

The mackerel has long been known in Scottish waters, and although there are substantial stocks it traditionally was of very minor importance compared with the herring. Indeed it was not until the herring stocks were reduced to a very low ebb in the 1970s that the full extent of mackerel stocks became apparent, when pelagic boats had to find an alternative fishery. As well as the North Sea stock, there is also a big West Coast stock which has become the subject of a major directed fishery.

The main mackerel stock is in West Coast waters. Here the main spawning grounds are in the deep water of the Celtic Sea to the south of Ireland in spring, and mackerel first join the spawning shoals at the age of two years. They feed in the winter on various life forms like shrimp and small fish on the sea bottom, but in their main feeding period in summer their food is from the zooplankton (Hardy 1959: 78, 79). A large part of this stock migrates northward after spawning and is the subject of the fishery off the West Coast of Scotland and between the Shetland Islands and Norway in the latter part of the year. However in recent years the pace and pattern of migration in northern waters has changed, and the shoals appear to be spending more time at the north end of this migration route, which means that they spend more time in what is now the Norwegian sector of the North Sea. The separate North Sea mackerel stock is smaller, and its spawning grounds are in the Norwegian Deep.

SHELL FISH

Although shell fish are over all much less important that demersal and pelagic species, some of them have a long-standing importance that in some cases extends from prehistory.

Obviously sessile species like mussels, oysters, limpets and whelks are easily gathered at the edge of the sea with no need for elaborate equipment, although in the deeper waters of estuaries they require dredges. In this group the oyster fisheries were long important, especially from the extensive beds on the Forth, although with intensified fishing in the 19th century they were almost eliminated.

Fisheries for crabs and lobsters have been prosecuted for centuries, and with the big interest in quality sea food of the modern period have become main commercial fisheries. These species characteristically live in shallow water on rocky coasts and are mainly exploited by small boats which can come close inshore. With the character of the Scottish coasts, these species are more plentiful on the West Coast than on the East, and are especially important on the islands of the Hebrides and also the Northern Isles. The lobster is much more valuable than the crab: it is found at depths between 5 and 40 fathoms, and in general the best fishing is around 12 fathoms (Fulton 1887: 195). In more modern times the nephrops, which lives on muddy bottoms all around Scotland, and is most abundant in the Moray Firth, has become the subject of a main directed fishery. It is generally best caught around dawn or dusk, and this is related to its habit of foraging for food in conditions of low light (Thomas and Davidson 1962: 14)

In all there is a varied resource base in the fish stocks in the waters around Scotland, and in the modern period this resource base has become more and more fully taxed. With the catching power now available in Scotland and elsewhere in the European Community and in Norway, the conservation of the resource base has become a major issue in fisheries.

REFERENCES

Annual Report of the Fishery Board for Scotland: Scientific Investigations (ARFBS.Sc.I.), 1883.

Coull, J. R. (1993) 'Beginnings of the Study of the Marine Environment in Scotland: Fisheries the Spur', in Dawson, A. H., Jones, H. R., Small, A., and Soulsby, J. A. (eds.) *Scottish Geographical Studies*, Departments of Geography, Universities of Dundee and St. Andrews, 258–266.

Fulton, T. W. (1887) 'The Scottish Lobster Fishery', ARFBS. Scientific Investigations, 189–202.

Hardy, A. (1956) *The Open Sea. Part I. The World of Plankton*, Collins, London.

Hardy, A. (1959) *The Open Sea. Part II. Fish and Fisheries*, Collins, London.

Hodgson, W. C. (1957) *The Herring and its Fishery*, Routledge and Kegan Paul, London.

Jones, R. (1962) *Haddock Bionomics II. The Growth of Haddock in the North Sea and at Faroe*, DAFS, Marine Research no.2, HMSO, Edinburgh.

Lamont, J. M. (1964) *Plaice Investigations in Scottish Waters. 1 Size-Composition of the Stocks 1910–1952*, DAFS, Marine Research no.1, HMSO, Edinburgh.

Rae, B. B. (1970) *The Distribution of Flatfishes in Scottish and Adjacent Waters*, DAFS, Marine Research no.2, HMSO, Edinburgh.

Rae, B. B. (1967) *The Food of the Cod in the North Sea and on West of Scotland Grounds*, DAFS, Marine Research no.1, HMSO, Edinburgh.

Raitt, D. F.S. and Mason, J. (1968) *The Distribution of Norway Pout in the North Sea and Adjacent Waters*, DAFS, Marine Research no.4, HMSO, Edinburgh.

Steele, J. H. (1961) *The Environment of a Herring Fishery*, DAFS, Marine Research no.6, HMSO, Edinburgh.

Thomas, H. J. and Davidson, C. (1962) *The Food of the Norway Lobster*, DAFS, Marine Research no.3, HMSO, Edinburgh.

Wood, H. (1960) *The Herring of the Clyde Estuary*, Scottish Home Dept. Marine Research no.1, HMSO, Edinburgh.

3

Fishing in Prehistory and Early History

In Scotland as elsewhere, the earliest evidence for human settlement and activity comes from archaeological evidence. In virtually all parts of the world, cultures dependent on hunting and gathering preceded those based on husbandry or farming, and this is the case in Scotland. The 'hunting and gathering' category itself includes a considerable variation in culture and economy, and in not a few cases fish provided an important part of the diet; and for the earliest settlers known in Scotland fish was a major food source. This is known despite the fact that conditions in general in Scotland are not favourable to the survival of bones on archaeological sites, as the predominant acid soils cause them to decay; and this is especially the case with finer bones, such as those of fish, in contrast to the more substantial bones of land animals.

Settlement must of course be located within reach of sources of drinking water; and where such sources consisted of streams or lakes there could be fish sources of food as well. In addition in many parts of the world peoples at the pre-farming level living in coastal areas have also used sea fish as food: the most easily exploited fish are obviously the sessile species like mussels and oysters, but there is also evidence over many thousands of years of free-swimming fish being exploited by a variety of methods, including trapping, spearing and hooking. There is a variety of direct and indirect evidence from various parts of the world for boats or rafts of some kind being in use from the Palaeolithic (or Old Stone Age) period, so that fishing is likely to have extended out from the shore from a remote period.

THE FIRST SETTLERS: THE MESOLITHIC

The archaeological evidence shows that Scotland was settled for several thousand years by hunting and gathering groups before the beginning of farming. Radio-carbon dating has shown that settlement began in the Mesolithic period around 7,000 BC, while the first evidence of farming, which brings in the Neolithic, is from around 4,000 BC. Known Mesolithic evidence is heavily concentrated around the coasts. Even the 7,000 BC date is seen as anomalously late in the context of North-West Europe, and has led to conjecture that earlier evidence may have been destroyed on coastal sites by a postglacial rise in sea level before that time (Woodman 1989: 20). The Mesolithic settlement evidence is typified above all by microliths (small blades) and is prominently located around the coasts and along the rivers. There is also a variety of other equipment, much of which appears to have been used for harvesting fish: included are bone and antler points and occasional hooks, harpoon heads and limpet scoops (Wickham-Jones 1994: 93–94). Despite problems of the survival of organic remains in Scottish conditions, considerable numbers of fish bones as well as the shells of limpets and other shell fish are known; in all it is clear that the early settlers depended to a great extent on fish for their sustenance. Evidently in many cases sessile shell fish (such as limpets and mussels) were of great importance in the diet, and indeed a feature that has been remarked in a series of locations are the refuse heaps termed 'shell middens' from the frequency of such shells in them. In an estuarine situation like that of the Forth, shell middens at places such as Inveravon and Polmonthill show the abundance of oysters, mussels, winkles, cockles and whelks available to Mesolithic peoples (Ritchie 1981: 16). It appears that such sites were particularly favourable for the requirements of early settlers, and the sheer volume of shells must have been built up over centuries. The mound at Polmonthill was built up against a bank behind the beach in the Atlantic climatic period, and is 170 yards long, about 25 yards wide and of varying height. It has been calculated that the number of oyster shells in it is between six and seven million (Stevenson 1945–46: 135–139). The sites with their shell middens on Isle Oronsay, the tidal island attached to Colonsay, have been known for over a century. While the bones of fish like cod and limpets had been recognised at an early stage, it took modern archaeological techniques, and in particular the use of fine sieving in the 1970s, to show the great importance in the diet of the commonest inshore fish, the saithe: the bones were too fine to be recognised by earlier investigators (Mellars 1979: 49). It is also the case that seal bones have been found in many coastal Mesolithic sites, and as well as giving a substantial amount of meat, seals would also have been valued for their skins. The exploitation of stranded whales is also testified at the head of the Forth estuary by the antler mattocks that have been found in whale carcasses (Ritchie 1981: 16).

While it may well be that Scottish Mesolithic settlers were in general offshoots from earlier groups in England, the situation is not uncomplicated. A part of the tool assemblages in Scotland have English antecedents mainly in the form of broad microliths, but many of the Scottish blades are of different and narrower styles (Wickham-Jones 1994: 47–50). A connection with Ulster has also been suggested for the settlements in the South-West, which have been interpreted as an offshoot of the Irish Larnian. The Ulster coast is easily visible from various places in South-West Scotland, and from several sites the distance across the North Channel is as little as 20 or 25 miles. No doubt future research will shed more light on the derivation of the earliest Scottish settlers. At any event a series of radio-carbon dates have now been obtained for Mesolithic sites: and although many of these are in the period between 4,000 and 3,000 BC (Wickham-Jones 1994: 103–104), they do in the case of the Lussa Wood site on the island of Jura go back to the period between 6,000 and 6,500 BC (Ritchie 1981: 13).

Population levels during the Mesolithic, and indeed for a long time afterwards, can only be conjectural. Early cultures are generally characterised by sparse populations and small group sizes, and by a degree of mobility; and this was the case with the Mesolithic settlers in Scotland. Although Mesolithic groups held the field for 3,000 years or more, the archaeological evidence is sparse. Much of it consists of debris that accumulated at settlement or camp-sites, and the most frequent are the 'lithic scatter sites', where stone blades and flakes have been found at the soil surface (Wickham-Jones 1994: 62–74), usually with no trace of any structures, and few such sites to date have been systematically excavated. In a number of cases however cave shelters are known in the South-West and in the Oban area of Argyll; and at Morton in Fife and Isle Oronsay in the Inner Hebrides hearths have been found with traces of structures which could have been dwelling huts or at least windbreaks. Various hearths have also been found, indicating cooking sites or camp sites. It has also been conjectured that the three linked stone rings found at Lussa Wood could be the earliest stone house foundations known in Scotland.

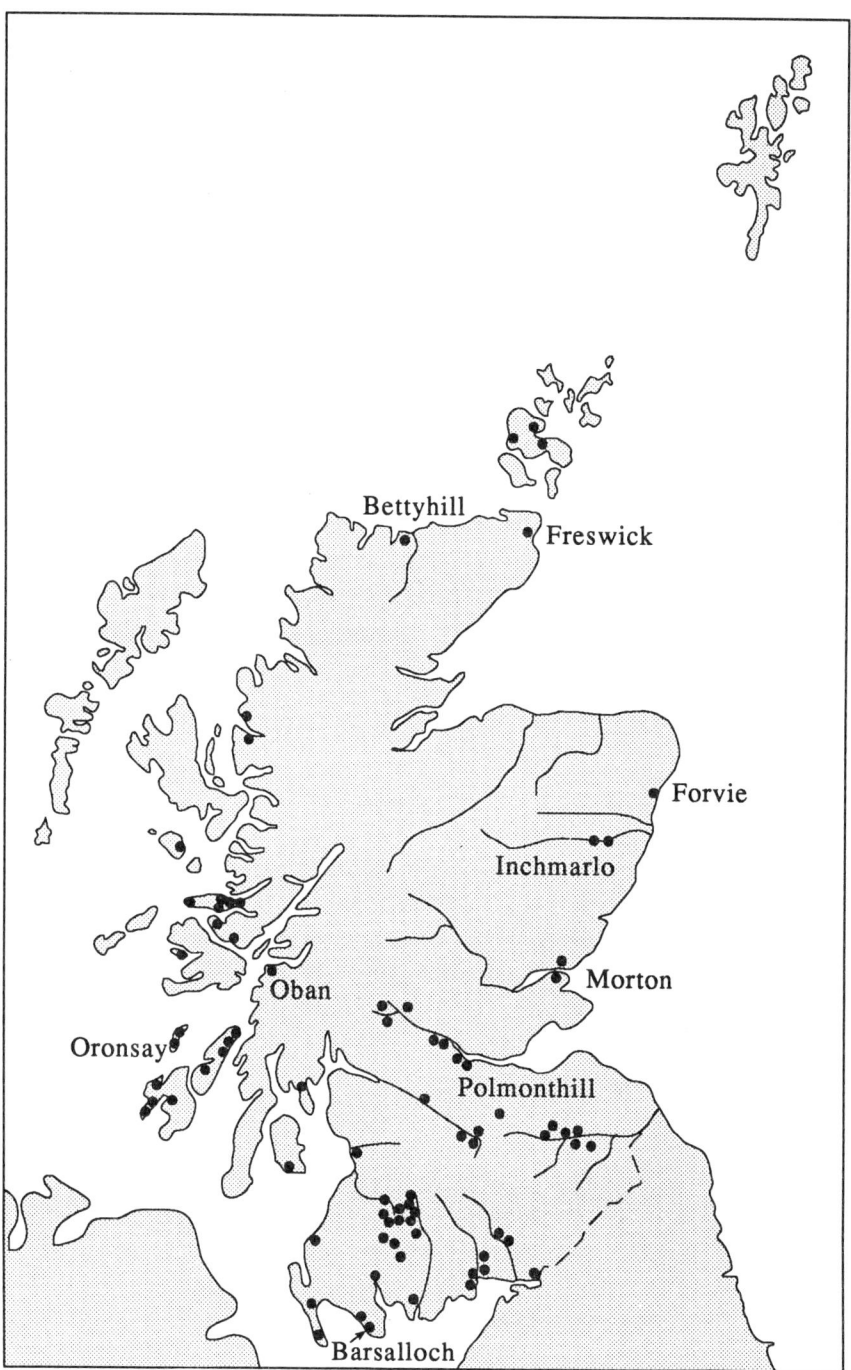

Fig. 3.1 Distribution of main Mesolithic sites in Scotland
(*After Wickham-Jones*).

The most prominent evidence of the presence of Mesolithic peoples is around the coasts of the southern parts of the country, especially in the South-West; but it is agreed that this represents the bigger scale of archaeological activity in the area as well as a possible greater abundance. The distribution also extends northwards along the coasts (fig.3.1), and has been found, for example in Fife, Aberdeenshire and Wester Ross. Sites have been found in Caithness, at Bettyhill on the north coast and at a number of places on the Orkney mainland (Wickham-Jones 1994:63). With the time span during which the Mesolithic peoples were unchallenged in Scotland, it is intrinsically probable that they reached all of the shores, although evidence in the north is still sparse, and the Shetland Isles present a special problem. Although Shetland is visible from high ground in Orkney, the passage from the one archipelago to the other would represent the longest open sea crossing in the country for primitive craft. Whether it was ever made by Mesolithic people may never be known: with the differential post-glacial changes in sea level which have occurred, the Shetland coast has undergone submergence since Mesolithic times, unlike the great part of the Scottish coast, and this precludes the discovery of intact sea-side sites should any have existed.

On occasion, the spade has yielded enough evidence to show some details of the life of Mesolithic groups. It is evident that on Oronsay that the more easily obtainable shell fish (here mainly limpets) constituted a main part of the diet, and provided at least two or three times the amount of food that other fish did. The indications are too that limpets were preferred to other shellfish such as winkles and whelks, because of their higher meat:shell ratio. Also different sites were occupied at different seasons: this is clearly indicated by the age-structure of the saithe exploited at different sites, as evidenced by the otoliths, or ear bones. With the different proportions of first and second year saithe age classes in different middens it has proved possible to postulate separate sites for summer, autumn and winter: and significantly the winter occupation site was at the most sheltered point on the island. The big concentrations of fish bones found in particular horizons and at sharply localised points in the middens suggest large numbers of fish being gutted after capture, and could also indicate efforts at seasonal preservation, presumably by smoking or drying. The predominance of saithe among the fish raises another question on the fishing methods used: although harpoons and bone hooks have both been found on Oronsay, these would not be particularly efficient fishing methods, and it has been suggested that for an abundant inshore fish nets might have been used. Nets are known in Mesolithic Denmark, and various types of small nets have been used in the Hebrides for catching saithe from unknown dates in the past (Baldwin 1882: 179–192). It is also possible that they might well have been taken in tidal traps like the yares known on many parts of the Scottish coasts till modern times (Mellars 1979: 48–57).

In some cases where more detailed investigation has been done, it is evident that particular sites could have sustained occupation and use over very long periods. On Oronsay radio-carbon dating shows sites occupied at a range of dates between 5200 BC and 3900 BC (Switsur and Mellars 1987: 140–143). In the case of Morton in Fife, detailed investigation was made of what were periodic camp-sites on a sandbank connected to the mainland at low tide and used over a period of some 1800 years from 6000 BC. This has shown the use of 40 different species of shellfish, although dominated by limpets and mussels: the total weight of shells in one midden was computed at 5,148 kg., or over 5 tons and the total number of shells at 10 million. Among fish bones at Morton by far the commonest were those of cod, a deep water fish that indicates that this community had boats of some kind for fishing offshore (Coles 1971: 343–356).

As well as the dominant coastal sites Mesolithic remains have been found in a variety of inland locations, most usually beside rivers: and it could well be that the most important of the river fish in later times, the salmon were a main item in the diet. It has been questioned whether the present picture of dominant coastal settlement could be unbalanced simply because Mesolithic remains at the coast are more obvious (Woodman 1989: 22). Inland remains are most numerous again in the south of the country in the valleys of the Tweed, Clyde, Forth and the rivers of the South-West; they have also been found on the Aberdeenshire Dee. Most often the signs are found on river terraces and on river-side bluffs. While the implications must be that fish were also exploited inland, from such sites settlers would have had a wider variety of game than at the coast; and in the later phases this would have been essentially forest game as the eco-system changed with ameliorating climate. The fact that at many of the inland locations Mesolithic evidence has been found below peat is related to general phases of climatic deterioration, especially in the cooler Sub-Atlantic period after the middle of the first millenium BC: this raises the question of how much more evidence has been concealed by the growth of peat and blanket bog in later millenia, and could lead to the inland evidence being anomalously sparse.

FISHING AMONG PREHISTORIC FARMERS

Even after the beginning of farming, there are indications that fish was often still one of the main sources of animal protein in the diet. However for centuries after the Mesolithic the degree of importance of fish in the subsistence of groups in Scotland is unclear, although fish bones have been found in sites of all ages from the Neolithic onwards. For the Neolithic period itself, evidence for fish in the diet has been most reported from the Orkney Islands: bones of bream, wrasse, conger eel and other species have been reported from Megalithic tombs like Midhowe and Knowes of Ramsay (Rousay) and Quanterness and the settlement sites of Knap of Howar (Papa Westray) and Skara Brae (Ritchie 1981:20, 29, 30, 37, 41). This is a reflection both of the relatively abundant remains of the Neolithic period in Orkney and the amount of detailed investigation the prominent monuments in the islands have received from archaeologists. It also suggests, however, that as a great part of the evidence for human presence in the Highlands and Islands is coastal, that fish also featured in the diet in other parts of the region.

With the accepted fluctuations there have been in climate in prehistoric times, the importance of fishing presumably varied with such things as the increase in peat formation and of difficulty in crop ripening, especially in marginal locations like Shetland. It has been suggested that the evident increase in fishing in the cooler Sub-Atlantic period in the second half of the first millenium BC, which is shown in the late Bronze Age in these locations by greater abundance of fish bones in settlement sites, is correlated with a greater emphasis on coastal settlement and more difficult crop ripening (Hamilton 1956:57: Elton and Baden-Powell 1936–37: 358). Subsequently the Iron Age evidence of the wheel house period suggests a renewed importance of cropping with grain storage pits, and a reduction in the number of fish bones has been taken to imply a reduced importance of fishing (Hamilton 1956:59). However the timing of this change has been challenged by evidence from the Western Isles, which show a more pronounced concentration of settlement at the coast in the Iron Age, and this appears to have been a reaction to a the deteriorating climate and the spread of peat. It has been deduced that this change also led to an increased degree of dependence on fish and shell fish (Armit 1992: 130).

Tracing developments in fishing after the Iron Age in Scotland is still much circumscribed by the very limited amount of archaeology done in Dark Age and later times. For Britain as a whole the archaeological record of fishing in the Medieval period is still recognised as very poor (Hutchison 1994: 129, 130). It is inherently probable that fish continued to be a major food source at the coast, and inland as well.

The main evidence available from the Dark Age is associated with the Norse in the Northern Isles. The impact of the Norse in fishing, despite their celebrated prowess in sea-faring, is less than clear in Scotland. The fact that excavated Norse settlement sites are still a rarity adds to the problem of any interpretation; and references to fishing in Shetland in saga material is so scant as to be effectively inconsequential. It was suggested by Hamilton after his excavations at the famous Jarlshof site in Shetland that the absence of fish bones in the early Norse period indicated that the community was primarily dependent on farming; and that fishing, as shown by the presence of line sinkers, only became important at a later stage (Hamilton 1956: 114, 157). However the absence of recognisable fish bones is scarcely a proof of lack of involvement in fishing, and even the artefacts interpreted as line sinkers could have had some other function, such as loom weights. On the other hand fish has been a staple item of diet in all recorded history in Shetland, and it is intrinsically probable that the scarcity of evidence for fishing at Jarlshof indicates a failure of evidence to survive, rather than an absence of the activity from a site that is right on the coast. In fact later excavations at Norse sites in Scotland have suggested that there was a great deal of fishing, especially in the later part of the Norse period. Sandwick on Unst has shown an 'overwhelming dominance of fish' when compared to Jarlshof (Bigelow 1985: 121); as well as remains of shell fish and bones of the 'piltock' (small saithe) which has always been much fished within known Shetland history, the big concentrations of bones of deep water species like cod and ling have led to speculation that by the 13th and 14th centuries there was some development of commercial fishing. Clear evidence of the importance of fish and shell fish has also come from several sites in Orkney such as Deerness and the Brough of Birsay, and in the case of Westness in Rousay there were a range of species of fish and shell fish: and prominent among the former were bones of the main commercial fish of later history, the deep water species of cod and ling (Kaland 1993: 311). In addition at Freswick Links in Caithness evidence has been found of Norse fishing on a scale that has led to conjecture that commercial fishing was also practised there (Batey 1987: 75).

It is a moot point how far the fishing practices identified in the prehistoric period continued through later history. By its nature, archaeology gives more attention to the ordinary and every day than does history, and much of the type of activity and equipment identified in the prehistoric period gets little if any recognition in historical records. The records available for especially the earlier historic period feature more the activity and fortunes of the leaders of society than of the general population. It is the case, however, that in many parts of the world strong elements of continuity in traditional folk culture have been identified between the prehistoric, historic and even modern periods. This has an inherent logic: once methods had been developed for utilising local resources using equipment made of local materials, there are good reasons for continuation within the context of subsistence economies. This does not preclude the possibility of innovation, but early methods could attain a degree of perfection that made improvement difficult and change rare. There are instances in which the use of fishing methods and equipment employed from time immemorial have indeed survived into the 20th century, and in some cases even till today. Fishing methods such as the use of tidal traps (the yares in Scotland) and fish spears were recorded in Wester Ross in the early 20th century (Bathgate 1948–49: 102). A variety of types of small net have been used into modern times for the catching of saithe inshore, often without the need of a boat. These included especially poke and hand nets: the former were vertically hauled on a rope and had the form of a bag net attached to a hoop; the latter was a bag net attached to a rod or pole (Baldwin 1982: 161–212); and on the Solway a variety of fixed nets, and the 'haaf net' which involves wading into the sea with a net mounted on a frame, were still very actively used in the 1950s (Kissling 1958: 166–174).

It is also striking that when detailed descriptions of the different parts of the country become available from the 16th century onwards, fish is often mentioned as a diet item, even in inland locations. While this is a long step forward from prehistory, elements in traditional cultures can be consistent over centuries or indeed millenia.

REFERENCES

Armit, I. (1992) *The Later Prehistory of the Western Isles*, BAR British Series 221, Oxford.

Baldwin, J.R. (1982) 'Fishing the Sellag: Hand Netting Traditions from Caithness, the Northern and Western Isles', in Baldwin, J.R. (ed.) *Caithness. A Cultural Crossroads*, Edina Press, Edinburgh, 161–212.

Batey, C.E. (1987) *Freswick Links, Caithness*. A reappraisal of the late Norse site in its context, BAR British Series 179 (i) and (ii), Oxford.

Bathgate, T.D. (1948-49) 'Ancient fish Traps or Yares in Scotland' *Proc. Soc. Antiq. Scot.* LXXXIII, 98–102.

Bigelow, G.F. (1985) 'Sandwick, Unst and Late Norse Shetland Economy', in Smith, B. (ed.) *Shetland Archaeology*, Shetland Times, Lerwick, 95–127.

Coles, J.M. (1971) 'The early settlement of Scotland. Morton, *Fife Proc. Prehist. Soc.* 37, 284–366.

Elton, C. and Baden-Powell, C. (1936–37) 'On the Relationship between the Raised Beach and an Iron Age Midden in the Island of Lewis, Outer Hebrides', *Proc. Soc. Antiq. Scot.* LXXI, 347–365.

Hamilton, J (1956) *Excavations at Jarlshof*, H.M.S.O., Edinburgh.

Hutchison, G. (1994) *Medieval Ships and Shipping*, Leicester U.P., London.

Kaland, S.A.A. (1993) 'The Settlement of Westness, Rousay' in Batey, C.E., Jesch, J. and Morris, C.D. (eds.) *The Viking Age in Caithness, Orkney and the North Atlantic*, Edinburgh University Press, Edinburgh, 308–317.

Kissling, W. (1958) 'Tidal Nets on the Solway', Scottish Studies 2, 166–174.

Mellars, P. (1978) 'Excavation and Economic Analysis of Mesolithic Shell Middens on the Island of Oronsay (Inner Hebrides)', in Thoms, L.M. (ed.) *Early Man in the Scottish Landscape*, Scottish Archaeological Forum 9, 43–61.

Switsur, W.R. and Mellars, P.A. (1987) 'Radio-carbon Dating on the Shell Midden Sites' in Mellars, P.A. (ed.) *Excavations on Oronsay: Prehistoric Human Ecology of a Small Island*, University of Edinburgh Press, Edinburgh.

Ritchie, G. and A. (1981) *Scotland. Archaeology and Early History*, Thames and Hudson, London.

Stevenson, R.B.K. (1945–46) 'A Shell-heap at Polmonthill, Falkirk', *Proc. Soc. Antiq. Scot.* 80, 135–139).

Switsur, W.R. and Mellars, P.A. (1987) 'Radio Carbon Dating on the Shell Midden Sites' in Mellars, P.A. (ed.) *Excavations on Oronsay: Prehistoric Human Ecology of a Small Island*, University of Edinburgh Press, Edinburgh, 139–149.

Wickham-Jones, C. (1994) *Scotland's First Settlers*, B.T. Batsford, London.

Woodman, P.C. (1989) 'A review of the Scottish Mesolithic: a plea for normality', *Proc. Soc. Antiq. Scot.* 119, 1–32.

4

The Development of Fishing Settlements

ORIGINS

Fishing has been prosecuted during history from a great number of places on the coast; and for several centuries a prominent part of the pattern has been specialised settlements, of which a main or entire function has been that of fishing. In addition there are a number of places of burgh status, particularly in the Highlands and Islands, that can be said to owe their existence to fishing: Lerwick in Shetland essentially owes its origins to the trade and service opportunities presented by the Dutch fishing fleets, while in the West Highlands Stornoway and Campbeltown owe their origins to Scottish efforts at fisheries development. While it is clear that fishing has had a high level of importance, especially in coastal locations, for thousands of years, in the great part of history the settlements involved have also had other means of securing the food supply; and for the great part of the time fishing has been united in varying degrees with farming. Indeed it was only with the development of a big commercial (as opposed to subsistence) component in the general economy that it was possible for some of the population to concentrate on fishing, and obtain the bulk of their food requirements in commodities like meal and butter, by exchange.

While it is possible to follow a broad trend of development, information on the early phases of most fishing places and communities is scanty. In many cases it is merely the name of a place that is recorded in some document which gives a date by which it is known to have been in existence. Even this is not uncomplicated, as there are places where the name which came to designate a fishing settlement had a wider or different earlier currency. There are various cases in which a place which came to be involved in fishing is first recorded simply as a place or district in the vicinity of the coast. In

Aberdeenshire Inverallochy and Boddam became by the 18th century substantial fishing villages, although their earlier significance was as estates with castles. In Berwickshire the district name of Coldingham was also used into the 19th century for the fishing village now known as St. Abbs. The development of separate fishing settlements at the local level can on occasion be effectively clarified when such names as 'fishertoun of . . . .' or 'seatoun of . . .' appear, as happened with, for example with Fisherton of Petty (near Inverness) and Seatown of Boddam (Aberdeenshire).

Recording can of course occur in both local and national records. At the local level the main basic sources are estate and burgh records. At the national level the Register of the Great Seal is particularly useful, with the manner in which places and pertinents are frequently itemised on various properties; and places can also be recorded in major national sources like the Exchequer Rolls, and the Privy Council Records. While there are fairly numerous references to fishings in such documents, the majority in fact refer to fresh water fishings; and even when 'fishertouns' occur it is possible for them to be at inland locations, although there must generally be an indication of an emphasis on fishing in the activity of a settlement. It is even possible that when a 'fish' element occurs in place-names it can be derived from words with no fishing connotation. The derivation of inland names like Fishrie and Fisherford in Aberdeenshire and Fishcross in Clackmannan is obscure, but is likely to incorporate a first element of some other derivation. Another series of documents which is useful in studying the development of fishing settlements is the early maps, including especially those of Pont and Blaeu; and on occasion these maps may actually have information on fishing actually entered on them: the Blaeu map of Shetland, for example records the great fishing effort of the Dutch.

While a variety of sources give some information on early fishing settlements, the circumstances of origin of many of them is obscure. While there are some exceptions, linguistic evidence suggests that in the big majority of cases the fisher folk speak the same basic dialect as their landward neighbours: this points to a common origin, the fishing population having become occupationally segregated relatively late in history. A possible source of origin of settlements was squatting by landless members of the population at the lower end of the social scale – the sort of circumstances that have given rise to not a few fishing settlements in different parts of the world.

There is evidence for the existence of fisher quarters of towns, like Fittie (or Footdee) in Aberdeen from the Medieval period, and at Ayr the fishing quarter was of unknown antiquity (Czerkawska 1975: 3). Newhaven and Fisherrow in the Edinburgh area are on record by the 16th century and Broadsea (Fraserburgh) by 17th century. As a general rule, even when living in close proximity with other members of the town population, the fisher folk were distinctive and largely separate communities right into the early 20th century. This applied even in relatively small towns like Nairn and Golspie (Dorian 1985: 1, 2), and it is only with the general lowering of social barriers in the present century that fishing families have become more dispersed within towns. Linked to this evidence for fishing communities attached to towns is evidence of early fish markets and marketing. A fish market is on record at Ayr as early as 1538 (Pryde 1937:80), and at Glasgow in 1596 (SBRS. 1876: 185): and records from the same century of personnel like fish gaugers show that there was also an organised trade in other burghs. In Edinburgh there was concern in this period that fish were being sold in Leith, forestalling the city market.

It is significant that not a few of the early towns (such as Berwick, Perth, Montrose, Aberdeen, Banff, and Inverness) were at river-mouth sites and had important salmon fisheries. The indications are that salmon fishing was particularly important in early commercial fishing, and may well have been a main reason for the development of distinctive fishing communities within urban society. It appears from records at places like Berwick and Aberdeen, salmon fishing in the estuaries could be operated by the same men who also fished in the adjacent sea.

As a general rule the main sequence of developments appears to have been that it was in the Medieval towns, with their greater populations and commercial opportunities, that specialised fishing communities first arose. With further development of the commercial economy specialised fishing spread into some rural coastal villages from as early as the 15th century and the development of non-urban fishing settlements accelerated by the 17th century. There are fishertouns recorded in the Register of the Great Seal from 1507 onwards. In 1507 one mentioned in the Inverness area (RGSS. II (1883): 663) may well be the Fisherton of Petty, while the one recorded in Kinneff (Kincardineshire) in 1541 (RGSS. III (1883): 557) could be Crawton; 'Michellis' (Muchalls) in Kincardineshire is specifically named in 1594 (RGSS. VI (1890: 62) and

Auchmithie (Angus) in 1608 (RGSS. VI (1890):754). However these, and doubtless other, villages were involved in a trade system in which towns were still the key points; and it was certainly an advantage to such villages to be near to main towns as a market outlet. For long fishing continued to be combined in varying degrees with farming in both town and country: and there is evidence for this in East Coast lowland areas, as well as in the Highlands and Islands. Crail was a burgh long involved in fishing and trading, and like other burghs it had arable land and a common muir; and in the 16th century land was being consolidated from run-rig (Murray 1964:85–87). The first feuars of Peterhead in 1593 were fishermen with small plots of about one-fifth of an acre who had rights to common pasture and peat cutting on the town moss (Arbuthnott 1815:18); the neighbouring burgh of Rattray, also involved in fishing at the same period, had infield, outfield and common land (Cumine 1887:1887–90:18); and in 1683 the fisher towns on the north coast of Aberdeenshire were reported as growing oats and barley (Garden 1907:141–142). The detailed siting of fishing villages is also a matter of significance. There are a number of cases where the site was actually some distance back form the shore, as happened at such places as Boatlea and Old Whinnyfold in Aberdeenshire (Summers 1988: 42), and Findon and Cove in Kincardineshire: and these contrast with the many examples where the settlement was as near the beach as possible to minimise the work of carrying fish and equipment between the boats and the houses. There is a strong presumption that in the cases where the settlement is back from the shore fishing was originally only part of the activity of a ferm toun, with the settlement site more centrally placed among the cultivated land. There are also occasional references from the 15th century onwards of holdings for fishermen which are in the form of crofts. Even with the reorganisation and specialisation that came with the Improving Movement, fishermen might retain small holdings. At Fishertown of Petty (Invernessshire), where the fishing opportunities were limited in the narrow firth, the fishermen depended as much on their crofts as on their catches in the 18th and 19th centuries (Bain 1925:126); at Torry (then in Kincardineshire) in the 1790s nearly all the 23 salmon fishermen still had land (OSA. VII:194); and at Whitehills in Banffshire as late as 1840 the men had 1.25 acre plots (NSA. XIII:237).

There are also indications, however that the tie between the fisher folk and the land was loosening in the 18th and 19th centuries. At Buckie, although the fishermen all had small holdings in the late 18th century, most of them had let the land to local crofters (Hutcheson 1887: 62); and at Inverallochy and Cairnbulg in Aberdeenshire in the 1790s the fishermen were conspicuous by being the only people in Rathen parish with no land (OSA. V: 15). In Ayrshire, the village of Dunure was built about 1800 in an area on the coast where previously there had only been a 'miscellany of crofts' (Czerkawska 1975: 3).

With the development of a definite emphasis on fishing, there was some adjustment in the desiderata for siting of settlements. The very general tendency was for villages to be sited immediately adjacent to the shore, and on rock rather than sand coasts; they were also where possible beside shingle beaches. Although many rock coasts may appear more forbidding, it was important to avoid sand foundations for houses. Also a boat pulled up on a shingle beach does not bed down like one on a sand beach, and loading and unloading from rock creeks, which bring deeper water inshore, is simplified. Thus despite considerable exposure, the preferred sites were those beside shingle beaches and rock creeks like Cairnbulg (Aberdeenshire) (plate 1) and Sandend (Banffshire) (plate 3) respectively. On a steep or cliff coast there was a tendency to build houses on small fragments of raised beaches at the cliff foot, as at Pennan (Aberdeenshire) and Crovie (Banffshire) (plate 2), despite the obvious danger from winter high seas. Where there was inadequate ground for house building at the cliff foot, there were many examples of villages built on cliff-tops from which the boats had to be reached by steep and often slippery paths. Such are St. Abbs (Berwickshire), Auchmithie (Angus), Portlethen (Kincardineshire), Whinnyfold (Aberdeenshire) (plate 4) and Portknockie (Banffshire): and the extreme case here is that of Whaligoe in Caithness, where to gain access to a creek in a bold cliff coast a zig-zag stair of 330 steps were actually cut in the cliff face (Wilson 1842, II: 190–191) (plate 6).

It appears that the Highlands and Islands, where fishing has into the 20th century been conducted from ferm towns and crofting townships, in effect preserves an earlier evolutionary stage. There was, none the less, a considerable body of opinion by the 18th century that the systematic building of fishing villages was the best method of fishery development in the Highlands, and John Knox, the prominent contemporary authority, suggested that about forty should be built at intervals along the coast between Arran and Dornoch. However only a handful were ever built, and a main reason for this was that Highland estates generally lacked the necessary resources (Munro 1989: 30–31). A partial exception was the island of Islay, where the Campbell landlords to a large extent copied the lowland model of reorganisation during the Improving Movement, and villages like Portnahaven and Port Wemyss were built for fishing (Storrie 1981: 187). The best examples of such villages in the West Highlands were Tobermory and Ullapool, built by the British Fisheries Society at the end of the 18th century; but like the majority of Highland villages they had no sustained success in their intended purpose. In South Argyll there were significant exceptions with successful settlements like Tarbert and Carradale, based on the Loch Fyne herring fisheries, in addition to such places as Inveraray and Campbeltown which acquired urban status. An isolated modern example of fishing village development in the Highlands and Islands is the village of Hamnavoe on the Shetland island of Burra, which is largely a spontaneous development from the later 19th century, as men from a crofting fishing tradition concentrated more and more on full-time fishing. However the general rule in the crofting area was that while fish always played a major part in the food supply of a population that was always predominantly coastal, they maintained an interest in the land, and in the Highland area crofter-fishermen are well recorded into the modern period, and indeed are still to be found in significant numbers today.

What evidence there is suggests that most fishing settlements were originally set up as daughter colonies of existing ones; and from the 17th century onwards there are not a few cases known in which the founders of settlements were induced by proprietors or factors to move to a new location. In the North-East, where fishing came prominently to the fore in the 17th and 18th centuries, there is a good deal of evidence for this process of proliferation of fishing villages. In Banffshire, Portknockie was founded by fishing families moving from neighbouring Cullen about 1677, Findochty by settlers from Fraserburgh (Aberdeenshire) in 1716, and Portessie by incomers from Findhorn (Morayshire) in 1727 (OSA. XIII: 401). In Kincardineshire there is a tradition that Johnshaven was started by fishermen moving from Peterhead before 1600 (Adams: pers. comm.), and in Angus, Ferryden has been attributed to the Scott lairds of Rossie recruiting fishing families from Banffshire about

1730. In the inner Moray Firth, the dialect differences which have been observed between the settlements of Avoch and Cromarty and their neighbours are thought to be associated with fishing families settling from the south of Scotland, and Avoch is known to be essentially a creation of the 18th century (OSA. XV: 625). Newtown of Ayr shows another unusual possibility: it was founded in the 1770s by incomers from the Pitsligo estate in Aberdeenshire, who owed their acquaintance with the Clyde to having been pressed on a man-o'-war (Czerwkawska 1975: 3). At a later stage in the early 19th century, fishermen were enticed from Ardersier in Nairnshire both to the rebuilt Burghead and the new planned village of Hopeman (3rd S.A. Morayshire, 1965: 133–134).

Although there are settlements involved in fishing on most of the Scottish coasts, the developments from the 17th century onwards have entailed that the leading fishing region with the greatest concentration of specialised fishing settlements is the North-East, as is indicated in fig. 4.1. Here on a length of coast of the order of 120 miles in length, there developed on the known record the remarkable total of 63 places involved in fishing: the majority of them date originally from the 17th and 18th centuries, and the preference for rock rather than sand sections of coast is obvious. the growth of these settlements is obviously linked to the development of a more commercial orientation in the local and national economy.

It is clear that in the better recorded 19th century there was a substantial increase in the fishing population, in harmony with broad national trends. Although some new settlements were founded, the main accent was on the expansion of existing places. In 1800 villages with more than 100 fishermen were unknown, whereas by 1900 not a few had over 300; and cases are known during the 19th century of fishing villages doubling their population in as little as 30 years (Gray 1978: 80). The mark of this trend is still obvious in fishing settlements today with the extent of 19th century housing, usually built on a grid pattern of streets.

In all, in the development of specialised fishing and fishing settlements, a broad trend can be seen. In general the first developments were in seaboard towns with the growth of fishing quarters, although the main activities of such towns were maritime trade and other central place functions. From the 16th century onwards there was also a proliferation of specialised settlements outside the towns, to the point that by the 18th century the main balance of commercial fishing effort was in effect rural as opposed to urban. This was

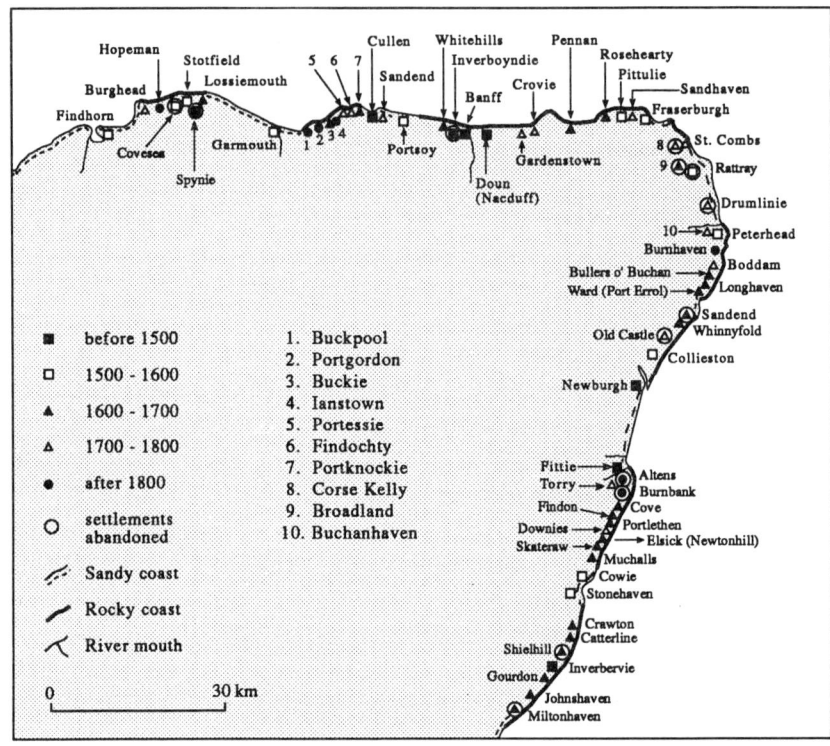

Fig. 4.1 Development of Fishing Settlements in
North-East Scotland, 14th–19th centuries.

succeeded by a period of further adjustments in the modern
period. Essentially the development of bigger scale operations from
bigger boats required harbours, and these could be economically
provided at relatively few places. At the same time maritime trade
also moved to bigger ships, and concentrated in a very few harbours
which had bigger hinterlands served by much improved systems of
overland transport. The harbours at which fishing now concentrated
were in the main seaboard towns for which the prime activity had
been trade, but which became mainly dependent on fishing: such
are Eyemouth, Anstruther, Peterhead, Fraserburgh and Wick. From
the late 19th century Aberdeen was to become a main anomaly in
this trend, in that it continued as a substantial trading port, while
also becoming the dominant white fish port in the era of steam
trawling. While varying numbers of the fishermen continued to live
in fishing villages, considerable numbers also took up residence in

the towns, and the big majority of the fishermen came to operate their boats from town harbours.

HOUSING OF THE FISHER FOLK

As is general with vernacular architecture in Scotland, very few examples of houses from before the modern period have survived among the fisher folk. What information there is suggests that the houses were often primitive and insanitary, and this can only have been aggravated by the fact that much of the necessary work of line baiting and gutting and cleaning fish was done in and around the houses. A late 18th century description of Fittie in Aberdeen speaks of comfortless low thatched houses, sparsely furnished and beset around with dunghills, fishing gear and drying fish (Morgan 1993: 15). It appears very general that traditional houses conformed to the vernacular norm in being earth floored, cruck framed and thatched roofed; and they would also have had peat fires in the middle of the floor. A custom which in places survived into the 20th century was that of sprinkling the floor with sand from the beach, which would become fouled with trampling and with work related to fishing; and at intervals it was swept up and replaced. With the low social status that the fisher folk often had, it may well have been that the construction of house walls would have often incorporated the use of materials like turf and peat as well as undressed stone.

While there are some examples of excessively small houses which could possibly be as old as 300 years or more, most of the picturesque old cottages for which many fishing villages are now noted are in fact products of the housing improvements of the 18th and 19th centuries, when the norm in simple Scottish housing became the two-roomed, gable-ended 'but and ben' with fireplaces in the gable-ends. With the fireplace removed from the middle of the floor, an early covering for the earthen floor was could be canvas, in the form of an old boat sail. At this stage, landlords often took the initiative in housing improvements, especially in new planned villages. There were also spontaneous improvements on the part of fishing families themselves, and in the middle of the 19th century the Fishery Board commented on the general trend of improvements in living conditions in fishing settlements (ARFB. 1855: 2). Most of the visible fabric of fishing villages and towns date in fact to the late 18th and 19th centuries, which was a general time of expansion, when most villages acquired enlargements laid out in orderly streets, some of which were on the familiar 19th century grid plan. Most prominent here is the construction of Pulteneytown, the extension to Wick on the south side of the bay built by the British Fisheries Society on a plan by Thomas Telford to cater for the rising herring fishery (Munro 1981:148–159).

In the Improving Movement the fishermen might be given premiums to build houses, as happened for example at St. Combs (Aberdeenshire); and the condition of this was that dimensions were stipulated, and the walls had to be of stone and lime (Summers 1988:22). It was also possible for the landlords to have improved houses built as an attraction to incoming fishermen, as happened at Buchanhaven, which is now part of Peterhead (Summers 1988:31–32). In the small coastal burghs of the Forth, and most especially in the East Neuk of Fife and in Newhaven there are the main exceptions to the general rule of single storey traditional fisher housing, with big numbers of old stone-built two-storey houses with pantile roofs, some of which are centuries old. Numbers of them have the traditional outside stairs to the upper floor, a style which was partly related to the narrower houses of former times, and in which the external stair effectively saved internal space on the ground floor. These reflect the long history of these places in fishing and other sea-faring, and speak of a greater level of past prosperity: and to this day the congested lay-outs of the East Neuk towns make them among the most scenic of traditional Scottish fishing settlements. There was also considerable building of two-storey fisher housing in the 19th century in main fishing towns like Peterhead, Buckie and Wick. In the later 19th century especially a prominent development in the main fishing towns was the development of special house styles, which as well as having living space for the family also had considerable loft space for storing herring nets and mending them. The building on a large scale of three and four storey tenement housing which occurred at Torry in Aberdeen at the end of the 19th century largely to house the families of trawlermen is in fact unique in housing for Scottish fishing families. When programmes of building council houses expanded in Scotland in the inter-war period, there were instances of these having special adaptations for fishermen in the form of net lofts, especially in the East Neuk of Fife.

From being concentrated in particular sections of towns, fishing families in the latest period have moved into more spacious houses and become much more intermixed with the general population:

this reflects the modern trend towards a less compartmentalised society, and the lessening of traditional social barriers as well as enhanced living standards.

TENURE AMONG THE FISHING POPULATION

The tenures of the fishing population in general have similarities to other elements of both the urban and rural populations, although they can show special features related to the fishing occupation. Where there was formal tenure, in the towns it appears that the general practice was for the fisher folk in royal burghs to rent their houses from urban landlords, and in the burghs of barony from estate lairds. In Aberdeen, although the fishing quarter of Fittie (or Footdee) is on record from the 13th century and there is specific reference to 'quhit fishearis' (white fishers) from the 15th century, records of the tenure of the fishing community are obscure until the 18th century, by which time there was a fishermen's friendly society, and one of the possibilities was for fishermen to rent houses from it (Morgan 1993: 15). When Peterhead was founded as a burgh of barony in 1593, there were fishermen who had feuar status among its first citizens (Skene 1952: 21–22); and in 1712 there is the record of the feuing of unused ground for fishermen's houses in the burgh of Nairn (Bain 1928: 248). Of all the fishing communities in towns in Scotland, undoubtedly the most prominent was that of Newhaven in Edinburgh, which for centuries had a major role in providing the capital's fish supply, and it appears to have had a special free status of which the origins are lost in history. Newhaven was in existence as a community by 1504 (McGowran 1985: 4) and actually featured in the early 16th century in the first building of a navy in Scotland; and with its development as a port it was taken over by Edinburgh in 1511, in case it should threaten the fortunes of Leith (Black 1951: 9–11,94). Subsequently the 'Society of Free Fishermen of Newhaven' was instituted, conceivably on some quid pro quo arrangement for the take-over: this society has a history of over four centuries, being known to have been in existence by 1572; it was incorporated by charter in 1573 (McGowran 1985: 203), and it was active in a number of ways. Its function as a friendly society underlined the hazardous character of the fishing occupation, and it also acted on behalf of its members in various disputes: these might be with other fishermen, with the town of Edinburgh or with other bodies.

Not unimportant in the towns were opportunities for other work involving boats, of which the special function of pilotage was most important, with local knowledge of the approaches and entries to ports being a big asset. For centuries pilotage was a regular function of the Newhaven men on the Forth (McGowran 1985; 42–44) and of the Fittie men at Aberdeen (Morgan 1993: 30–31); at Montrose the men from the neighbouring fishing villages of Ferryden and Usan acted as pilots (Ouchterlony 1907: 41–42); and fishermen at other places such as Arbroath (Hay 1887: 377) and Peterhead (NSA. XII: 369) did similar work. Not unnaturally men from fishing families also engaged in other seafaring occupations: throughout history men with seafaring experience found employment on trading and other vessels, and there was always some two-way movement between fishing boat crews and those of other craft. Trading ships tended to involve rather less discomfort than fishing boats and gave more predictable returns, although they could involve being away from home for protracted periods.

While the fisherfolk in towns evidently were able to share in the experience of feudal times that 'town air makes free', they appear to have been under the control of lairds longer in villages on estates. The Pre-Reformation situation outside the towns is seldom clear, but it is known that in Easter Ross the fishermen of the village of Hilton supplied the Abbey of Fearn with fish and lived rent-free (Ash 1991: 156): presumably this was not entirely an isolated instance with an ecclesiastical landlord, although it is scarcely likely to have been typical with lay landlords. It was not automatic that the fishermen would have control of their own affairs after the Reformation, and indeed with the growth of the commercial economy they could fall within the control of tacksman or merchants. In 1636 in Montrose there is the record of fish curer David Newton owning seven fishing boats, and forbidding the crews to sell any fish other than to himself at the risk of the penalty of £6 (Scots) (Adams 1993: 226). It is clear that the control of lairds was getting less with the economic and social changes which were in train by the 18th century: and although the greater freedom that the fisher folk were to acquire was not associated with the legal changes that were specific to a group like the coal miners, they were not unaffected by the main currents of social change. The manner in which from the 18th century and even earlier they moved between estates in starting new villages (as mentioned above) also shows an important element of freedom, but it was not universal:

there is extant an agreement of 1713 among proprietors in Easter Ross, which specifically prevented the fishermen leaving one master for another (Ash 1991:160); and as late as 1805 the Earl of Northesk was able to obtain a legal decision giving him the right to look on his white fishers as serfs or thralls, following an attempt by a family to leave his village of Auchmithie for the town of Arbroath (Hay 1887:446). Also, in the leading fishing settlement of Buckie, the fishermen did not break with the laird until 1821, after which they began to build their own boats (Hutcheson 1887:19), rather than have the laird supply them periodically as part of a wider lease arrangement.

The usual early expedient in fishing villages appears to have been a type of steelbow tenure, with rents being paid jointly for house, land and boat: in the North-East especially there is much evidence of this. Examples are extant of multiple tenancies being given to groups of fishermen who could constitute boat crews wholly or in part, and it was claimed in the early 19th century that this type of arrangement had operated 'from time immemorial'; and instances of it going back to the early 17th century were quoted (Bell 1812:54). In 1681 at Rosehearty, Lord Pitsligo provided boats for crews of six men and undertook to replace the boat every five years (Taylor 1982:3). It was also possible for single tenancies to be given, at times with the specific stipulation that the lessee recruit a boat crew, as happened in the mid-17th century on the Panmure Estate in Angus (SRO. GD 45/18/257; GD 45/18/296); and as early as 1579 at Eyemouth there was a lease in a feu charter to a man and his co-fishers for half of a boat, along with five houses and gardens (MacIver 1906:267). On the Newtarbat estate in Easter Ross in the 18th century, the landlord provided houses and gardens along with boats and equipment, and it was the practice to replace the boat every seven years (OSA. VI: 425); and at Fishertown of Petty at the same period, the fishermen were stated to hold tenancies 'in a peculiar manner' in that the headman or skipper paid the rent to the landowner for the entire crew (Bain 1925:126). In effect, the implication appears to be of a type of arrangement with a tacksman with sub-tenants: one which no doubt became more common with time, as it did with farming tenures. In the 1720s at Johnshaven the fishermen were advanced £5/12/- every sixth year for a new boat, and paid an annual rent of £2/10/- (Adams 1991: 30). There are signs of an evolution towards a system in which the fishermen supplied part of their own capital requirements by the end of the 18th century at Buckie, which was certainly one of the foremost fishing places in Scotland by that time. The cost, fitted out, of the six-man boats used was about £24, and the proprietor gave a crew £11 towards the purchase of a new boat every seventh year. The men bound themselves to continue in the boat for seven years, and they were also charged an annual rent of £5/3/3 and six dried cod; for the smaller boats, which cost about £12, the proprietor played no part (OSA. XIII: 402).

The earliest known rents could have a considerable element in kind as well as money. Also there are records of teind fish being payable, and in the 16th century at both Anstruther Easter and Anstruther Wester the lairds took one tenth of all fish landed as teinds (Watson 1986: 22). At Eyemouth there is the record of teind fish payable in 1628 (MacIver 1906: 268); and in the 18th century, when more records become available, there are records at places like Newhaven (Black 1951: 20–23) and Fraserburgh (Cranna 1914: 59) of teind fish having been paid, although by that time the practice appears to have generally lapsed. Rents in kind were of course mainly paid in fish; at Cullen, for example, in 1642 each skipper was bound to give the laird 'two dissoun fishes' for each day spent fishing (Cramond 1888: 41); and in the 18th century on the Newtarbat estate one fifth of the catch was paid as rent, although this was becoming converted to money (OSA. VI: 425). However at Cowie (Kincardineshire) in the 18th century rents in kind were limited to 100 haddocks and 3 large cod from the skipper for the year, along with a pint of oil from each man (Christie 1974: 13): this suggests that these had something of a vestigial or token status by this time; and the same would certainly apply to the six dried cod paid annually in the 18th century at Buckie (Hutcheson 1887: 17), quoted above. As late as 1840 at Pennan in Aberdeenshire the six boats paid a rent of £20 and some dried fish in annual rent (NSA. XII: 269). However the fact that rents could also include items like bere or meal confirms the that the early fishing villages also had a stake in the land: at Fishertown of Petty (Inverness-shire) in 1769 rents were still in bere, meal, oats and straw, and the fishermen depended as much on their crofts as on the proceeds of fishing (Bain 1925: 126).

On various estates it is clear that the lairds and their factors saw the leases to fishermen as part of a wider function in the working of the estate, beyond providing supplies of fish and making money rent payments. There could be special service provisions, and here the availability of boat men who could use their craft to transport a variety of heavy or bulky commodities was especially important before overland transport improved during the modern era. On the Panmure estate it could be part of their terms of lease that they transported coal and other materials for the laird (SRO. GD 45/18/296; GD 45/18/313), and at Buckie the 'binnage' freights for which the fishermen were liable involved the transport of building materials, including stone and timber (Hutcheson 1887: 19). At Cowie (Kincardineshire) in the 18th century the fishermen were also bound to do harvest and peat cutting work for the laird (Christie 1974: 13); and in the same period in Easter Ross it was usual for fishermen also to provide services including shearing sheep, making hay and carrying peats (Mowat 1981: 49). At Dunure in Ayrshire there was in the early 19th century a more spontaneous link-up between fishermen and farmers, where the farmers supplied the carts to transfer mussels transported from the northern Clyde lochs to mussel scalps at the shore, and in exchange the fishermen supplied seasonal harvest help (Czerkawska 1975: 10).

With the general expansion of fishing, it is not surprising that in the 18th and 19th centuries the general tendency was for the organisation of it to get into more specialised hands. At Broadsea (Fraserburgh) in the late 18th century all boats were supplied by a tacksman, who had rented the fishing from the laird: each boat was bound to give the tacksman a sixth part of its catch, and the men were bound to stay in the boat for a fixed term (Cranna 1914: 56). Merchant firms also could step in, and a firm that was to have interests in a number of places in the north were the Falls of Dunbar: they were for decades the providers of capital and the merchants at Avoch in Easter Ross, and after they went bankrupt in 1788, the Northumberland Fishing Company took over and continued in their place (Mowat 1981: 49).

Despite the various rental arrangements that are on record, it is noteworthy that right until the late 19th century, by which stage many of these settlements had expanded to have populations numbered in hundreds, many of the people were without proper legal title to their houses. This had some remarkable effects in economic terms, and may be illustrated by the fact that in the Aberdeenshire village of St. Combs in 1840 the total annual rental value of houses and gardens for the whole village was £61/4/- at a time when the annual proceeds from fishing was computed at £1567 (NSA. XII: 229). An inquiry conducted in 1884 by the

Fishery Board for Scotland showed that this continuing situation had grown up by a combination of benevolence (or inaction) on the part of the superiors, and trust on the part of the fishermen, but it had become a liability to the fishermen themselves. Although they gained some advantage from being liable for no taxable subject apart from the ground their houses stood on, they were in the possession of assets which could not be the subject of security or credit. On the Banffshire coast, which had the greatest concentration of fishermen in the country, there was property in fishermen's houses to the total rental value of £55,000, which with proper legal title, would have allowed the ready borrowing of £35,000 for fishermen for such purposes as improving their boats or taking part in harbour improvement schemes. The situation in Banffshire was also typical of that in other parts of the North-East and of the islands (ARFBS. 1884: xlvii–l).

In all, it is possible to distinguish a general trend in the tenure arrangements of the fisher folk, which is related to wider socio-economic developments, and also to the increasing scale and specialisation of the fishing enterprise. At the earliest phases small groups of fishing families had small holdings, and paid a single rent for house, land and boat; and this rent was paid in a combination of kind, services and money direct to the laird or his factor; probably too this was most often in the form of a joint tenancy with a boat crew. With the passing of time, and as the scale of operation expanded, the money component in the rent became more and more dominant; and increasingly the fishermen dealt with tacksmen or merchants rather than lairds or factors. In time, while the fishermen continued to pay rents to estates for their houses, the fishing operation became unconnected with the lairds. However, the full modernisation and regularisation of the tenures for fishermen was long delayed. It was still the case in 1891 that many a fisherman had no documentation for his house beyond the receipt he had for the price he had paid for it, and was technically a squatter (ARFBS. 1891: xix).

Shetland was something of an anomaly (see ch. 6) in tenurial arrangements. Here there was very little local commercial market, in contrast to the situation in most of the country, by which the fisher folk disposed of most of their fish by direct sale to consumers. Instead in the 'Shetland method' the fishermen were bound to lairds and merchants in a truck system, by which virtually all the commercial catch was exported, and participation in fishing for the landlord was a condition of holding house and land.

A FAMILY BUSINESS

As in so many traditional trades, fishing formerly depended very much on collective family effort; and there is considerable evidence that many fishing settlements began with a nucleus of a handful of families. This is clear for the late 17th century for several villages for which the population is recorded in 1696 in the Aberdeenshire Poll Book. It is also significant that the populations of many fishing villages are even now dominated by very small numbers of surnames, which derive essentially from the founding families when the settlements were much smaller. Thus Boddam (Aberdeenshire) has its Cordiners, Sellars, Hutchisons and Stephens; Gardenstown (Banffshire) its Watts, Wests and Wisemans; and Auchmithie (Angus) its Cargills, Spinks and Swankies. The use of 'tee-names' to designate particular individuals and families, which is a very common characteristic of fishing communities is partly related to this limited number of surnames, although in traditional society it was of course a characteristic not limited to fishing situations.

The fact that the actual work of fishing was done virtually exclusively by the men is in danger of obscuring the fact that the women and other members of the family played essential roles. Historically very little organised information was ever collected on the work of the women, and much of what is known depends on folk memory, although up to the last generation this was lively and vivid. In the most recent period there have been some systematic studies of fishing communities by social anthropologists and sociologists, which have gone some way to emphasising the family role, even if such studies are not primarily historical.

There is ample evidence of the great work the women in baiting lines for inshore fishing, and this was a practice which continued into the inter-war period of the 20th century in some places, and in the case of Gourdon (Kincardineshire) persisted into the 1970s. Before the rise of trawling as the main method of white fishing in the late 19th century, there was an immense aggregate effort in baiting lines, and this very generally also involved the collecting and 'shelling' of mussels. To the women too fell much of the work of gutting and cleaning the fish; and in addition that of salting and spreading on shingle or stone beaches (or sometimes trestles) the part of the catch which was to be cured before sale. There are various records too of the wife wading into the sea carrying her husband on her back to the boat, so that he would not get his feet wet at the start of a voyage. How widespread this practice was is difficult to know, and it may well have begun in times before there were reliable watertight seaboots; and there are certainly records from the late 19th century of young boys going to sea barefoot. Probably better known is the role of the women in disposing of the catch. Formerly, and to some extent until well into the 20th century, fish could be disposed of by the fishwife carrying her creel of fish to the customers. This meant going to town markets where these were within access: and also from many coastal villages fish wives went inland to the farms and crofts of the rural areas, where the fish might be disposed of by barter for such products as butter, meal and potatoes. It appears to have been not infrequent for the fishwives to make return journeys of ten to twenty miles or even more for this purpose. It was frequently said that it was essential for a fisherman to marry a woman from a fishing community, so that his wife would have the necessary skills, and also be inured to the toil that was inescapable. Until relatively recent times it was virtually unknown for there to be intermarriage between fishing and other communities; and even when fishermen did start on occasion to take wives from outwith their own community, their partners most commonly came from other fishing communities. This trend was related a good deal to the development of mobility in the fisheries (ch. 12), which gave greater opportunity for men and women from different fishing communities to meet.

As the fisheries expanded in the modern period, parts of the work of cleaning and curing fish for market came within the purview of specialist merchants and curers. Some of these came from the fisher folk themselves, and their hired help often came from them too; and this did give some scope for women to become money earners to aid household finances. It was also possible for part of the work of line baiting to be on a paid basis (e.g. Hay and Walker 1985: 29).

When the major effort of the fisher folk became progressively more directed to the herring fishery during the 19th century, family effort continued but was considerably transformed. In the herring fishery the work of gear maintenance became substantial: for long the nets employed were fully as valuable as the boats themselves, and wear and tear of nets was on a scale that necessitated a major effort in mending between seasons; and the great part of this fell on the women folk. This was done largely by fishermen's wives and daughters, although there could be some paid help. There was also a major source of employment in the herring curing yards: here the main tasks were the gutting of herring and the packing of them into barrels; and here also the women of the fishing communities were the main source of the work force.

Even now fishing still tends to be a family occupation, and many of the Scottish fleet are still family boats, and there is also some survival of family businesses in fish merchanting and processing. While company ownership did make an appearance on a major scale with the development of steam trawling from the 1880s, that has now substantially lapsed in our modern era of 200-mile limits. The survival of family enterprise in fishing is partly related to fishing being an industry of limited importance in the modern economy; but it is also related to the special nature and traditions of fishing itself. In the catching sector especially, there has always been resistance to any development of, or take-over by, companies.

REFERENCES

Adams, D. (1993) 'Abundant with all kinde of fishes: Sea fishing before 1800', in Jackson. G. and Lythe, S. G. E. *The Port of Montrose. A History of its harbour, trade and shipping*, Hutton Press and Georgica Press, Tayport and New York, 225–236.

Adams, D. (1991) *Johnshaven and Miltonhaven. A Social and Economic History*, Chanonry Press, Brechin.

Annual Report of the Fishery Board (ARFB.) 1855.

Annual Reports of the Fishery Board for Scotland (ARFBS.), 1884, 1891.

Arbuthnott, J. (1815) *An Historical Account of Peterhead*, D. Chalmers and Co., Aberdeen.

Ash, M. (1991) *This Noble Harbour. A History of the Cromarty Firth*, Cromarty Firth Port Authority, in Association with John Donald, Edinburgh.

Bain, G. (1928) *History of Nairnshire* (2nd edn.), Nairnshire Telegraph Office, Nairn

Bain, G. (1925) *The Lordship of Petty*, Nairnshire Telegraph Office, Nairn.

Bell, R. (1812) *A Treatise on the Election Laws*, A. Constable and Co., Edinburgh.

Black, R. M. (ed.) (1951) *Society of Free Fishermen of Newhaven. A Short History*, Society of Free Fishermen, Newhaven, Edinburgh.

Christie, J. (1974) *The Empty Shore. The Story of Cowie, Kincardineshire,* Stonehaven.

Cramond, W. (1888) *The Annals of Cullen* (2nd edn.), Buckie.

Cramond, W. (1893) *The Making of a Banffshire Burgh. Early History of Macduff,* Banffshire Journal Office, Banff.

Cranna, J. (1914) *Fraserburgh Past and Present,* Rosemount Press, Aberdeen.

Cumine, J. (1887–90) 'The Burgh of Rattray', *Trans. Buchan Field Club* 1, 114–121.

Czerkawska, C. L. (1975) *Fisher Folk of Carrick,* Molendinar Press, Glasgow.

Dorian, N. C. (1985) *The Tyranny of Tide. An Oral History of the East Sutherland Fisherfolk,* Karoma, Ann Arbor, USA.

Edwards, D. L. (1921) *Among the Fisher Folk of Usan and Ferryden,* Advertiser Office, Brechin.

Garden. A. of Troup (1907) 'An Account of the Northside of the Coast of Buchan' in *MacFarlane's Geographical Collections* II, T. and A. Constable, Edinburgh, 133–143.

Gray, M. (1978) *The Fishing Industries of Scotland 1790–1914. A Study in Regional Adaptation,* Oxford U.P., Oxford.

Hay, E. and Walker, B. (1985) *Focus on Fishing. Arbroath and Gourdon,* Abertay Historical Society Publication no.23, Dundee.

Hay, G. (1899) *A History of Arbroath,* T. Buncle and Co., Arbroath.

Hutcheson, G. (1887) *Days of Yore,* Buckie.

Bishop Leslie (1578) 'History of Scotland', in Brown, P.H. (1893) *Scotland before 1700 from Contemporary Documents,* David Douglas, Edinburgh, 113–183.

List of Pollable Persons in the County of Aberdeen in 1696 (LPPCA.) (1844) Spalding Club, Aberdeen

McGowran, T. (1985) *Newhaven-on-Forth. Port of Grace,* John Donald, Edinburgh.

MacIver, D. (1906) *An Old Time Fishing Town: Eyemouth,* James McKelvie and Sons, Greenock.

Morgan, D. (1993) *The Villages of Aberdeen. Footdee,* Denburn Press, Aberdeen.

Mowat, I. R. M. (1981) *Easter Ross 1750-1850: the Double Frontier,* John Donald, Edinburgh.

Munro, J. (1982) 'Pulteneytown and the Planned Villages of Caithness', in Baldwin, J. R. (ed.) *Caithness. A Cultural Crossroads,* Edina Press, Edinburgh, 130–159.

Munro, J. (1989) 'The Planned Villages of the British Fisheries Society', in Smith, J. S. and Stevenson, D. (eds.) *Fermfolk & Fisherfolk,* Centre for Scottish Studies, University of Aberdeen, 50–62.

Murray, J. E. L. (1964) 'The Agriculture of Crail, 1550–1600', *Scottish Studies* 8, 85–95.

New Statistical Account (NSA.):- XII, 213–236 (Lonmay Parish). XII, 258–274 (Aberdour Parish). XII, 344–396 (Peterhead Parish). XIII, 220–244 (Boyndie Parish).

Ochterlony of Guinde (1907) 'Presbitrie of Brechin' in *MacFarlane's Geographical Collections*, II, T.and A. Constable, Edinburgh.

Old Statistical Account (OSA.):- VI, 15–20 (Rathen Parish). VI, 417–435 (Tarbat Parish). VII, 194–217.(Nigg Parish). XIII, 392–427 (Rathven Parish). XV, 610–639 (Avoch Parish)

Pryde, G.S. (1937) Ayr Burgh Accounts 1534–1624, *Scottish History Society*, Constable, Edinburgh.

Registrum Magni Sigilli Scotorum (Register of the Great Seal of Scotland, or RGSS.). Vols. II (1424–1513) (1882), III(1513–1546) (1883), VI (1593–1608) (1890), ed. Paul, J. F. and Thomson, J. M., H. M. Register House, Edinburgh.

Scottish Burgh Records Society (SBRS.)(1876) *Extracts form the Records of the Burgh of Glasgow 1573–1642*

Scottish Record Office (S.R.O.) *Panmure Estate Papers GD* 45/18/276; 45/18/296;45/18/313.

Storrie, M.C. (1981) *Islay: biography of an island*, Oa Press, Port Ellen, Islay.

Summers, D.W. (1988) *Fishing off the Knuckle – the Fishing Villages of Buchan*, Centre for Scottish Studies, University of Aberdeen.

Third Statistical Account (T.S.A.)(1965) *Morayshire*, Oliver and Boyd, Edinburgh.

Taylor, J. (1982) *Rosehearty: Its History as a Fishing Town*, Rosehearty Heritage Society.

Watson, H.D. (1986) *Kilrenny and Cellardyke. 800 Years of History*, John Donald, Edinburgh.

5

The Early Herring Fisheries

INTRODUCTION

The greatest fish stocks in the waters around Scotland were always those of the herring. It is inherently probable that the earliest fisheries for them were at various parts of the coast for local use, but when they do enter the historical record it is primarily in a commercial context. While there were organised commercial fisheries for herring in Scotland and elsewhere from Medieval times, the full potential for herring fisheries was not appreciated until the great rise of the Dutch North Sea herring fisheries from the 15th century. For long the main Scottish herring fisheries were concentrated in inshore waters in the southern parts of the country, but they also extended to the West Highland coasts by the 15th century; and here one of the incentives was that there was less competition from the fishermen of Holland and other European nations.

None the less the greatest potential always lay in the open sea fisheries of the North Sea, and these became a major source of wealth for the Dutch, and excited the envy and emulation of all North Sea countries. Nowhere was the feeling on this as keen as in Scotland, as the great part of the Dutch fishing effort was conducted off the Scottish coasts, especially in Shetland waters. A consequence was strained relations between Scotland and Holland, as the Scots took various measures aimed at restricting the Dutch, while trying to rival them.

Great as was the potential of the herring fisheries, they were always prone to sizeable fluctuations; in addition, as a fatty fish they were particularly prone to rapid spoilage after catching, and necessitated the mastery and monitoring of the technique of salt curing in barrels: the Dutch were the leading exponents of this, but it was for long a major problem in Scotland.

EARLIEST DEVELOPMENTS

The earliest evidence for herring fisheries is in Medieval documents, although there must be a high probability of fisheries on some scale being much older. The main gear with which herring are known to have been fished is nets, which is the best means for catching such a shoaling species; and while nets are known in southern Scandinavia from Mesolithic times, the very general absence of evidence of herring in prehistoric coastal sites has led to the conclusion that their fisheries originated later. A main reason for this has been suggested as the considerable extra labour needed in making nets was not warranted in small coastal communities (Clark 1952: 56, 89–90). While some signs of prehistoric herring fishing have been detected in Scandinavia, it has been generally accepted that the Dark Age is the most likely time for them to have started on any real scale. By the Medieval period they are known at various places and in a number of countries around the North Sea. In Scotland the right to fish for herring on the Firth of Clyde is specified in a charter given by King David I to the Abbey of Holyrood in the 1138 (APS. vol. 1 (1124–1423), 1844: 368). Herring also feature specifically in the Parliamentary records as early as 1240, when the dues payable for herring entering into trade in the burghs are itemised at the rate of 4d per last on fresh herring (Mitchell 1864: 132, 133). However the main early source in which herring are recorded is the Exchequer Rolls, where they are fairly prominent from the 13th century onwards. Herring from Inverness, Crail, Ayr and Dumbarton are mentioned as early as 1266 (ERS. I (1878): 19). It is clear, however that the Firth of Forth was the main location of herring entering into commerce: Crail was particularly important as a centre of activity in the trade, and it is repeatedly mentioned in the 14th century: and nearly all the coastal burghs on the Forth were recorded as being active also. It is evident too that by the 15th century there were also important fisheries on the Firth of Clyde, with Dumbarton and Ayr being prominently involved. There are indications from the earliest records that part of the catch was cured, with mention of salt for curing, 'red herring', and 'herring houses', which are most likely to have been curing or storage premises.

There are various parts of the Scottish coasts where herring come close to the coast in their life-cycles and migrations; and there are also local herring stocks of more limited dimensions. The abundance of food contained in such a shoaling fish is inherently likely to have stimulated fisheries for it, although what part of the catch went to local use must for long be a matter of conjecture for want of information. When the situation becomes clearer in the 17th century, there was a substantial consumption around both the Firths of Forth and Clyde, and local use appears to have accounted for the greater part of the catch; and this may well have been the case for several centuries previous.

The earliest records from the 12th century show that rights to herring fishing were vested in the Crown, and herring was recognised, along with salmon, as the most important fish species. While direct information on the activity of fishing is scarce, the records of trade, particularly as shown in customs, give some gauge of its importance, and in time these are supplemented by a variety of other sources. The descriptions and collections of information from around 1600 onwards, as part of more organised attempts at national stock-taking and in connection with map-making, are also useful and give additional glimpses of some places; they also on occasion give some detail of fishing itself. In the fisheries on the Forth, there is the record of the granting of a right to fish herring in the 12th century by David I to the Abbey of Holyrood (Cochran-Patrick 1892: 71); and by the 15th century there is also evidence of activity on the Clyde (Cochran-Patrick 1982: 71–72). These sheltered firths allowed the exploitation of the fisheries inshore on a seasonal basis, and their importance was certainly enhanced by their location at either end of the most densely populated central belt of the country.

The main gear used in the fishery was evidently nets, although how they were deployed in early days is less clear. As well as nets, even for a soft-mouthed species like the herring, there was some use of hooks to catch them, and this continued in some measure into the 19th century. The nets are most likely to have been made of linen, and may well have been made by fishing families in fishing settlements, as again occurred into the 19th century, before the innovation of factory-made nets. The method of drift-netting, in which nets are put in the sea hanging down from the surface (fig. 5.1), would fairly certainly have been used to some extent, as it was in general the main method used until relatively recent times. However, in shallow water close to the coast, it could have been supplemented by the use of fixed or anchored nets. There could also have been the use of sweep-nets, or seines, as were used in later

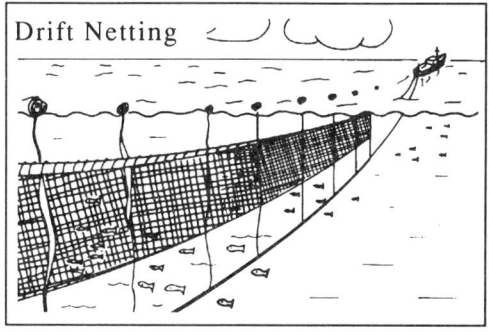

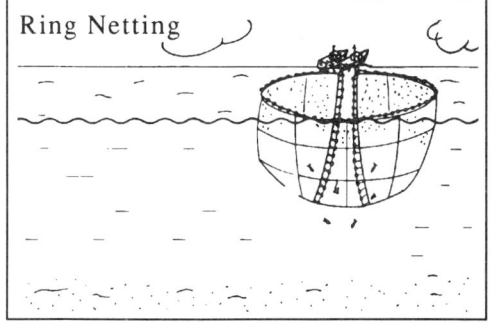

Fig. 5.1 Main methods of herring fishing.

times on the Norwegian fjords, or for that matter on the Clyde. With the strongly seasonal regime of the fisheries, it is very likely that many (or most) of the personnel involved were part-time, and would have been occupied at other times of year in other activities. As well as other fisheries like salmon and white fish, these activities could also have included various other occupations. To judge from the limited information available, and from what happened in later herring fisheries in several countries (including Scotland itself), those involved may well have been mainly men from farms and ferm-touns, especially if they had limited land. However the fact that the main herring fisheries were in the autumn must have brought them into some conflict with the demand for harvest labour. When the situation starts to become clearer from the 15th century, it is evident that although the scale of operation of the individual boat was inevitably limited, there could be very considerable aggregate effort of hundreds of boats directed at the fisheries in the season. The fact that in herring bumper catches were always a possibility must have been a very considerable incentive when the mass of the population was poor.

Herring entering into trade was usually taxed as a source of revenue. How far it was possible to administer and collect the duty imposed from 1424 of one penny on every thousand fresh herring sold in the country is doubtful. However it is clear that duties imposed at the same time did yield some significant income, at least to the burghs on the Clyde: these duties were at the rate of four shillings for each last of 12 barrels cured by Scotsmen, and six shillings when cured by strangers; and fourpence per thousand red herring (Cochran-Patrick 1892: 71–72).

While the herring fisheries were for long an important source of wealth, there were formidable and persistent problems in organising them to the best effect for commerce. Inshore herring fisheries are particularly irregular, because of variations in the path of migration as well as in abundance, and catching capacity with the boats and gear available must have been limited. However, the great problem for a nourishing but fatty fish which easily spoiled after catching was to have it properly cured for out-of-season use and for transport. Traditional curing techniques included both the preparation of white and red herring. Curing of white herring involved salting direct into barrels, the herring being packed in layers with intervening layers of salt; and for best keeping quality it would have involved gutting before packing, although this appears often to have been omitted. Preparation of red herring, on the other hand, entailed less use of salt, but meant prolonged smoking over wood fires for a month or more before packing in barrels; and while this was done in Scotland, red herring were in fact more a product of the English East Anglian autumn fishery, where the herring were less fat.

The variability of the fishery posed a constant challenge in gearing the numbers of barrels and amounts of salt to the catches; and when big peaks occurred in the landings, curing often had to be done in great haste. This was one of the reasons for the apparently frequent failure to gut the herring before curing; and even as late as the 19th century it was known for herring to be gutted without the use of knives, but by opening the bellies with the hand. There were further major problems in getting barrels and salt of adequate standard for curing. While the Convention of Royal Burghs tried to stipulate the use of barrels of standard size, both size and quality of construction were variable. In addition, it was necessary to import the barrel staves and hoops from Scandinavia, and this was another problem in maintaining adequate supplies. For salting herring it was always desirable to use the superior quality solar salt from the southern European countries, rather than the native Scottish product which was made by coal-fired evaporation at the coast, and was often contaminated with various impurities. While some solar salt did reach Scotland from the 14th century onwards (Ewan 1990:89), supplies were scarce and notably irregular, trade flows often being interrupted during international power and naval struggles.

It was above all in the organisation of curing that Scotland, and other European countries, for long lagged behind the Dutch.

INSHORE HERRING FISHERIES IN THE 17TH AND 18TH CENTURIES

As in so many fields in Scotland, information on the herring fisheries expands from the 17th century, and considerably more detail becomes known of their scale and importance. While the main locations of activity continued to be the more densely populated southern parts of the country, there was increased interest in them in the north, much of it associated with initiatives from established places on the Forth and Clyde.

It was claimed that the great 'Lammas' drave on the Forth in the 17th century gave great employment to fishermen and shore workers on both sides of the firth, particularly in Dunbar and Pittenweem. This fishery appears to have been in origin mainly a 'ground drave', in which nets were attached to posts in shallow water at the coast, and the fishery lasted only a few days in the year (Gray 1978:18), but it also involved drift nets by the early 18th century. Sir Robert Sibbald's 'Discourses' document from 1701 is very valuable for the details it gives of this time (Sibbald 1701: 50–52). It lists six places on the south shore and ten on the north which were much involved in the fishery; and it is apparent that the main catching power was on the Fife side, where the bigger boats used were more than double the size of other inshore boats and had crews of seven or eight men, compared with the four- to six-man crews in other parts of the country. With the evidence from earlier times, it appears that this was the part of the country that had long been in the van in herring fishing. Evidently there had been sufficient success at the fishery to generate the extra capital to get bigger and better boats. It is probable too that these were the boats which are recorded elsewhere as fishing at the North Isles and in the Minch at other seasons (ch. 13). At the drave there were 104 of these boats manned by a total of 746 men, and 'countrymen' joined regular fishermen in making up crews during the season; and this supplementing of regular crews in the seasonal labour-intensive fisheries is also attested in various other herring fisheries. There was also a greater but uncertain number of smaller boats which also took part in the fishery.

Details are also given for the bigger boats of the share system of dividing up the proceeds from the fishery and this is an important precursor of the systems used in later times. Expenses were first deducted, and the money was then divided into 20 shares. Two shares were allotted to the teind, each crew member got one share, and a share was allotted to each of the nine nets, while the boat owners got the remainder, i.e. two or three shares. This system is typical in the high proportion of the returns allocated to the gear, which is a reflection both of its high relative value and of the high rate of depreciation. The proportion of the proceeds given to the boat owners appears relatively low, but to judge from the practice of later times may well be misleading. The regular fishermen may have owned the boats wholly or in part, as they regularly did in later times, and they probably owned the gear: hence their income from the fishery would be in several components, as in addition to their labour share they would have stood to make additional earnings in proportion to their ownership stake in boat and gear.

There was a high level of consumption of fresh herring around the Forth, and fresh herring were also sent in bulk as far as Newcastle. There was also much curing for home use and for export. Adair recognised that the early season herring were better suited to smoking than to pickle curing; and he estimated the total amount shipped at 1200 to 1600 lasts (*c.*14,000 to 19,000 barrels) annually, in addition to the high local consumption (Adair 1707: 37). One merchant estimated the total profit in a year form the fishery as high as £40,000 stg. (Sibbald 1701: 49): it was such returns that no doubt stimulated Sibbald to be forward-looking and comment on the need for proper 'work houses' for the curing of herring, refining of salt, and making nets, sailcloth and cordage; and of the need for 'magazines' for the proper storage of all of these (Sibbald 1701: 61). While such advances were eventually made, they had in the main to wait until the 19th century.

The fishery on the Forth went into decline in the 18th century, and was at a low ebb between for half a century from about 1725.

The primary cause of this appears to have been a period of reduced size or changed migrations of the herring stock: this is the sort of fluctuation that is better known in the coastal herring fisheries of Scandinavia. There were also economic readjustments after the 1707 Union, and the low period in the Forth herring fishery has also been attributed to the imposition of the tax on imported salt after 1707. In addition, the rise of the fisheries on the Clyde may have helped depress it by generating more competition on the markets, especially the Irish one. The Irish market had risen to a leading position as an outlet for Scottish herring, and the Clyde was more conveniently located to export to it. It is possible that the Fife fishermen, some of whom had already developed a significant degree of mobility in fishing herring elsewhere, tried the summer fishery elsewhere at this time during this period of problems and readjustment. An increase in interest in herring fishing reported form this time on the Moray Firth, and by tradition derived from Fife, may well be connected.

The Clyde was better for inshore herring fishing than the Forth, with a more reliable herring stock and a greater area of sheltered water on which to conduct the fishery: and Loch Fyne especially was a main place of operation. It does, however, appear to be later in time in coming to the fore, and this is in accord with the main concentrations of population and the leading towns being in the east of the country until the 17th century. Even so by the 15th century there were burghs on the Clyde, including Irvine and Greenock, which played a key role in the organisation of the fishery, and it involved a major effort in coastal communities generally during the season. While behind the Forth in standards of boats and equipment, there were by the late 17th century up to 600 (Sibbald 1701: 49) or even 900 boats (Adair 1707: 43). In the light of the known exaggeration in data on the herring fleets of the time, including those related to the Dutch, such general figures may well be over-estimates. Even so, they clearly indicate a very considerable fishery, and it is also recorded that the boats typically had four-man crews, each one providing an equal share of nets. While the share system is described in less detail than that on the Forth, it appears that there was a share allocated to the boat, and there were cases where the skipper got three shares from the proceeds, in virtue of owning the boat and providing the provisions. It was also stated that in a good fishing the earnings after expenses could reach £12 to £15 stg. to the share (Sibbald 1701: 49). This for the time represented a spectacular level of income, and it no doubt gave a great impetus to the growth of the fishery. The main centre was Greenock, and in addition to a great consumption of what was a main food staple in the counties adjacent to the Clyde, this gave a cure of 2,500 lasts (30,000 barrels) for export. At the end of the 18th century there could be at least 500 boats in the season on Loch Fyne alone, but the crews were still of four men (OSA. V: 292). In fact the main initiatives in the move to larger boats for the boat fishery were taken on the East Coast, where there was a greater advantage of having boats which could work on the open sea in more difficult weather.

Herring appear to have been fished sporadically on other parts of the coast than the Forth and Clyde, but with little sign of a sustained and regular organisation of curing. The main instance of this was the Moray Firth coast, where there is a tradition that Fife men introduced the technique of drift-netting for herring before 1700 (Hutcheson 1887: 34).

In the late 17th and early 18th centuries the town of Cromarty had a prominent phase of prosperity in the time known locally as the 'herring drave' (Miller 1828: 4–5). This was based on the summer inshore fishery, and although the fishery failed before 1730, evidently due to changes in the migration pattern of the herring, it was also taken up in Nairn (Bain 1828: 246). There was also a winter fishery on the inner Moray Firth by the later part of the 18th century (Pennant 1771: 178). Here Avoch in Easter Ross emerged as a main centre, and boats also came from a range of other places; in 1795 it was recorded that the fishery commanded the attention of 60 to 80 boats (OSA. XV: 628). The men in Avoch itself also engaged in the summer herring fishery on the Caithness coast, and they needed two sets of nets, as the winter herring were smaller, and had to be taken with a smaller mesh. The fishery developed on a bigger scale at Buckie (Hutcheson 1887: 18) where the men operated in four-man boats with one net each, but in a good herring season might make as much as they did in the rest of the year from line fishing. It did however involve considerable extra labour, the nets being spread out to dry almost every day to minimise deterioration from the repeated exposure to salt water (Hutcheson 1887: 18, 35). As an inshore fishery, the extent to which it was prosecuted probably varied with the abundance of herring in individual years.

It was after 1770, when incoming merchants from the Firth of

Forth triggered growth in Caithness (Gray 1978: 29) that the herring fishery took firmer root in the Moray Firth area; and Caithness for a whole century after this was to be the leading area in the country and attracted boats from much of the Scottish coast. The increasing interest and tempo of activity from the 1780s made a big impact on the fishing settlements of the Moray Firth. In the Buckie area especially, where previously the herring fishery had only been prosecuted in the late summer after the cod and ling fishery, it now became a key part of the annual round for a fleet of 30 of the bigger six-man boats of *c*. 10 tons which had been built up for line fishing (OSA. XIII: 400–402). In the 1790s there was already a fleet of 200 boats working at Wick in the herring season (OSA. X: 9); and the growing success of the fishery is attested by the total of between 200 and 300 herring boats with seven or eight man crews recorded in the Moray Firth area in the early 19th century (Parliamentary Papers 1803: 294–295): there was sufficient profit to allow investment in the bigger class of craft which was a signally important factor in the expansion of the fishery. The Old Statistical Account for Rathven parish, which includes Buckie and neighbouring settlements, gives important extra details of the organisation of the fishery in Caithness. It is evident that there were incentives given by the curers who effectively controlled the fishery, and there were two alternative types of arrangement. The one which gave more security to the fishermen gave each boat a bounty of £8 for engaging with the curer, together with 5/- arrival money at the start of the season and another 5/- at the conclusion; and 2/- was given for the 'Saturday pint' as well as a bottle of whisky a day. However under this arrangement there was apparently no obligation on the curer to pay a fixed or minimum price for the herring delivered to him. The other type of arrangement, which gave each boat 10/- per barrel for its herring along with the daily bottle of whisky (OSA. XIII: 403), obviously gave the fishermen greater incentive to catch more, and must have been an attraction to stronger-going crews. It is noteworthy that when the Caithness fishery began its vigorous expansion from this time, it was this second type of arrangement (or engagement) that became its main organisational basis. The regular inclusion of whisky in bargains with curers allowed crews in effect a dram a day in work that even in summer involved considerable cold and discomfort.

EFFORTS AT EXPANSION OFFSHORE

Small open boat fisheries were part of the pattern of operation in herring fisheries well into the 19th century, but there were ambitions from at least the 15th century to operate on a greater scale. The great precedent and model for this was the Dutch, and for long it was accepted that the best prospects lay in fitting out bigger boats to carry stores of barrels and salt as well as nets, and operate offshore to seek out the herring in deeper waters. There were aspirations both to rival the Dutch in the North Sea, and to develop the herring fisheries of the West Highlands and Hebrides, where there was less challenge from the Dutch and other continental nations. Envy of the Dutch led to direct conflict with them as well as with the English, and there was the attempt in the 17th century to lay national claim to the fisheries off the coast; as well as reserving all bays and firths for home fishermen, this also asserted claim to all fishing within a 14-mile limit (Cochran-Patrick 1892: 74). The impetus to exploit the Minch fisheries has also been seen as the main motivation for the increased effort to extend more effective national control to the West Highlands (Grant 1930: 540–542).

Dutch success in the herring fishery was in fact only part of their performance as a leading maritime nation, which had major interests in international trade and also was a leading power in whaling. In the herring fishery, their pre-eminence was dependent on a series of inter-related factors. There was in the first place the technique of catching herring with the drift net in the open sea by decked busses which could seek them out wherever they happened to be in a particular season; added to this was an expertise in curing, which involved gutting the fish immediately on catching, and packing them in standard barrels with solar salt. In organisation there was a legally defined season beginning on 24th June, and this prevented the herring from being caught too early in the year: in the early season they were unsuitable for salting, before they were full from feeding. From the start in late June off the Shetland Islands fishing moved in a general southerly direction as the season advanced, and it culminated with the autumn fishery in the southern bight of the North Sea (Beaujon 1883: 357–358). In addition there was a government system of protecting the herring fleet at sea with escorts, and inspection of the quality of the cure, which stipulated the use of barrels of standard size and of appropriate amounts of salt. There was an integrated system of marketing in the main consuming area of the North European Plain, which involved the organised obtaining of market intelligence, and the gearing of supply to demand (Unger 1980: 243–279). In contrast Scotland in the European context was for long a small, poor and backward country which lacked the technique, capital and organisation to rival the Dutch; and although there were periodic efforts to promote and develop the deep sea herring fisheries, in the main they had restricted success.

For long the Scots and others were dazzled by the success of the Dutch, and much of the literature produced in Britain related to the Dutch fishery much exaggerated the size of the Dutch herring fleet and their catches; and right until the early 19th century the main thrust of attempts to develop the fishery focussed on catching by busses and on-board curing. There were also various attempts to recruit skilled Dutch fishermen to help stimulate the industry in Scotland. The development of a buss fishery became a longstanding national preoccupation and repeated attempts were made to promote it, although they of course varied in intensity in view of other national priorities and exigencies at different periods. The main impetus for the bigger-scale ventures came from the burghs, as is clear in a number of enactments of the Scots Parliament. The first recorded legislation to this end appears to date from 1471, when it was specified that certain lords and burghs should provide ships and nets for the fishing (ALCBS. 1910: 34), and measures to promote and regulate the herring fisheries thereafter came at relatively frequent intervals. The Convention of Royal Burghs was the leading national body engaged in promoting the fishery: it was in a key position, as all trade in herring, at home as well as abroad, was officially in the hands of registered merchants. In 1493 the emphasis on bigger boats was specified when it was enacted that all burghs should fit out boats of at least 20 tons, in number according to the burgh revenues, and that able-bodied idle men be pressed to crew them. Evidence to show how effective such measures were is scarce, but repeated attempts to enforce them into the 17th century suggest that success was limited. None the less the expansion of the deep sea herring fishery was a continuing preoccupation; and what effort there was in the earliest phases seems to have come mainly from the burghs on the coast of the Forth, although burghs on the Clyde also fitted out busses to fish in the north-west. Certainly by 1600 from the East Neuk of Fife, which was to be the main seat of the fishery, fleets of half-decked 'crears' were sailing to fish herring round the Western and Northern Isles (Watson 1986: 33–34), and there are also various other references to other boats going to the north-west from towns on the Clyde. A passing mention in the Aberdeen regi-ster of births states that a fleet of a hundred sail arrived at the town's roadstead in January 1587 on its way back from Loch Broom (*Analectica Scotica*, vol. I 1832: 286), and this suggests that the scale of effort even in the 16th century could be relatively substantial. How these boats compared with the Dutch busses is unclear, but the crears seem to have been something in between the usual Scottish inshore boats and the Dutch busses; and it is on record that they also joined the small craft in the seasonal inshore fishery on the Forth.

The main meaning of the word 'crear' is 'a small trading vessel' (DOST., vol. I, 1937: 735), and this suggests that these may have been vessels for which trade was also an important function. The amount of effort put into fishing may well have had an inverse relationship with trade opportunities: trade in general gave more predictable returns than fishing, although fishing must always have had the attraction of the occasional bumper season. Trade was generally at a low ebb in the winter part of the year, but it was pos-sible at that season to fish herring in the Minch, and this might well explain the visit of a hundred boats mentioned above to Aberdeen in 1587. The importance of the dual function of fishing and trading for Scottish vessels is corroborated by a study at the Norwegian end of the timber trade, which was particularly important in Scottish overseas commerce in the 16th and 17th centuries. Although the Scottish boats involved were more numerous than those from any other European country, they were conspicuously smaller, and many were also used for fishing: many of the Scottish boats were of less than 30 tons, and some were even as small as eight tons. It is also significant that the item most represented in the cargoes to Scotland was barrel hoops which would have been used mainly in fish curing; boards as well as beams were also included and presumably went partly to supply barrel staves; and the main desti-nation was the Fife ports, which played a leading role in fishing (Lillehammer 1986: 101–105). The fact that these crears were able to operate both at fishing and trade represented an element of diversification which must have helped to maintain a continuing sea-faring tradition.

Various of the enactments from at least as early as 1487 refer specifically to the herring fisheries of the West Coast (ALCBS. 1910: 44). Here the main fisheries were at first in the Clyde, and already in 1555 the West Coast burghs had long fished for herring (ALCBS. 1910: 80); and there was concern about the fisheries in the 'west sea' and those of the north and west in 1587. There were additional problems attached to developing a fishery in the north-west: as well as being sparsely populated, remote and Gaelic-speaking, the area had almost no towns to function as commercial centres. With the commercial fishery as a monopoly of the royal burghs, development of the fisheries of the West Highlands had the additional complication of being perforce by remote control; and the evidence shows the concern of the Lowland burghs, especially those on the Clyde to keep the trade in herring in their own hands.

How much success there was in these early attempts to develop a buss fishery is unclear, although by the 17th century there seems to have been a relatively large number of busses of some kind involved on the West Coast. About 1630 it was claimed that there were over 200 'cowper' boats: while these were big boats, the fact that they were dependent on up to 1500 or more small catcher boats to provide them with herring (NLS Library (Edinburgh) MSS. 31.2.16) shows what was to be a persistent weakness in the organisation: the fishery was still essentially an inshore one, and there was little attempt to develop the Dutch type of open sea oper-ation in which the busses themselves did the fishing. It may also be questioned whether the numbers of boats involved in Scotland were as large as claimed: in the absence of any central record, such numbers could be impressionistic. It is also recorded in the same manuscript that this fishery was directed at supplying the home market, rather than being much of a source of export earnings.

There is however a description from the early 17th century of the manner of operation of 'busses' based on the Forth (Knox 1785: 189–190), which does indicate a more competent level of operation and which made better use of available opportunities. Whether these 'busses' were different from the crears mentioned above may be doubted; but it is clear that these vessels had evolved a pattern of working that effectively maximised the fishing oppor-tunities around Scotland. Indeed it anticipates in several important respects the pattern of operation in the 19th century when Scotland became world leaders. It entailed a whole-year cycle, which effec-tively began in the month of March, when the busses set out for two-month trips to the Orkneys. The spring has always been a time when herring are scarce, and at this season these boats caught her-ring mainly as bait for fishing for cod, which were caught by hand line, gutted and salted on the boats, and dried on Orkney shingle beaches before being taken back to the home ports. Cod always had higher unit value than herring, and in spring it was more profitable to use what herring were available as bait. The boats then in early June took aboard herring nets, salt and barrels, and set out to fish at Shetland; and if the fishing went well they might return to port with all their barrels full, and take on fresh nets, salt and barrels to continue fishing until the end of July, when they returned home and took part in the home herring season at the entrance to the Forth in the company of a big fleet of small boats in August and early September. When the home season was over, they went to the herring fishing on the West Coast lochs, which lasted till Christmas. After this the busses were laid up for over two months for maintenance; in these first months of the year the men did spend some time in lining for white fish in small boats, but a major concern is almost certain to have been work on their herring gear: herring fishing always involved much wear and tear of gear, and the nets were frequently as valuable as the boats from which they were employed. It is recorded that the busses changed gear between their different fishings, and while no doubt some net mending was possible during and between seasons, experience of later times suggests that there would have been a considerable back-log by the New Year. Possibly most significant, the unknown author of this document asserted that 'by this practice the men became the most expert fishers in Europe'. While this claim is no doubt extravagant in face of the pre-eminence of the Dutch at the time, it does accord with fishermen on the Forth acquiring a leading position in Scotland, and a degree of mastery in the fishery which was to be a key element in the eventual success of the fishery.

While Scotland had something of an established interest in the herring buss fishery, there was evident and official dissatisfaction with the limited achievements in it, and from the 17th century onwards fishery development was increasingly seen as a task for specialised bodies committed to it. In the pursuance of this there were several attempts to found companies dedicated to the purpose in the 17th and 18th centuries (Dunlop 1978: 7–14): and there were also other organisations, including government boards, which sought to foster the fishery. There was a general increase in the tempo of the development efforts in the post-Restoration period from 1660, in broad accord with advances in a number of other fields.

From the late 17th to the early 19th centuries, the promotion of the fishery was seen as a proper field for government financial incentives. While there was always considerable debate on the best form of incentives, policy was for long dominated by the objective of increasing exports: this was done both by giving export bounties and by exempting from the heavy salt duties salt which was to be used for curing herring for export. In the second half of the 18th century especially there was to be an added momentum to the development efforts, and the continuing debate on the best methods of promotion is reflected in government reports and other sources. One of the by-products of the introduction of government incentives in the form of export bounties, together with the tonnage bounties to fit out busses, was the production of official reports, together with the gathering of a considerable amount of statistical data in satisfying government requirements of accountability: these give information on a new level of completeness and detail.

In the 17th century there were two major efforts to generate a greater measure of success through the setting up of Royal Fishery Companies, which were to have trade monopolies an incentive and protection for expansion. The first of these was the outcome of an Anglo-Scottish Commission and was floated in 1632. It was grandiose in its objectives, and envisaged the construction of 200 new boats of from 30 to 50 tons which would fish on a year-round basis, the spring time in between herring seasons being filled by trips to the lines for white fish (Elder 1912: 36–37). Such a plan was obviously intended to emulate the Dutch method of operation: it was unduly optimistic on raising capital, and the Civil War in the middle of the century supervened. It was clear within a few years that very little had been achieved and that the company was deep in debt. Later, after the Civil War the Royal Fishery Company of Scotland was instituted in 1670, but also had little success and was wound up in 1690 (Elder 1912: 108–115). In fact both of these companies were badly under-capitalised and never got properly started; and the background troubles on the international scene, with friction between Britain and Holland and the war between Holland and France gave added problems. It also appears that what busses were acquired were none the less built not in Britain but in Holland, and manned mainly by Dutch fishermen (Jenkins 1927: 87).

After the Union of 1707, there was a fresh initiative for economic development in Scotland, although for the fisheries there was in effect to be a full century more of groping for a suitable policy to promote them. There was a will to provide government funds in the form of underwriting of approved ventures, investment bounties for the fitting out of bigger boats, and export incentives in the form of barrel bounties; but there was also the big complication of the tax on imported solar salt to protect the home (especially the English) salt industry.

An important step was the setting up in 1727 of the government Board for Manufactures and Fisheries, which by the terms of the Act of Union had £6,000 per year to dispose of; and it was resolved form the start that the main bulk of these funds would be used to promote the herring fishery and the linen industry (Hamilton 1932: 79). This board then took over from the Convention of Royal Burghs the control of curing and packing herring. Although the Board for Manufactures and Fisheries gave premiums to reward successful fishing ventures from 1727, this did not stimulate any immediate success. By mid-century there were arguments being advanced to provide greater incentives to encourage the fishery: the grounds of these were that the Dutch were now at a low ebb following war losses, and that a deep sea fishing fleet was necessary to provide a reservoir of trained seamen for the navy.

A result of this increasing pressure was that arrangements were made to provide herring busses with duty-free solar salt, although these were to generate a tangle of complex regulations which were to limit their effect. A more substantial move was made in 1750, when it was decided to award tonnage bounties of 30/- per ton for the fitting out of busses, and at the same time to establish a new company which would have capital aid from the government, but which would not have a monopoly in the fishery: this was the Society for the Free British Fishery. While many new busses were built for the Shetland and Yarmouth fisheries, it appears that the Society abused the government grants it had been given, while the outbreak of the Seven Years War in 1756 also hampered progress; and this proved to be yet one other effort which had little long-term effect. However the subsequent government decision to raise the level of bounty from 30/- to 50/- per ton in 1757 did encourage a flurry of private investment, especially in the Clyde area of Scotland, and there were over 200 busses operating in the last 30 years of the century. There was an expansion in the number of busses to a peak of 294 in 1776, after which the outbreak of the American War of Independence disrupted the important West Indies market; and it also led to increases in the cost of necessary materials and to changed priorities for shipping. In addition there were problems in the arrangements for paying the bounties, and after the raising of them to the level of 50/- per ton, in only two years was the full amount paid in Scotland (Dunlop 1978: 10–11).

PROMOTION OF THE 'BOAT' FISHERY

However, as a result of hard-won experience there were adjustments to the system of incentives late in the 18th century, which subsequent developments showed to be important, and indeed decisive. A new act of 1786 reduced the tonnage bounty to 20/- per ton, while giving busses a bounty of 4/- per barrel for herring satisfactorily cured. More important, for the first time it brought the 'boat' fishery within the orbit of incentives by offering for it a bounty of 1/- per barrel; and in 1795 this was raised to 2/- per barrel. In 1787 the importance of the small boats was further recognised by the sanctioning of a practice that had hitherto been illegal – that of selling their catches to busses (Hamilton 1963: 120, 117), and this was of particular value on the West Coast. While the rate of outlay for the 'boat' fishery was effectively less than half that for the busses, it is evident that the new incentives gave it a marked impetus, and the more complete information gathered before the end of the century shows that it was definitely dominant: in 1798 285 busses produced a total of 84,942 barrels, while an undisclosed number of craft in the boat fishery yielded 134,172 barrels (RRBF. 1798: 372–375).

While there were obvious results of the bounties, there was always controversy as to their efficacy. As well as the abuses and inefficiencies recognised by an authority like John Knox (1785: 52-54), they were subjected to the scrutiny of no less a figure than Adam Smith. In a cogently argued passage, he showed the limited effects that the buss bounties had had in stimulating the fishery. He focussed especially on the eleven-year period from 1771 to 1782, by which time the system was past any trial period. Herring caught by the busses was costing the government well over their sale value per barrel, and in an oft-quoted utterance, he stated that 'it has . . . been too common for vessels to fit out for the sole purpose of catching, not the fish, but the bounty'. He also claimed that the encouragement of the buss fishery had been at the expense of the boat fishery, which he said had been ruined. The boat fishery was also the main supplier to the home market, and the buss bounties had done no good to the many common people who depended on herring for much of their diet (Smith 1896, II: 20–25). The voice of Adam Smith was no doubt influential in the 1786 decision to extend the barrel bounties to the boat fishery.

In fact, despite the proven success of the Dutch, there was always some support for the encouragement of the boat fishery. As the various attempts to promote the buss fishery continued to give disappointing results, the arguments in favour of the boat fishery were reinforced. The main force of the argument was that whereas the Dutch had perforce to operate at hundreds of miles form base because of the location of the herring, the Scots could catch herring close to their own coast, which made possible the much simpler and cheaper expedient of on-shore curing. There were always limits to the amount of barrels and salt which even decked boats could carry, and also to the amount of the eventual catch they could take aboard and store; but these constraints scarcely applied to curing stations on shore.

By the late 18th century, there were spontaneous initiatives and innovations on the part of merchants and fishermen, in addition to government incentives which were starting to generate a new dynamism in the boat fishery. Prominent here was the crews from Buckie and from Avoch who were going to the summer fishery at Caithness for six weeks, where they had the security of engagements with curers.

Another element in the strategy of fishery development was the founding of special settlements for the purpose, particularly on the West Highland coast. This area was almost wanting in centres of commerce, or indeed in population concentrations beyond the size of ferm-touns. Fishing was seen from an early date as a means of regional development, and there was an obvious need for specialised fishing villages, and for towns which could allow trade to develop. Early efforts in this direction were the founding of the royal burghs of Stornoway, Campbeltown and Fort William by James VI. In the case of Stornoway there was also the attempt to bring in people of proven expertise from Fife and from Holland to act as catalysts for development, but these swiftly came to naught and evidently produced more friction than development (Dunlop 1978: 7). There were to be other new settlements in other period, where the general rationale was to select sites adjacent to good fishing grounds. The principle was most prominently articulated by John Knox in the late 18th century, when he suggested the building of about forty villages, equipped with storage and curing sheds, on the coasts of the North-West Highlands (Knox 1785: 145–146). Knox's thinking must have been influenced by the contrasting situation on virtually all the Lowland coasts by this time, where specialised fishing settlements provided the bulk of the fishing effort. Although the British Fisheries Society from 1786 did set out to build fishing villages, and there were also attempts to operate on a similar way on individual estates, in virtually no case were such settlements successful in any sustained way in the development of the herring fishery. As well as facing frequent shortage of capital and a variety of problems of organisation, such attempts were set against a background of stressful social change in the Highlands which made many fear for their security of tenure on the land; and rather than commit themselves completely to the uncertain yields of the herring fishery, there was a general wish to have crofts to cultivate.

The whole principle of setting up specialised villages encountered a major and sustained problem of organisation in the irregular behaviour of the herring shoals, which were prone to frequent different lochs and parts of the coast in different years: this made it a problem to have stocks of barrels and salt at the right place at the right time. What effort was mounted on the West Highland coast was nearly all in small open boats, rather than the bigger boats which could seek out the herring offshore and could have more dependable results. Even when such bigger vessels from outside were sent to the West Coast, the general practice was for them to use small boats for the work of fishing, supplemented by the native small craft. In effect they were depot ships, and had little advantage in operation over the boat fishery.

THE EXPORT TRADE

Inevitably the best available index of the changing importance of the herring fishery over time are the figures for export; although not always consistent or complete, these do give a picture at the national scale that was never available for the part of the catch consumed in Scotland till the end of the 19th century. The data that are available are somewhat varied in form: the earliest are mainly measures of value, but from the 17th century onwards there are statements of quantity, the earliest of which are generally in lasts. Although the last is a unit that varies, especially in relation to the type of commodity measured, its general equivalent in modern measure has been stated at two tons; and in the case of herring it has been given as 12 barrels (OED. VIII 1989: 671). While the export statistics show fluctuations which must reflect the interplay of variations in catches, home demand and market conditions abroad, they do in the long term show a general upward trend indicative of expansion. At the same time it is probable that up to the late 18th century the bulk of Scottish herring found their way on to the home market, and there is consequently a major unknown in the scale of the fishery.

By the late 15th century the customs receipts for herring exports were valued at £390, when they constituted 11.5% of the total value of exports: they were the second item in order of importance after wool and hides, and were comfortably ahead of the better known salmon on the list (Grant 1934: 69). While there was a high amplitude of fluctuations in herring exports, the general level in the late 15th and early 16th centuries was around 150 lasts (1,800 barrels); and after 1540 there was a considerable increase in the level over the rest of the century to a level around 500 lasts, and in the peak year of 1541 it even reached 1,500 lasts, or 18,000 barrels (Guy 1986:79). While this higher level of exports must have added to the prosperity of the merchants involved, it is apparent that it also produced tensions, with the measures that were enacted in the effort to guarantee that home requirements got preference over exports (Samuel 1918: 87–89). Figures for the actual value of herring exports for the years 1611–1614 show an annual value of £100,000 (Scots), which was double the value of the better known salmon exports (Mar Report 70–74); and in the same century before the Civil War around 400 lasts of Scottish herring were passing annually through the Sound to markets in the Baltic (Sound Toll Registers).

Available data for the remainder of the 17th century suggest that the 1610–1630 period was something of a high point in the export trade. It was subsequently damaged by the upheavals of the Civil War, and the exports in 1669 were less than half of those of 1611–1614, although the second Royal Fishery Company had been given export bounties to promote the trade. With the general economic upturn after the Restoration there was a recovery, and in the closing years of the century a total of 4,500 lasts (54,000 barrels) were reported as being regularly exported from the Forth and Clyde between them (NLS. Ms. 33.5.31). Despite such achievements and the continuing encouragement given by government in the 18th century, in the period between 1750 and 1800 exports varied between 15,000 and 70,000 barrels (Cochran 1985: 45).

While Scottish fish exports reached most of the countries on the western seaboard of Europe and extended into the Mediterranean, there were dominant trade flows and these changed over time. An important direction of movement was into the Baltic, to reach the main international market on the North European Plain, although here there was direct competition from the Dutch, and there could also be competition from the Norwegians and Swedes. It is clear that the Dutch dominated the trade and supplied the premium quality product, while the Scots and others were not sufficiently well organised to guarantee the quality of the cure, and they supplied a lower value sector of the market.

The old Scottish ally of France was also a main market, and by the late 17th century was the leading single overseas destination, especially for the herring cured on the Clyde. In the 1680s this export was estimated as amounting to 2,500 to 3,000 lasts per year (Smout 1963: 222), and it was a major disruption to the Scottish trade when in 1689 France prohibited the import of salt herring (RCRBS., vol. IV: 260). It was at this juncture that it fortunately proved possible to increase the export to the Baltic, due to the difficulties experienced by the Dutch in their conflict with France. It is noteworthy that the Clyde ports of Glasgow and Greenock, as well as the East Coast ports, were now participating in this Baltic herring trade (Riis 1992: 62). This was also aided by the national decision to introduce export bounties on herring in 1699: in the first decade of the 18th century almost half the herring passing through the Sound came from Scotland, and the volume could reach over 1000 lasts annually (Smout 1963: 223).

While the premium markets were always on the European continent, there were other markets for Scottish herring in Ireland and the Caribbean by the 18th century, and these became the main outlets, although they too had no guaranteed stability. These market sectors were very largely supplied by herring caught on the West Coast: this was a combination of small boats operating on the Clyde, and busses working in the West Highland lochs. While exports conspicuously fluctuated there was a general level of around 30,000 barrels from 1750 onwards, and in some years before the end of the century they topped 100,000 barrels (RRBF. 1798: 129, 209). By the second half of the 18th century Ireland was the main single market, and was generally preferred over that of the more distant Jamaica and the other West Indies; there was a tendency for these markets to vary in complementary directions, with high exports to Ireland corresponding to low exports to the West Indies. This was not, however, solely related to Scottish production: although the Dutch production was by this time a fraction of its peak, the second half of the 18th century was the great time of the Bohuslän fishery in Sweden, and as well as dominating the Baltic market for half a century, it was also a competitor on the Irish and plantation markets. Inevitably this complicated the pattern of trade flows, but in aggregate Scotland was at last becoming a main competitor on the scene; and this was to become increasingly clear in time with its advance to international leadership which was to come in the next century. The essential basis for this had developed in Fife and Banffshire, in both of which there was the established use of bigger boats with eight or nine man crews, and the operation of these was linked to shore-based curing: and curing and merchant expertise developed on the Clyde was also to play a prominent part. It was in shore-based curing above all that the Scots had the decisive advantage over the Dutch method of operation; and when an effective government organisation was eventually mounted to superintend and guarantee the standards of curing, this proved the key to success.

REFERENCES

Acts of the Parliament of Scotland (APS.), vol. 1 (1124–1423), (1844), H.M. Register House, Edinburgh.

Adair, J. (1707) *A Short Account of the Kingdom of Scotland and the Firths, Roads, Ports and Fishings about the Coast*, National Library of Scotland (NLS.) Ms.19.3.28.

Analectica Scotica, ed. Maidment, J. (1832), vol.I.

Beaujon, J. (1883) *The History of the Dutch Sea Fisheries*, London.

Ancient Laws and Customs of the Burghs of Scotland (ALCBS.), vol.II (1910), Scottish Burgh Records Society, Edinburgh..

Clark, J.G.D. (1952) *Prehistoric Europe. The Economic Basis*, Methuen, London.

Cochran, L.E. (1985) *Scottish Trade with Ireland in the Eighteenth Century*, John Donald, Edinburgh.

Cochran-Patrick, R.W. (1892) *Medieval Scotland*, James Maclehose and Sons, Glasgow.

Dictionary of the Older Scottish Tongue (DOST.), vol.I (1937), Scottish National Dictionary Association Ltd., Edinburgh.

Elder, J.R. (1912) *Royal Fishery Companies of the Seventeenth Century*, Aberdeen University Press, Aberdeen.

Ewan, E. (1990) *Townlife in Fourteenth-Century Scotland*, Edinburgh University Press, Edinburgh.

Exchequer Rolls of Scotland (ERS.), Vol. I (1264–1359), ed. Stuart, J. and Burnett, G., (1878), H.M. Register House, Edinburgh.

Grant, I.F. (1930) *The Economic and Social Development of Scotland to 1603*, Oliver and Boyd, Edinburgh.

Grant, I F. (1934) *The Economic History of Scotland*, Longmans, Green and Co., London.

Gray, M. (1978) *The Fishing Industries of Scotland 1790–1914. A Study in Regional Adaptation*, Oxford U.P., Oxford.

Guy, I. (1986) 'The Scottish Export Trade, 1460–1599' in Smout, T.C. (ed.) *Scotland and Europe 1200–1850*, John Donald, Edinburgh, 62–81.

Hamilton, H. (1963) *The Economic History of Scotland in the Eighteenth Century*, Clarendon, Oxford.

Hamilton, H. (1932) *The Industrial Revolution in Scotland*, Oxford U.P., Oxford.

Hutcheson, G. (1887) *Days of Yore*, Buckie.

Jenkins, J.T. (1927) *The Herring and the Herring Fisheries*, P.S. King and Son Ltd., London.

Knox, J. (1785) *A View of the British Empire, more especially Scotland (3rd edn.)*, London.

Lillehammer, A. (1986) 'The Scottish-Norwegian Trade in the Stavanger Area in the Sixteenth and Seventeenth Centuries', in Smout, T.C. (ed.) *Scotland and Europe 1200–1850*, John Donald, Edinburgh.

Lythe, S.G.E. (1960) *The Economy of Scotland 1550–1625*, Oliver and Boyd, Edinburgh.

Mar Report (1904) Historical Manuscripts Commission Report on the Manuscripts of the Earl of Mar and Kellie, preserved in Alloa House.

Martin, A. (1981) *The Ring-Net Fishermen*, John Donald, Edinburgh.

Miller, H. (1829) *Letters on the Herring Fishery in the Moray Firth*, Inverness Courier, Inverness.

Mitchell, J.M. (1864) *The Herring, its Natural History and National Importance*, Edmondston, Edinburgh.

National Library of Scotland (NLS.) *Discourses Anent the Improvements may be made in Scotland for Advancing the Wealth of the Kingdom in these parts* (Ms 33.5.31), by a Well Wisher to his Country. Attributed to Sir Robert Sibbald.

Old Statistical Account (OSA.):- V, 287–308 (Inveraray Parish). X, 1–33 (Wick Parish). XIII, 392–427 (Rathven Parish). XV, 610–639 (Avoch Parish).

Oxford English Dictionary (OED.) VIII (2nd. edn.), 1989.

Parliamentary Papers (1803) *Reports from Committees of the House of Commons*, vol.X.

Pennant, T. (1771) *A Tour in Scotland*, vol. I, London.

Records of the Convention of the Royal Burghs of Scotland (RCRBS.), William Paterson, Edinburgh.

Report Respecting the British Fisheries (RRBF.) (1798), Evidence and Papers.

Riis, T. (1992) 'Long Distance trade or Tramping: Scottish Ships in the Baltic, Sixteenth and Seventeenth Centuries', in Smout, T.C. (ed.) *Scotland and the Sea*, John Donald, Edinburgh.

Samuel, A.M. (1918) *The Herring; its Effect on the History of Britain*, John Murray, London.

Scottish National Dictionary (SND.), vol. 4 (1956); vol. 6 (1971).

Smith, A. (1896 edn.) *An Inquiry into the Nature and Causes of the Wealth of Nations*, George Bell and Sons, London.

Smout, T. C. (1963) *Scottish Trade on the Eve of the Union 1660–1707*, Oliver and Boyd, Edinburgh.

Sound Toll Registers (Öresundstoldregnskaberne), Rigsarkivet, Copenhagen.

Unger, P.W. (1980) 'Dutch Herring, Technology, and International Trade in the Seventeenth Century', *Journ. Econ. Hist.*, 40, 243–279.

Watson, H.D. (1986) *Kilrenny and Cellardyke. 800 Years of History*, John Donald, Edinburgh.

6

The Early White Fisheries

In Scotland the term 'white fish' has a long established use as including the demersal species such as cod, haddock, saithe, and ling, and also flat fish species such as sole, plaice and skate. Bones from most of these species have been identified at archaeological sites of a variety of ages, and there is sufficient evidence in the records to show that fisheries for them were important by Medieval times, although they get considerably less recognition than herring and salmon. The main reason for this is that they entered less prominently into commerce. They excited fewer comments from politicians, travellers and other observers, and made less mark on national affairs. They also involved less spectacular peaks of activity than did the herring and salmon fisheries, and showed lesser variations in abundance. However there is good reason to think that this combination of circumstances has contributed to their importance for the generality of the Scottish population being inadequately recognised.

There is, however, evidence for the long-standing importance of white fish. The wealth of dialect terms in different parts of the country for the different species is significant, and many of these are known to be centuries old; and these include words to designate sub-species and fish at different stages of growth (Watt 1989: 5–9). The sporadic references from early travellers and chroniclers also suggest that white fish were a very common diet item around the coasts. In the Exchequer Rolls from the 13th century onwards there are sporadic references to 'mullones', generally accepted to be dried cod in a series of East Coast burghs, together with early references to 'hard fish', which most probably means the same. There are various other fragments of evidence which attest to the early importance of white fish: at Montrose, for example the town hospital, which was founded at some date before 1245, had the teinds of the local white fish boats (Adams 1993: 225). When the situation becomes clearer from the 17th century onwards, what is very prominent is that around the coast of the most populous parts of the country were a series of fishing communities in villages and within towns for the majority of which the major activity was white fishing. In the 17th century there are around fifty cases known in which, in the saisines, rights to white fishing at the coast were conveyed along with lands (Stewart 1869: 21). While there is some doubt on the early legal status of white fisheries, the indications are that they were never in practice an exclusive right; and what was specified in the saisines was in fact a right to certain houses, yards and crofts, along with the boats from which fishing was conducted, rather than the fishing as such (Bell 1812: 53–54). The specification of such rights in the 17th century, but not earlier could well be indicative of an expanding fishery in which coastal landowners had an interest: there is certainly evidence for growth in the fishery at this time.

White fishing appears to have started largely for local consumption and local markets, and included trade by barter. It had an important role in the food supply of the coastal zone, although over time it entered increasingly into longer distance trade. In the case of the Shetland Islands, white fisheries from the 16th century were the main basis of the commercial economy (Smith 1984: 15 ff.), and the main markets were on the continent; in this respect the early modern economy of Shetland, while on a smaller scale, compared with those of North Norway and Newfoundland.

The methods used in catching white fish through the great part of history are likely to have been dominated by hook and line. There were, however a great range of methods employed in different places and circumstances, some of which did not even involve the use of boats: there was also for centuries some use of leisters (fish spears), traps and nets and other gear in shallow water along the coast. Such methods as setting baited lines at a low tide to be hauled at the next low tide, and wading into the sea with the 'half net' (Symson 1684–92: 79–80, 116) were recorded in Galloway in the 17th century; also recorded was the collecting of sand eels in great quantity at low spring tides from sand banks in the Dornoch Firth (Anon 1908: 101–102). There were a variety of types of net used in inshore waters, especially for the taking of the saithe, the most abundant white fish species near the coast; and these included dip and sweep nets (Baldwin 1982: 164, 191). These methods go

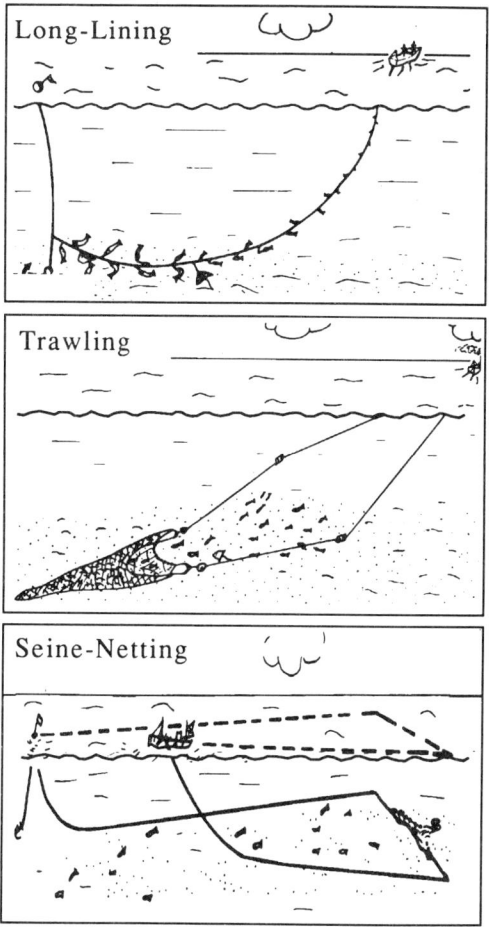

Fig. 6.1 Main methods of white (or demersal) fishing.

back an unknown distance in time, and continued in use into the 20th century.

In the broader European context, the use of hooks is known to go back to the Stone Age, with the important advance to barbed hooks taking place by the Bronze Age (Clark 1952: 85–89); and early lines were made of materials like nettle fibre and lime bast. Evidence for early fishing equipment in Scotland is very scarce, and little has been reported before the sinker stones known in Viking times in Shetland. Within known history, lines were made of hemp or linen, and hooks of iron, and later of steel. In any case, a more substantial issue is the type of lines used. Early lining would almost certainly be hand lining, with one or two hooks on the end of the line. However, the great part of lining within known history was long-lining, which involves a line being laid on the sea bottom, with hooks at intervals, anchors at the ends, and floats from the ends of the line to show their position. The main types of long line traditionally were two: there were 'small lines', which had relatively small hooks at intervals of three or four feet, were usually baited before putting to sea, and were used near the coast for fish such as haddock, sole and codling. While recorded practice varied around the coast, these were generally used in 'strings' on each of which were a hundred or more hooks (Mather 1969: 6). In contrast was the 'great line', with bigger hooks at longer intervals (say six feet) which was used offshore for bigger fish, especially cod: and it was usually baited aboard during the process of shooting. It was general practice with both small and great lines to shoot miles of line at a time. Also by the 18th and 19th centuries it was common for the two types of line to be used by the same men at different seasons, and these seasons could vary on different parts of the coast. Great lines were used in trips further out to sea for bigger fish, at distances of up to 15 miles or more; small lines, on the other hand were used in the inshore zone, and the main species caught was haddock. Generally in the North-East the main time for great-lining was from April to June (Arbuthnott 1815: 38–39); but on the Forth great lines were used form August to December, while small lines in the first half of the year prior to the summer herring season (Washington 1849: 60).

In small line fishing, there was a great deal of effort by the women folk, aided by the children, in collecting bait and putting it on the hooks before the men went to sea: this work regularly involved the daily baiting of thousands of hooks by the individual woman. By far the most common bait was mussels, which were collected from the scalps at the shore in the early morning. A remarkable illustration of the work involved here is the fact that in the 1790s in the Aberdeenshire parish of Slains, the school roll fell markedly in the main fishing time of the summer, as there were so many children of over six and seven years from the fishing villages of Collieston and Old Castle who had to help then with line baiting (OSA.V: 284). In Carrick the women collected the mussels at times between 2 a.m. and 6 a.m., depending on the state of the tide; and a woman might 'shell' (i.e. take from the shell) as many as 2,400 mussels in a morning (Czerkawska 1975: 10). In Arbroath the shelling part of the work began about 4 a.m., and the total task involved in getting the bait and putting two mussels on each of the 1,300 to 1,400 hooks took a total of about ten hours (Hay and Walker 1985: 29–30).

The history of the long line in Scotland, and indeed in Europe, is obscure. It is known that the Dutch were using long lines in offshore fishing by the 17th century, which indicates that they had sufficiently mastered the technique of manoeuvring a ship under sail well enough to handle gear of which the position in the sea was essentially fixed, and did not drift with the ship as did the drift net which the Dutch developed for their main herring fishery. It is considerably simpler to manoeuvre a boat to a long line with oars, and it is intrinsically probable that the long line has a longer history of use in inshore waters. It was certainly much used in Scotland by the 18th century, and by that time boats would often literally shoot miles of line at a time; and the technique was already likely to be centuries old. There could be some corroboration of this in the claim by Hugh Miller in the mid 19th century that the methods of fishing with which he was familiar had been practised with no real change for many generations (Miller 1844: 330).

In the detailed work of fishing, the fact that the great part of the work was done within sight of the land allowed the sighting on land marks to fix position at sea with some precision. In the North-East, for example, the bearing of hills like Mormond and the Bin Hill of Cullen over various coastal features were used. The most remarkable case of this which has been documented is that of Fair Isle: here for the fishing of the recognised grounds around the island (which is itself less than three miles long) over 80 examples of 'meads' (or sightings) with cross bearings are on record (Eunson 1961: 181–198). It was also possible in inshore fishing for particular sea areas to be reserved for particular boats or crews, as was recorded for example in Shetland (March 1971 I: 37); and at Fraserburgh the right to fish on particular banks was recognised as being the prerogative of particular crews (Cranna 1914: 59). It was known in Barra for there to be run-rig at sea, whereby different boats took it in turns to fish particular banks on different days. These are examples of local administrative practice which are very likely to have featured widely.

THE DEVELOPMENT OF
COMMERCIAL FISHERIES AND TRADE

Trade records do contain some references to white fish from Medieval times onwards: and references occur fairly regularly in the Exchequer Rolls from the 13th century onwards. The most prominent early records are to trade in 'mullones', 'stockfish', and also to 'haberdynes' or 'abirdenes', all of which have generally been thought to be dried and salted cod fish. The cod as a big fish is one of the best fish available for the purpose of drying and salting without the need for barrels, and gives a relatively big weight for the work involved. It appears, however that these references do not refer exclusively to cod. Various other species, including especially ling, but also saithe (or coalfish), hake and haddock could also be so preserved. This is evident when the record becomes clearer in the 17th and 18th centuries, and they could also have entered into the earlier trade. There are various references to white fish among early chroniclers, and Bishop Leslie in 1578 in listing the fish commonly eaten in the country included turbot, flounder, plaice and cod (Leslie 1893: 140)

It is clear that for centuries the export trade in cured white fish was a small fraction of that in both herring and salmon. The Exchequer Rolls show that there was a limited and irregular export by the 15th century and Pedro de Ayala recorded the export of stockfish in 1493; however the 1471 customs receipts show the value of the export to be less than 10% of either that of herring or salmon (Grant 1934: 69). The export data show big year to year fluctuations in cod exports between 1460 and 1600: they ran at a general level of around 20,000 fish annually, but good years could rise above 50,000 fish, and in the period between 1540 and 1580 could exceed 80,000 (Guy 1986: 78).

The circumstances under which commercial white fishing was organised appears at the earliest stages to have been from the fishing quarters of towns, such as Newhaven (Edinburgh), Fittie (or Footdee) in Aberdeen or Broadsea in Fraserburgh. In such locations, fishermen obviously had a ready market on their doorstep, and this produced the inevitable problem of the control of the trade which town merchants wanted to retain by insisting that fish be sold only at the market cross. It is clear that there were attempts to forestall this, as was recorded for example in the 17th century at Montrose (Adams 1993: 226). However as time went on non-burghal fishing settlements grew up or were instituted on coastal estates; and fishing was to become most frequently prosecuted from rural settlements. These included specialised fishing villages from at least the 16th century.

Although the main areas of early commercial white fishing are likely to have been in the south of the country on the Firths of Forth and Clyde, there was certainly some activity elsewhere, as is shown in the Exchequer Rolls and various town records. In the 16th century, for example, trade in cod, ling and skate is frequently recorded in the records of Inverness (HVAL.1911: 67). The indications are that by the 17th and 18th centuries there was a change in regional balance with the North-East with its many (and proliferating) fishing villages becoming an area that, as well as supplying its own needs, had an expanding external trade. This was mainly to the Midland belt, and an example of the mechanism of the trade is shown in the contracts that George Beattie, a Montrose merchant, had in the 1720s with fishermen at Findochty and Cullen in Banffshire: he was to take all the cod and haddock that they sold over a period of five or more years (SRO. RD 2/125). However, the Forth was the main destination of this trade, but it also reached the Clyde; and to a significant extent it penetrated into England, especially to the London market. There is sufficient evidence in the Old Statistical Account to show that there was by the late 18th century a substantial and regular trade in cured fish from the North-East in a southern direction: the villages of the Banffshire coast were collectively exporting fish to the value of thousands of pounds, and Collieston in Aberdeenshire was also prominently involved in addition to the burghs of Peterhead and Fraserburgh (Coull 1969: 30). As early as the mid-18th century, records of substantial shipments of dried codling from Johnshaven to the Forth suggests that the Angus villages were also much involved in this trade (Adams 1991: 29). On the West Coast too there is evidence that in the early 18th century, the annual export of cod from Gairloch and Lewis was relatively substantial (HVAL.1911: 67). The indications are that such trade links, to which the substantial productive effort catered, had often been initiated from the south of Scotland. Merchants from the burghs on the Forth were especially prominent and the commercial networks were extending outwards to the west side of the Moray Firth and into Orkney. In the North-East however there was the beginnings of a response within the region whereby merchants in the area were starting to get a significant share of the trade.

In addition, from the burghs there were from at least the 18th century ventures with bigger boats to cod fishing on the West Coast: here the main effort came from burghs like Greenock and Campbeltown on the Clyde, but Stornoway merchants also played a part; there was participation too from East Coast burghs like Fraserburgh and Anstruther, and in the case of Peterhead smacks went in addition as far as Iceland. Such ventures were in the main stimulated by the export bounties being paid on cured fish; and over most of the country such efforts were subordinate to local fishing to supply the home market.

An important and largely distinct part of the national pattern were developments in the Shetland Islands, which led to them becoming the main seat of production for an export to Europe which extended into the Mediterranean. This contrasted with the West Highland coasts: here, while the inroads of commerce had begun by the end of the 18th century at a few points – notably Barra, Lochs (Lewis) and Gairloch (Wester Ross) – white fishing continued to be dominated by small scale local subsistence.

SHETLAND: THE 'HAAF' FISHERY

The Shetland Islands are the most prominent case in Britain of a land base of limited productivity adjacent to productive fishing grounds, and is comparable on a smaller scale of the type of juxtaposition that is better known in Norway, Iceland, Newfoundland and Labrador.

In these remoter fishing regions around the North Atlantic, the establishment of market links with more populous areas has been crucial to the growth of commercial fisheries, and characteristically the original initiatives in establishing such links came from more developed external areas, with merchants going out to the periphery in quest of trade. It has been shown that fish from Shetland were finding their way into North European markets by the late Medieval period through visiting ship merchants coming from North Germany. In the rather better known period between 1550 and 1710, visiting merchants traded for fish, along with 'wadmel' – cloth, at places all around the archipelago: these merchants were dominantly Germans from Bremen and Hamburg, but also included men from Scotland and England (Smith, H.D. 1984: 7–25). They were involved in fish curing, and operated from May to August; and they have been credited with showing the Shetlanders the method of fish curing (Smith, R. 1986: 84–85).

One of the most distinctive things about Shetland is that when foreign merchants withdrew in the changed situation of the 18th century, the local lairds took the initiative in developing fisheries as the main basis of the commercial economy. The withdrawal of the continental merchants was itself a matter of controversy. It has been seen as part of changed trading relationships after the 1707 Union, and in particular as a consequence of the imposition in 1712 of a tax on imported solar salt, which was needed for curing. However an important part of the background to the withdrawal of the merchants is a previous legacy of mounting problems in Shetland in the 17th century, which here was 'one long crisis' (Smith, B. 1992: 99): there was the collapse of the structures of government, as well as repeated harvest failures and epidemics. The severity of the position is seen not only in the problems of the tenantry, but also in the fact that several landlords went bankrupt. It was in this atmosphere of problems and recession that the German merchants withdrew.

There was a hiatus following the withdrawal of the German merchants in the late 17th century before the local landlords began to promote commercial fishing, but the systematic development of this can be seen from about 1720 (Smith, B. 1992: 101). The indications are that the responses of the Shetland landlords to the export bounties on cured fish given from 1727 were important. While the promotion of the fishery shows a rational appreciation of available resources, it also put the lairds in a powerful position in the organisation of production, and what came to be known as the 'Shetland method' has ever since been a major field of debate. It essentially involved a system by which tenants were bound to fish for their landlords, usually as a condition of holding house and land: they were crofter-fishermen, for whom the land provided a substantial part of the home food supply and for whom fishing was the main commercial activity. The term 'fishing tenures' has been used to designate this landlord-tenant relationship, and it encapsulates an important part of the Shetland system, but it does not in itself include another important part of it, that of the control of the terms of trade. The landlords had a monopoly position in deciding the prices tenants got for their fish, and also those at which they bought their fishing gear, meal and other stores. It was reported at the end of the 18th century in evidence to a Parliamentary Committee that the landlords met before the start of the fishing season to fix in concert the price of fish, and that they had an apparent profit on the fish of 100% (RCHC. 1803, vol. X: 26, 29). The landlords' case was that the fishermen got various other advantages like low rents, but it would be difficult to dispute that in their position of control the landlords did minimise the risks to themselves, and perforce put more risk on the fishermen, who lacked the resources to cope with uncertainties and fluctuations, and often became saddled with debt.

Previous to the lairds taking control of the organisation of the fisheries, they were largely confined to inshore operation within five or ten miles of the shore, and were mainly prosecuted from fourern (four-oared) boats in day trips. However by the middle of the 18th century it was becoming more common for fishermen to venture further off-shore in sixern (six-oared) boats, and to make two trips per week. While oars were regularly used, it is likely that the sixerns also had sails. Fishing was by long line, and sixerns would shoot up to six or seven miles of line at a time. The main fish sought were cod and ling, and Shetland was to become one of the main European suppliers of cured ling: the other main suppliers to the market in cured big fish were Norway and Newfoundland, where the production was emphatically dominated by cod. One result of this greater emphasis on off-shore operation was that the main 'haaf' stations were established in peripheral locations within the islands (fig. 6. 2), and the most prominent seats of the fishery were in the north-west part of the islands, which were within about thirty miles of the deeper water of the continental edge, where the best ling grounds are. Fethaland at the northern tip of the Shetland Mainland was the greatest of the haaf stations, although Stenness and Gloup were also of first rank. As well as the catches from the haaf stations there was also a significant continuing output from inshore grounds, especially from the Burra Haaf in the south-west and the Fetlar Firth in the north-east.

The shore stations were located at bays with shingle beaches, on which boats could be drawn up when not in use, and on which fish could be spread to dry after being split and salted (plate 10). Shingle beaches were a natural hard surface which allowed water to drain from the fish after washing and air to circulate around the fish to promote drying. It was usual for the fishermen to concentrate on the work of fishing, while the splitting and salting on shore was the

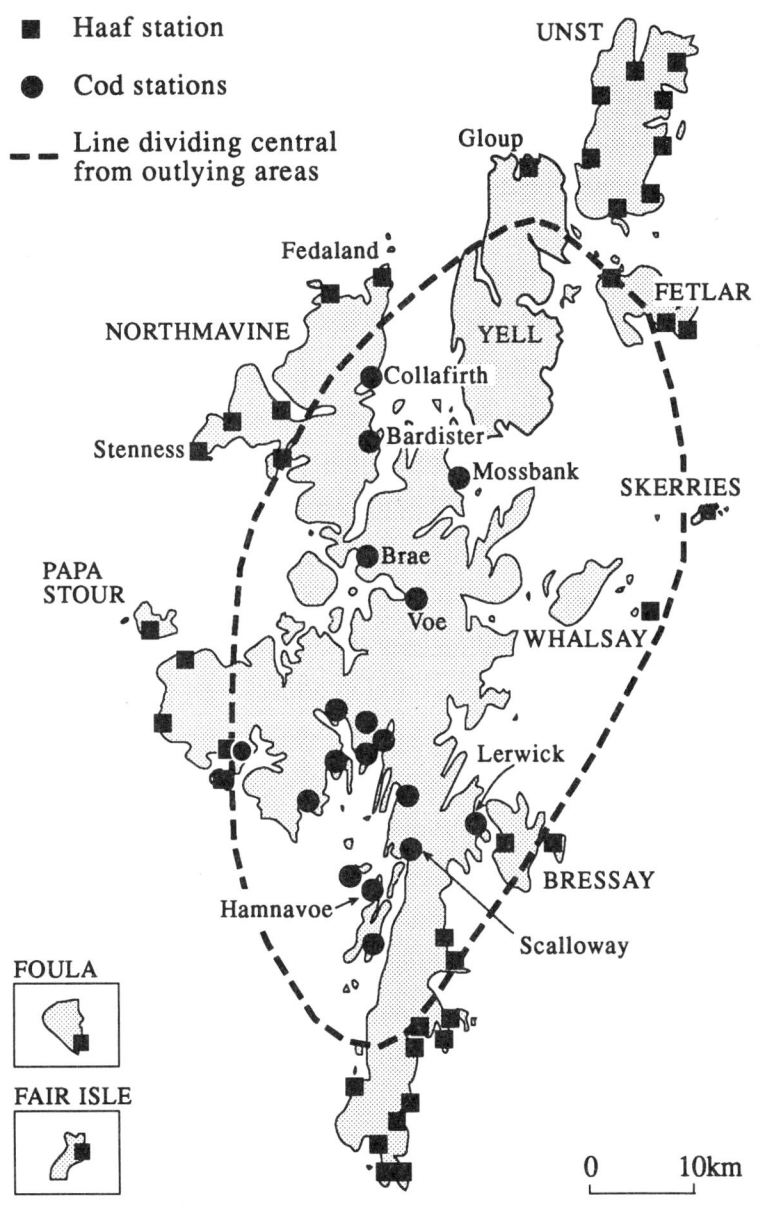

Fig. 6.2 'Haaf' stations and cod bases in Shetland, 18th and 19th centuries (*After H. D. Smith*).

work of old men and boys. The locations of the main haaf stations made it necessary for the majority of the fishermen to live during the two-month summer season in 'lodges' or huts; and with the value of timber in the Shetland situation these were roofed only during the summer when they were in active use (Edmondston 1809: 236–240).

There was also a significant trade in fish oil; and although this was made in part from boiling the livers of ling and cod; but the main source of this was the fishery for saithe, and here there was something of a sub-specialisation with this fishery being most active at the south end of the islands in Dunrossness, and in the area of the Fetlar Bight to the north-east.

The Shetland system was no simple stereotype, as was recognised in the best known contemporary commentary record of it by Edmondston (Edmondston 1809:232-259). There was significant variation in relative levels of rent and in the extent of compulsion exerted on the fishermen to sell their catches to their landlord. It was claimed that when the export was in the hands of foreign merchants the lairds, by allowing 'booths' (storage buildings) and curing operations on their land, got an income that was only one third of that of the merchants (RCHC. 1803 vol. X: 63). The assumption by the landlords of control of the fishing itself inevitably gave them a much bigger share of the profits. Especially in the early phases, transactions were largely conducted by entries in the ledgers of the lairds' factors, with very little money changing hands. It was the practice for the fishermen to be bound to give the lairds the entire proceeds of the summer fishing; there was also a nominal rent in butter and cash, and on some estates the landlords were also bound to hand over part of their farm produce, especially of their animals (Smith, B. 1992: 96). With the dominance of the haaf fishery in the economy and with its main bases around the edges of the archipelago, one of the results was that the peripheral parts of the islands came to be dominated by a few large estates (Smith, H. 1984: 120).

While a considerable part of the economy functioned as paper transactions in estate ledgers, it was necessary to buy salt for curing from outside Shetland. Boats also had to be acquired, and these were bought in kit form from Norway; and although the sixerns used in the fishery were less expensive than decked boats, they were none the less a considerable capital item. The merchant James Hay stated the cost of such boats in the late 18th century at 'from £5 to £8 stg. according to size' (RCHC. 1785–1801, X: 26); and Edmondston gave the value of a sixern fully equipped for use (with mast, sail, oars and lines) at £26.7.6; and of this the lines, which could extend to a total length of seven miles, was virtually half the total (Edmondston 1809: 246). At this time the typical rent for a land holding was around £3 (Scots) (Smith, B.: pers. comm.). While boats and gear could last for a run of years, this made the value of the land rent only a small fraction of that of the boat and gear. Boats were characteristically owned in shares, but with some variations in the arrangements: there were normally six shares, and it was possible for a crew of six each to have one; but sometimes the skipper had two shares, and the landlord might have one or more shares. It was also the case that many boats were hired, at an annual charge of about one third their value (Edmondston 1809: 235, 247). The relative value of land and boats indicates that there was a considerable distortion built into the system, and it was the very general perception of the fishermen that they were oppressed. It is scarcely surprising that there was also a persistent clandestine element in the trade, with small merchants buying fish and also oil directly from the fishermen at higher prices than those offered by the landlords. However this clandestine element tended to be limited as the landlords had various sanctions at their disposal, and in the extreme case could evict the fishermen.

The system was not of course static in time: it was to be the main basis of the islands' economy for the best part of two centuries. There was over the period a considerable population increase, and the landlords have been accused of operating what amounted to a population policy in the development of the fishery. This is another debatable issue: although the population was certainly increasing in Shetland, it was also increasing in most of the rest of Scotland where the nature of the economy was widely different from Shetland. However the landlords evidently did make some use of the traditional custom that no young man could take a wife until he had possession of a piece of land on which he could settle, and they made way for earlier marriages by sub-dividing holdings. Certainly the available demographic data, though not precise, indicate a remarkably rapid rise in numbers in the second half of the 18th century, just at the time that the haaf fishery was becoming firmly established. There was an increase of 5,000 or 6,000 (i.e. between 30% and 40%), and this contrasts forcibly with an increase of around 100 in the neighbouring archipelago of Orkney, where the base population was of a comparable size and which also at the time had a major economic development in the rise of the kelp industry. The fact that in Shetland agricultural holdings were very generally reduced below a size that could provide a family subsistence inevitably gave the landlords a strong lever in enforcing participation in fishing.

Shetland was partly the victim of being in an outlying position with a small population, and to this day around the North Atlantic there are similar cases where the volume of trade and traffic militates against competition and works to the disadvantage of the producer – in this case the fisherman. The lairds in Shetland have been generally accused of oppression; and in defending the system, it was usual for the lairds to stress the difficulty of organising production on any other basis in a remote location with a limited population. While in Britain it was in Shetland that this type of system was best developed, it was not unknown elsewhere; in the latter 18th century in Lewis there were cases where participation in fishing was a condition of holding land, and the land rent was low while the landlord had a commercial monopoly on the fish (Knox 1784: 339). It was also observed that in the Western Isles the people also suffered from the lack of an open market, but the scale of trade was so small and activity so dispersed that there was little scope for improving the system (RCHC. 1803, vol. X: 33).

Whatever the socio-economic context of its production, 18th century records show clearly that the export trade in cured white fish from Britain was dominated by the Shetland Islands, and the more detailed records of the second half of the century show a general level of exports of 500 to 1000 tons, of which around 80% came from Shetland. It is noteworthy too that from a main market orientation in the traditional direction of North Germany in the mid–18th century, there was an increasing emphasis on the Mediterranean market, which was the most important in Europe for cured cod and ling. While there were big lairds like Gifford of Busta and Henderson of Gardie who were directly involved in all phases of the trade from production through to continental marketing, it became characteristic that an expanding trade got more and more into the hands of specialist merchants and agents (Smith 1984: 61–74): although the periphery lagged in various respects in its commercial and social development, it was not isolated from main economic trends.

MODERNISATION OF THE WHITE FISHERIES
UP TO THE 1880S

With the modernisation of the British and Scottish economy, a gathering momentum of expansion in white fisheries can be seen in the 19th century. Basic factors were the rise in average living standards, and an expanding population which could be reached by fresh fish by improving transport: as well as the turnpike roads which were widely used by the early 19th century, there was the rapid spread of the railway network from about 1840. The overall result was a strong and sustained market expansion; and there was a response in the increase of fishing effort. This included both an intensification of inshore working and an expansion of offshore fishing. Even so, this continued to be dominated by the hook and line method until the latter part of the 19th century, when there was the major innovation of trawling. While it is not possible to gauge with precision the scale of expansion before comprehensive statistics of the Fishery Board begin in 1887, there are many indications of growth in the catch of fish that was disposed of in both fresh and cured states.

In the 1820s the expansion of traditional cured fish for the market was stimulated by the granting of bounties by the Fishery Board (fig.6.3). Evidently the great response that had been triggered in the herring fishery by the award of bounties led to a parallel extension of bounties to cured cod, ling and hake. While there is clear evidence that this stimulated expansion, the Board's own data show that the cured sector here was only a minor proportion of the trade: in an estimate made in 1843 it was found that in addition to a production of 4,641 tons dry cured and of 5,123 barrels, there were at least 20,595 tons which were otherwise disposed of (ARFB. 1843: 2). Most of the catch was in fact going to the fresh market, while the bounties were encouraging dry curing, which took considerably more effort and time. Although all such bounties were abolished in 1830 with the moves towards free trade, with improving communications there was sufficient market demand for production to expand.

Related to the award of bounties, the Fishery Board from 1821 onwards recorded that part of the catch of the bigger round white fish which was cured: from 1821 to 1830 a bounty of 4/- per cwt. was given on cured fish which passed a quality inspection by fishery officers. While the bounty applied officially to 'cod, ling and hake',

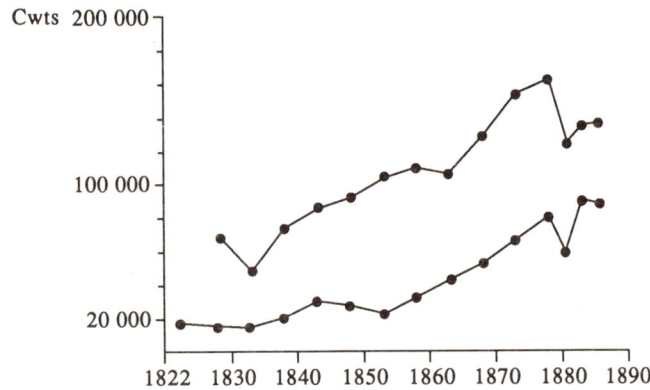

Fig. 6.3 Production and export of cured cod, ling, and hake
by 5-year means, 1822–1886.

other species were also in fact accepted and these included saithe
and tusk. Cured white fish continued to be inspected for export by
fishery officers between 1831 and 1850, and statistics of production
continued to be gathered after official inspection stopped. However
there is a minor complication in following the trend of Scottish production in that the bounties covered also cured fish exported from
England until 1849. This came from landings recorded at St. Ives
in Cornwall, and also from catches of smacks (mainly from Whitby)
which took part in the cod fishery. While the salt cured production
was important, it came largely from peripheral areas outwith the
market range of fresh and smoked fish. Although it was an activity
all around the coast, for the areas of the East Coast where the main
concentrations of fishermen were based it was a subsidiary activity.

One of the results of the government bounties granted in the
1820s was the expansion of the offshore fisheries for cod with
decked smacks. Such vessels, as well as catching fish themselves,
also bought it from open boat fishermen, and for them this could
function as a market link. The big majority of the smacks were in
fact Scottish, and in Scotland the big majority were from Shetland.
The relatively big scale of investment involved might appear anom-
alous in a poor marginal group of islands: the numbers in Shetland
relate significantly to the limited investment opportunities there
were in these islands, while there were alternative outlets for investment in most maritime communities on the Scottish mainland.
There had actually been an earlier beginning to this fishery in Shetland, and a number of factors contributed to it. As well as being

encouraged by the government bounties, it could increase the output of the islands' export staple of cured white fish; it gave another outlet for activity for the ships that island merchants needed for trade; and for fishermen it gave an opportunity for employment without the degree of bondage and control by landlords that was usual in the 'haaf' fishery. Not least, it could provide a useful cover for the smuggling in which some of the merchants are known to have been engaged (Smith, H. D. 1984: 104–112).

The fisheries by smacks were in part conducted on banks off the Scottish coasts, especially around the Northern and Western Isles; but smack fishing introduced a new element with the capability of going on longer voyages. The smacks from Shetland in the second half of the century made their main effort on the banks at Faroe, Iceland and even Greenland, where catch rates were higher (Goodlad 1971: 132–144). The pattern of bases of smacks at Shetland was complementary to that for the open boat 'haaf' fishery: in contrast to the peripheral pattern of bases for the latter fishery, smacks needed sheltered voes (bays) in which to anchor, and they operated from a central area within the islands (fig. 6.2); also much of the capital for them came from the estates and merchants of this area (Smith, H. D. 1984: 119–120). Operating at such distances inevitably gave additional problems in curing the catch: gutting and splitting had to be done on board and the fish wet salted for temporary preservation, to be followed by full drying on return to base, before onward dispatch to final destinations.

The production of cured white fish increased substantially in the 19th century (fig. 6.3). While the trend was inevitably irregular, it is noteworthy that with the bounty incentive in the 1820s output doubled during the decade from 2,500 to 5,000 tons: and although after the withdrawal of bounties in 1830 it dropped back to the 1820 level, this was followed by renewed expansion from the mid-1830s; and this effectively justified the official view that market demand in itself would provide sufficient incentive. From the middle of the century annual production was regularly over 5,000 tons, and by the peak year of 1878 output exceeded 9,000 tons. By this time an additional factor had come into play, with increasing home market demand for the part of the catch which could be landed fresh. This was essentially due to improving overland transport by railway, and it gave an outlet for species like halibut and skate, which were more difficult to cure than cod and ling.

As with the herring trade, it was the official aim to promote the export sector, and for cured white fish the Spanish market was especially important. Growth in what was a very competitive export market was in fact slow until the mid-1850s, although the rise thereafter meant that from the later 1870s over half the production went abroad; and despite the run-down of line fishing from the 1880s as steam trawling expanded, production and export both continued high until the end of the century. Curing provided an outlet for surplus trawl as well as line catches. Although the home market was now dominated by fresh fish, there was a continuing market for cured fish on the continent, where economic development at this stage was slower, and where there was an established taste for cured cod and ling in increasing populations.

The location of cured production continued to be significant: Shetland was always the leading district by a comfortable margin. In the 1820s the archipelago accounted for nearly half the total, and the broad tendency thereafter was for its proportionate share to decrease as production expanded. Production was important too in Orkney and in the Hebrides, especially in Lewis, Barra and Islay; and the North-East, Fife, and Eyemouth were also regular contributors. In effect the main seats of production were the places from which open boats could most easily reach areas of deeper water. Although the Northern Isles dominated the smack fishery, Greenock and Ayr on the Clyde, with their maritime traditions and access to urban sources of capital, also participated in this sector of production. Throughout this time the home market was dominant, but there was a continuing export market which included other countries in Europe as well as Spain, while there were also exports to Ireland and to the Americas.

Despite the increased effort and catch in the bigger white fish, the greater developments in Scotland in the 19th century were actually in the inshore fisheries. The outstanding change of the period was the expansion of the market for fresh and lightly smoked fish, to the point that it greatly overshadowed the salt cured market. A major response to the increasing market opportunity was the expansion of the fishery for haddock (Gray 1978: 88–90), which had always been the main species exploited by inshore line boats. As well as being distributed fresh, there was a great expansion in the market speciality of the 'smokie' or 'finnan', the latter originally named after the village of Findon in Kincardineshire. This began in the North-East by the fisher families themselves smoking the fish, which as well as giving a distinctive flavour acted as a short-term preservative, and was well established by the 18th century. Smoking over peat fires was usual in the rural areas, but in towns there was some use of oak chips (Arbuthnott 1815: 40). With expansion the practice was taken up along most of the East Coast, and specialist curers came in on the trade and came to account for most of the output. This method of light curing, which took only a few hours, effectively widened the market range, even in the days when transport was by the fishwife and by horse and cart. With the spread of the railway network the towns and cities in general came within the market orbit, and Glasgow especially became a main market. Even a centre like Peterhead which, with its position on the North-East shoulder was well placed for the cod fishery and had a long history of involvement in it, had substantially redirected its efforts to haddock fishing in both summer and winter by 1840. Although Peterhead was by this time prominently involved in the rising herring fishery, it was calculated that the white fisheries were yielding over £4,000 annually, or more than double that of the herring fishery (NSA. XII: 380).

The development was stimulated by a strong price rise, indicative of a market-driven demand backed by rising purchasing power, and between the 1840s and the 1880s the indications are that the going price rate increased about three times (Gray 1978: 89). The one attempt made by the first Fishery Board to estimate the size of the white fish catch was in 1871, when the value of the smaller round fish (mainly haddock) was £264,595, or 55% of the total; this compared with £206,201 for the bigger round fish (mainly cod and ling); and the remainder, valued at £12,280 was made up of flat fish (ARFB. 1871: 5).

In the longer term there was also in the latter 19th century the extension of the haddock fisheries to peripheral areas like Shetland and Lewis: by this time the mainland markets could be reached via regular steamer connection. Such areas, were however inevitably at a measure of disadvantage with extra marketing costs and greater difficulty in the fish reaching destinations in the best condition.

SOURCES OF BAIT

Supplies of bait were always required for line fishing, and as activity expanded the availability of bait became more and more a matter for concern. Some variety of baits were used, and those which were employed depended mainly on local availability, but also to some extent on the species of fish being sought: in general order of preference the main baits were mussels, clams, lug-worms and limpets (Fenton 1992: 139, 140). For the inshore fisheries, which continued to dominate the supply of white fish until the advent of trawling, by far the most important source of bait was mussels. Scotland overall was fortunate in its supplies of mussels: the mussel has a considerable tolerance to water of different levels of salinity. Because of this, in addition to many small beds (or 'scalps') around the coasts, there were extensive beds in all the firths and in many of the estuaries. It was the beds in the Montrose Basin and in the Firth of Clyde below Dumbarton which came to be recognised as the best and most productive. The importance of mussel bait was underlined by a three year study done at Eyemouth in the 1880s, which showed that the weight of mussels needed for bait was between 85% and 90% of the weight of fish that were caught (RCSMBB. 1889: iii).

For centuries there was apparently little problem in getting bait anywhere on the coast, although its importance was great enough for rights to it to be itemised along with lands in various charters. There is evidence, however, that before the middle of the 18th century mussel bait was starting to acquire a scarcity value: the town of Tain was starting to make money by the 1730s in encouraging Moray Firth fishermen to come and collect them from the extensive beds in the Dornoch Firth (Munro and Munro 1966: 81). There could also be considerable effort put into ridding the mussel beds of the mussels' main predators of starfish and whelks by women and children collecting them, as happened at the Rossie beds in the Montrose Basin (RCMBB. 1889: 79). By the second half of the century it was becoming clear that along much of the coast the available supplies of mussel bait were becoming seriously strained. It had already got to the point that in various places fishing was being to some degree restrained by shortage of bait. At Collieston in Aberdeenshire for example, which was near to one of the better sources of bait in the Ythan estuary, the local tacksman in the 1840s was preventing the fishermen going out more than twice a day because of the numbers of mussels they were using: however local fishermen got mussels at a preferential rate, paying £3 per year, or, if they were over sixty, £2 (NSA. XII: 595). The demand for bait rose steeply in the second half of the century: it was recognised as late as 1889 that most of the fishermen of Scotland, now numbering some 50,000, used mussel bait for at least part of the year

(RCSMBB. 1889: iii). In response to this expanded demand there was some increase in alternative types of bait such as lug worm or butcher's offal; but much more important was an increasing trade in bait from one point on the coast to another, and this was facilitated by improving transport. The main sources for this expanding trade were the main firths and estuaries: it was an active trade with boat cargoes from at least the early 19th century, and in the second half of the century it increasingly went by the more rapid expedient of rail transport. For the most active area of the North-East, mussels might be conveyed from the Findhorn estuary, the Montrose Basin, the Cromarty and Dornoch Firths, and the Firths of Tay and Forth; and by the 1880s they were coming in quantity from Port Glasgow on the Clyde. The bait could be used directly, or transported for temporary 'storage' to local mussel scalps.

Mounting concern about the supply of mussel bait in time led to official concern, and in the Act of 1868 there were provisions for the Fishery Board to administer the exploitation of the mussel beds, as for want of regulation they were 'threatened with extinction' (ARFB. 1871: 6). Action was, however less than immediate, probably because the acute controversy of the effects of trawling on the line fisheries for two decade occupied centre stage (ch. 9). After the Fishery Board was reconstituted in 1882 it was given the power in Scotland in 1885 to regulate shell fisheries, which had previously been the responsibility of the Board of Trade. The new board's reports make it clear that the issue of the supply of mussel bait continued to be of major concern in the closing years of the 19th century, despite the rapid domination of the white fish market by trawl catches.

Official reports in 1887 and 1889 revealed a situation in which many of the mussel beds had deteriorated badly through over-exploitation, and in which line fishing could be seriously hampered for lack of bait. The once important beds at Tain in the Dornoch Firth, for example, were now almost useless; and the harvesting of about 30,000 tons of mussels from the Clyde beds in ten years to 1889 had ruined the most extensive beds in the country (RCSMBB. 1889: x, vi); they had previously been largely protected from use by their distance from main fishing settlements.

The traditional situation in which there had been adequate supplies of bait everywhere, with fishermen being able to use it without restriction, was now changing. While there was little restriction on nearly all the West Coast outside the Clyde in the 1880s, the intensity of use had led to problems almost everywhere on the East Coast, where the great part of the white fish landings were made. In addition to the problem of getting bait there was also concern as to its quality: there were various complaints about undersized mussels, and of empty shells which had been dredged up along with full ones. There were also various restrictions on taking mussels, many of them of recent origin, and the price of them at between £1 and £2.50 per ton had risen several times in a single generation. Many of the main beds were worked or administered by lessees, and the rate at which mussel beds had appreciated in value was highlighted by the one hundred times increase in letting value over an 80-year period to £500 annually of the beds on the Sands of Dun in the Montrose Basin (RCSMBB. 1889: iv). In addition there were on the Tay estuary fishermen whose full-time livelihood was the taking of bait mussels. From numbers of places in the North-East there were by the latter 19th century expeditions for mussels to such West Coast destinations as Glencoe, Loch Hourn, Ballachullish and Loch Snizort (Fenton 1992: 143): this was an indication that the issue was becoming desperate.

It was obvious that more management of the beds had become necessary, and by the 1880s there was in some places a reaction to the deteriorating situation with local management arrangements. There were preferential rates for mussels for local fishermen at St. Andrews, Findhorn, and Ferryden along with Usan. Also a local firm had bought and was administering the main beds of the Montrose Basin: they were able to guarantee good quality large size mussels, and although the price was high at £1.75 to £2 per ton, the demand always exceeded the supply. After 1885 the Fishery Board was given the power to restrict the exploitation of defined scalps to named persons, and to impose restrictions on the use of scalps to protect them from injury (ARFBS. 1886: lxv). A bill was also put through Parliament in 1894 for the better regulation of clam and mussel beds. In addition to being given power to acquire scalps and beds, the Board was charged with drawing up a schedule of all mussel and clam beds and fisheries: this resulted of a list of 72 of them, the big majority held by Royal charter (SROAF 7/135). They were widely different in size, and the biggest extended to thousands of acres; and it was a recognised principle that local fishermen should have a primary interest in them.

The concern about the supply of mussel bait and the somewhat belated effort to conserve and improve them in the late 19th century represented a reaction to pressures that had been building up for decades. From the 1880s inshore line fishing went into decline in face of the great increase in the supply of trawl caught white fish, and the use of mussels went down from around 200,000 cwt. in the mid–1880s to well under 100,000 cwt. by the end of the century. The growth in the late 19th century of the fishery for cod with anchored nets in the nearshore zone, especially in the Moray Firth, represented a reaction both to the increasing scarcity of bait, and a new opportunity in an area now officially banned for trawling. By the inter-war period of the 20th century, the supply of mussel bait was a minor issue.

Bait for the offshore line fishery, in which the hooks were bigger and more widely spaced, was not an issue of anything like the importance that it was in the inshore fishery. While mussel bait might in earlier times be used in this fishery, by the later 19th century the general practice was for lines to be baited as they were shot: in the Shetland 'haaf' fishery the usual bait was piltock (small saithe) or haddock (March 1971, I: 40), while on the mainland the bait was mainly cheap fish like herring, mackerel or squid, especially the former. Bait could be taken aboard before the start of a trip, fresh or salted. It became the usual practice for such line boats to carry a few herring nets to catch bait, and they were often independent in their bait supply.

REFERENCES

Adams, D. (1993) 'Abundant with all kinde of fishes: sea fishing before 1800', in Jackson, G. and Lythe, S. G. E., *The Port of Montrose. A History of its harbour, trade and shipping,* Hutton Press and Georgica Press, Tayport and New York, 235–236.

Adams, D. (1991) *Johnshaven and Miltonhaven,* Chanonry Press, Brechin.

Annual Reports of the Fishery Board (ARFB.), 1843, 1871, Edinburgh.

Annual Report of the Fishery Board for Scotland (ARFBS.), 1886.

Anon (1908) 'The Description of the Province of Sutherland' (undated description) in *MacFarlane's Geographical Collections* III, T. and C. Constable, Edinburgh, 96–110.

Arbuthnott, J. (1815) *An Historical Account of Peterhead,* D. Chalmers and Co., Aberdeen.

Baldwin, J.R. (1982) 'Fishing the Sellag: Hand Netting Traditions from Caithness, the Northern and Western Isles', in Baldwin, J.R. (ed.)

Caithness. A Cultural Crossroads, Scottish Society for Northern Studies, Edina Press, Edinburgh, 160–212.

Bell, R. (1812) *A Treatise on the Election Laws*, A. Constable and Co., Edinburgh.

Clark, J.G.D. (1952) *Prehistoric Europe: the Economic Basis*, Methuen, London.

Coull, J. R. (1969) 'Fisheries in the North-East of Scotland before 1800', *Scottish Studies* 13, 17–32.

Coull, J. R. (1989) 'Fisherfolk and Fishing Settlements', in Smith, J.S. and Stevenson, D. (eds.) *Fermfolk and Fisherfolk*, Aberdeen University Press, 26–49.

Cranna, J. (1914) *Fraserburgh Past and Present*, Rosemount Press, Aberdeen.

Czerkawska, C. L. (1975) *Fisher Folk of Carrick*, Molendar Press, Glasgow.

Edmondston, A. (1809) *A View of the Ancient and Present State of the Shetland Isles*, John Ballantyne and Co., Edinburgh.

Eunson, J. (1961) 'The Fair Isle Fishing Marks', *Scottish Studies* 5, 181–198.

Fenton, A. (1992) 'Shellfish as Bait: the Interface Between Domestic and Commercial Fishing', in Smout, T.C. (ed.) *Scotland and the Sea*, John Donald, Edinburgh, 137–153.

Goodlad, C. A. (1971) *Shetland Fishing Saga*, Shetland Times, Lerwick.

Grant, I. F. (1934) *The Economic History of Scotland*, Longmans, Green and Co., London.

Gray, M. (1978) *The Fishing Industries of Scotland 1790–1914. A Study in Regional Adaptation*, Oxford UP., Oxford.

Guy, I. (1986) 'The Scottish Export Trade 1460–1599', in Smout, T. C. (ed.) *Scotland and Europe 1200–1850*, John Donald, Edinburgh, 62–81.

Hay, E. R. and Walker, B. (1985) *Focus on Fishing. Arbroath and Gourdon*, Abertay Historical Society Publications, no. 23, Dundee.

Hepburn, A. (1906) 'Description of Countrey of Buchan, Aberdeenshire, 1721' in *MacFarlane's Geographical Collections*, I, T. and C. Constable, Edinburgh, 38–45.

Highland Village Association Ltd. (HVAL.) (1911) *Home Life of the Highlanders 1400–1746*, Robert Maclehose and Co. Ltd., Glasgow.

Knox, J. (1784) *A View of the British Empire, more especially Scotland* (4th edn.), London.

Bishop Leslie (1578) 'History of Scotland', in Brown, P. H. (1893) *Scotland before 1700 from Contemporary Documents*, David Douglas, Edinburgh, 113–183.

March, E. (1971) *Inshore Craft of Britain*, I, David and Charles, Newton Abbott.

Mather, J. Y. (1969) 'Aspects of the Linguistic Geography of Scotland III: Fishing Communities of the East Coast' (Part 1), *Scottish Studies* 13, 1–17.

Miller, H. (1844)'Report by the Commissioners for the British Fisheries of their Proceedings 1842' (unsigned) *North British Review* 1, 326–365.

Munro, R.W. and Munro, J. (1966) *Tain through the Centuries*, Tain Town Council, Inverness.

Mussel Grounds in Tidal Waters and Drying Grounds for Fishermen's Nets (1887), HMSO., Edinburgh.

Neish, R. (1952) *Old Peterhead*, P. Scrogie, Peterhead.

New Statistical Account (NSA):- XII, 344–396 (Peterhead Parish). XII, 589–598 (Slains Parish).

Old Statistical Account (OSA.):- V, 275–286 (Slains Parish)

Reports from Committees of the House of Commons (RCHC.), vol. X (1785–1801).

Report of the Committee on Scottish Mussel and Bait Beds (RCSMBB.) (1889), HMSO., Edinburgh.

Scottish Record Office (SRO.) AF 7/135. (Schedule of Mussel and Clam Beds).

Scottish Record Office (SRO.) Register of Deeds (RD) 2/125.

Smith, B. (1992) 'Adam Smith's Rents from the Sea: Maritime Sharecropping in Shetland', in Smout, T. C. (ed.) *Scotland and the Sea*, John Donald, Edinburgh, 94–113..

Smith, H.D. (1984) *Shetland Life and Trade 1550–1914*, John Donald, Edinburgh.

Smith, R. (1986) *Shetland and the World Economy. A sociological history of the 18th and 19th centuries*, University of Edinburgh Ph. D. (unpublished).

Stewart, C. (1869) *A Treatise on the Law of Scotland relating to the Rights of Fishing*, T. and T. Clark, Edinburgh.

Symon, A. (1907) 'A Large Description of Galloway' (1684–92), in *MacFarlane's Geographical Collections*, T. and C. Constable, Edinburgh, 51–128).

Washington, J. (1849) 'Report' in Parliamentary Papers, 51: i–xxiii, i–73).

Watt, R.A. (1989) *A Glossary of Scottish Dialect Fish and Trade Names*, Department of Agriculture and Fisheries for Scotland, Scottish Fisheries Information Pamphlet no.17, HMSO., Edinburgh.

7

The Herring Fishery in the Nineteenth Century: The Rise to Pre-eminence

ORGANISATIONAL DEVELOPMENTS: ESTABLISHMENT OF THE FISHERY BOARD AND THE BREAKTHROUGH TO SUCCESS

The various efforts which had been made over centuries to capitalise on Scotland's main fishery resource, the herring, finally came to full fruition in the 19th century. This was related to a combination of circumstances, but could be summed up in the experience previously gained in organisation (especially in the latter 18th century), and the general economic advance that was bringing Britain to the forefront in the world commercial system. In direct terms it was obviously related to the setting up of the Fishery Board as a promotion agency in 1809, which can be judged as an essential success in developing a system of quality control in herring curing, as well as stimulating general improvement in the various aspects of the trade. Also important after the end of the Napoleonic Wars was a period of general peace and substantial political stability which facilitated great expansion of trade in Europe and beyond. The overall result was a sustained expansion, and from a level of production of around 50,000 barrels annually at the start of the century it passed a million barrels in the 1880s, and was to touch two million at the peak (fig. 7.1); by the latter 19th century the levels of production were several times any ever achieved by the Dutch. The great rise in the herring fishery was to change the balance of effort among Scottish fishermen. While other fisheries for white and shell fish continued, as the 19th century progressed the main tendency was for the herring fishery more and more to dominate the scene, although the other fisheries were still prosecuted to varying extents outside the herring seasons. The main

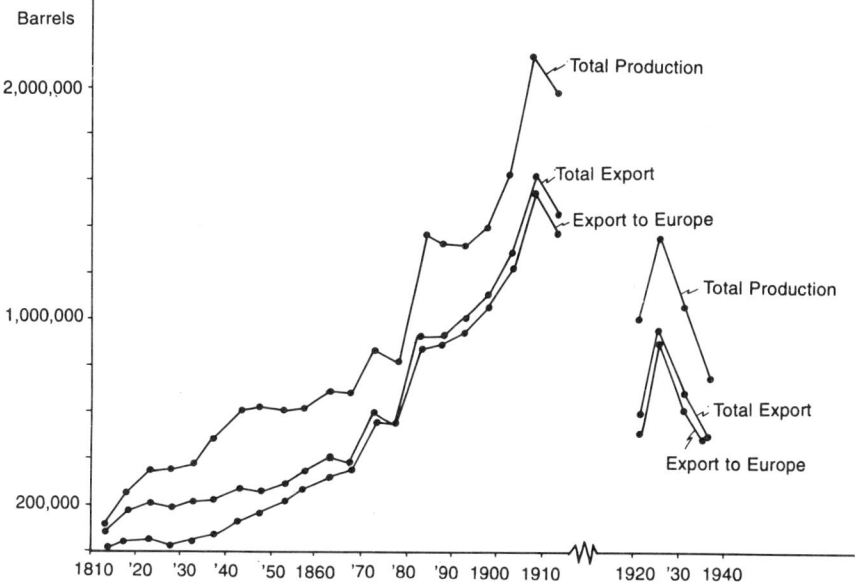

Fig. 7.1 Herring cure and export, 1809–1939,
by 5-year means.

seats of the fishery were emphatically on the East Coast and the fishing was dominated by East Coast boats, although the West Coast was also to play a distinctive part. This was partly through a considerably separate development on the Clyde, but also to the West Coast playing host to East Coast boats for part of the year (ch. 13).

In the 18th century, it was the accepted orthodoxy that government intervention and encouragement was proper for the promotion of new developments. For the herring fishery there had been a lively debate on the best means of encouragement, and the main development efforts were to take the form of the granting of tonnage bounties for the fitting out of big boats to fish in the open sea in the Dutch manner; and this was supplemented to an extent by export bounties. There was, however also a considerable body of opinion which felt that, with Scotland's position near to the herring grounds, more emphasis should be given to the open boat fishery, which could be developed with much less capital expenditure; and this implied that the main challenge was to organise a system of oversight of curing on shore which would guarantee the quality of the product. The linking of the'boat' fishery to onshore inspection of curing by fishery officers spread around the coast was in fact to be vindicated as the main key to success.

The granting of a bounty of 1/- per barrel for herring cured on shore had in fact been begun as a secondary element in policy in 1785, and by the early 19th century had contributed to a significant element of success, especially with the expansion of the fishery on the Caithness coast. While the Fishery Board from its inception gave an enhanced emphasis to the boat fishery by giving a bounty of 2/- per barrel on herring satisfactorily cured on shore, in its early phases it still wished to promote the offshore fishery. In addition to giving tonnage bounties to busses, it also aimed to develop the 'coast' fishery: this was to involve boats in the 15–30 ton range, which had to be at least half-decked, and which were to fish on the open sea and cure their catches on board (ARFB. 1810: 10). Herring caught by boats on the coast fishery also qualified for the barrel bounty, and in addition herring for export qualified for the excise bounty of 2/4 per barrel. Although the export bounties were discontinued in 1815, at the same time the level of bounty on the cure was raised to 4/- per barrel, at which level it was to continue for 11 years, before being phased out between 1826 and 1830. In effect the barrel bounties were successful in pump-priming with the establishment of a substantial boat fishery by 1830, the products of which supplied the home market and gave a considerable surplus for export. Although complaints were not wanting with the phasing out of the bounties as the *laissez-faire* ideas on economics became accepted orthodoxy, the fishery continued to grow. The new economic climate was not in any case wholly disadvantageous, as the burdensome salt duties were also repealed in 1824.

However, there was an adjustment in priorities by the Fishery Board in the light of the limited success in its efforts to promote the coast fishery. After it became clear by the early 1820s that the development of the boat fishery was leaving that of the coast fishery far behind, it was decided from 1824 to divert the £3000 from the annual budget which had been ear-marked to encourage the coast fishery to other uses. These were mainly to aid the construction of harbours (ch. 16), but also for a run of years to help poor fishermen in repairing damaged boats.

As well as promoting increase in production, the Fishery Board was much concerned, especially in the early phases, to improve the handling and curing of the herring. At the beginning they laid it down that for catching nets with a mesh size of at least one inch from knot to knot be used, to prevent the catching of undersized and immature herring. They did, however, have to allow a few years grace for the wearing out of gear already in use. To reinforce this it was stipulated in 1816 that each barrel contain at most 1250 herring. To improve quality there was also the encouragement of the shaking of herring from the nets at sea, rather than after the boats returned to land.

The Board was also concerned to stipulate regulations for barrels, so that they should be leak-proof and sufficiently robust to withstand transport: in 1815 it was ruled that barrel staves be a minimum of 0.5 in. thick, and they also moved to eliminate the use of undersized barrels, as the big majority were less than the standard capacity of 32 gallons (ARFB. 1829: 3).

More important still was the improvement of curing practice, and for a start it was insisted that the herring be cured within 24 hours of landing, and that they should not be allowed to lie in troughs open to the weather. Gutting of the herring was also promoted as a means to a better and more lasting cure: this had often been omitted, especially when landings were heavy. Encouragement was also given that the gutting should be done by opening the belly of the herring with a knife for the extraction of the gut, rather than by the cruder method of using the finger. A most important element of policy which took some time to be accepted was the proper sorting of herring for packing: this involved the separation of full from spent (spawned) herring, and also the separation of adults from the younger 'matties' (or 'matjes', a term borrowed from the Dutch); and it also involved the discarding of any fish that were broken or were in any other way unsatisfactory. Basic too was the use of adequate amounts of clean salt. When gutted and packed, the barrels were allowed to stand for ten days to 'pine', before being filled up with more herring and salt, and the lids put on to seal the barrels. To qualify for the bounty they had to be inspected on a sample basis by the fishery officer, after which they were branded with the crown brand as a guarantee of quality. Initially there was slow progress in getting the principle of sorting accepted, but from the early 1820s, Banff became known for its improved practice (RBPFSJ. 1856: 7), and the ensuing more ready acceptance on the continental markets led to a fairly rapid spread of the practice.

There were various other matters on which the Fishery Board exercised jurisdiction, and one of the most important of these was marine superintendence. With big numbers of boats involved in the fishery, and boats often moving from one part of the coast to another, there was always the danger of conflicts of interest between fishermen; and this was not limited only to Scottish fishermen but included those from other parts of Britain and from continental countries. With boats often in relatively close proximity putting out trains of drift nets which could extend for half a mile or more in a fishery that essentially operated at night, there was obviously scope for gear damage and indeed collisions of vessels. For superintendence the Board operated two fishery cruisers, and in time it also brought in other provisions, the most important of which was under a new act in 1868 in the registration of boats, and the accompanying painting of distinguishing numbers on hulls and sails (ARFB. 1868: 4).

A matter in which the Board helped inshore fisheries at minor cost was in the provision of communal village mercury barometers in the times when household barometers were rare or unknown. By 1868 there were 34 of those (ARFB. 1868: 50), and they were of service to fishermen, especially in the prediction of sudden weather changes.

As well as the encouragement given by the Fishery Board, herring curers as a group also played a key role in the expansion of the fishery. Although by the 18th century there were fish curers operating along much of the Scottish coast, the main experience and expertise in herring curing was still in men on the Forth and Clyde. It was curers from the Forth who, originally coming for the cod and ling fishery to Caithness, saw the potential there for the herring fishery; and they induced fishermen to prosecute it by giving price guarantees under the engagement system. Curers from these traditional areas, either directly or through their agents, continued to play an important role in the expanding fishery, although local curers also became main actors on the scene. While curers generally had one main base, the bigger curers tended to spread their interests to several. In the case of James Methuen, the most prominent of them all in mid-century, his yards were established all over Scotland and he extended his operations into eastern England and the Isle of Man; and in aggregate he had thousands of boats fishing for him (Bertram 1865: 259, 260). The curers were the organisational pivots of the fishery and the 'engagement' system originally developed in Wick at the end of the 18th century was to prove one of the two main organisational pillars which gave an important element of guarantee and stability in a fishery in which there was always a big element of uncertainty. The other pillar was the crown brand on the barrels, which gave a government guarantee of proper curing.

While there was some flexibility within the engagement system, at its core was a price guarantee by the curer for the herring caught during the summer season of about two months which began in July and extended into early September. In giving this the curer obviously had to anticipate the market, and there was an inevitable element of speculation in his calculations. The herring crew were given a 'complement' or upper limit to the amount of herring to which the agreed price applied, and this was most often 200 crans (*c*. 35 tons). It was in fact exceptional for this amount to be reached during most of the 19th century, so that the guarantee usually applied to the whole catch. Moreover it was often possible for a crew to dispose of over-complement herring, although usually at less favourable rates. The characteristic pattern was for curers to show a very considerable consensus on the price offered per cran, but various other elements might enter into their agreements with fishermen. Very often a bounty was given at the start of the season, and this became an important way of attracting the best crews; there could be various other perquisites offered by the curer, such as lodging on shore, free cartage and net ground; the latter allowed nets to be spread to allow rain to wash the salt from them.

The organisational basis of the fishery did change in time, and once it was established on a growth trajectory there was considerable debate about what should be the proper role of government, especially as free trade became established orthodoxy. It was decided to phase out the bounties, which had effectively primed the pump for growth, between 1825 and 1830: and although curers and fishermen understandably complained, the momentum of growth was maintained. After the abolition of the bounties there was also a prolonged debate on whether branding also should be phased out and quality control effectively left to the market. However there was continuing support for branding from various groups in the trade (RB.FSJ. 1856: 5–7), and the service of inspection and branding continued to be provided at government expense. Understandably after the phasing out of bounties in 1830 there was a considerable reduction in the proportion of barrels presented for branding; it was eventually decided in 1858 that while branding would still be undertaken by fishery officers, that it would now be subject to branding fees of 4d. per barrel (ARFB. 1858: 2): this once again caused a reduction in the proportion of barrels branded, and this was partly now because there were bigger curers who were prepared to market their cure on the strength of their own reputation.

The other main organisational pillar of the engagement system continued as a main feature of the fishery until the 1880s, by which time it was on a much expanded scale. It was not entirely inflexible: the main variant was the herring price, which varied from season to season according to the market; the other significant variable was in engagement terms in any one season as curers were in competition to get the best crews. Even so it was its insufficient flexibility that was ultimately to lead to its removal from the 1880s.

IMPROVEMENTS IN THE FLEET AND ITS GEAR

Curers were a main source of credit for fishermen in the acquiring of better boats and gear. The progressive improvement of the fishermen's capital equipment was in fact basic to the growth of the fishery. Even before the advent of the Fishery Board, there was sufficient success in the fishery to encourage at main centres like Wick, Buckie and Anstruther the use of bigger boats than were normally used for the inshore line fisheries. With the developing success of the boat fishery, especially on the East Coast, the use of these bigger boats became fairly general, largely through the curers who stood behind the fishermen. At the same time there continued to be some use of smaller craft during the summer season. A strong tradition of reinvesting the profits from the fishery developed among the fishermen; but although there was a slow, if accelerating, increase in the size of boats they were predominantly of open construction until around 1880, despite the efforts of the Fishery Board from before 1850 to encourage the use of decked craft as a safety measure. Although decked boats eventually became accepted, the early ones generally had a low rail, and it is scarcely surprising that the fishermen felt safer in their proved open craft.

In the earliest phases of the boat fishery, craft of some range of sizes was used, although few of them were over about 25 feet in length: they were powered by oar or sail, and they could be pulled up on the beach on return to shore if necessary. However by the 1820s fishermen were acquiring bigger sail–boats specially for the herring fishing and these were 30 feet and more in length; and the need for harbourage was a strong influence in centralising the main effort at the few suitable harbours. By mid-century leading fishermen might have boats of 40 feet, and soon after some were built having a small forecastle forward with bunks; and the first fully decked herring boat was built in 1855 (ARFB. 1855: 2). It was not to be until the later 1870s, however, that new boats were being generally built fully decked and of lengths up to 50 feet, and in the 1890s boats of over 60 feet made their appearance. The increase in carrying capacity of boats was of course far greater than the increase in length, and the cost increased from c £46 in the 1830s (SROAF 34/1: 29) to £500 or more in the 1890s, while the increase in crew strength was from four or five to eight men. With the increase in size of boat there was of course increase also in the size of mast and sail, and often the installation of a second mast. By the late 19th century the weight of mainmast, boom and sail could be two or three tons, and presented a major task to raise by block and tackle. This had to be done regularly at sea, as mainmast, boom and sail all had to be lowered when the boat was lying at her nets after shooting, to prevent excessive rolling.

An important consequence of the trend towards bigger boats was an enhanced need for harbours (ch. 16). Although till late in the century herring boats were commonly pulled up at the end of the season, this was a major effort, and there was a need for piers where herring could more easily be landed and for harbour basins where boats could lie safely. This became a major factor at centralising operation at main harbours, and in the leading area of the fishery, the open East Coast, effort prominently concentrated at major harbours like Wick, Fraserburgh and Peterhead. However, partly because of the congestion that became endemic at these ports, smaller centres continued to be involved in the trade throughout the century.

The increase in boat size is only part of the story of success, however, as expansion of catching power was largely achieved by parallel and equally important improvements in gear: indeed for most of the century the value of nets and gear was actually in excess of that of the boats. At the start of the 19th century, boats are quoted as using up to a dozen nets which were home made of linen or hemp. The improved boats of the 1830s had trains of about 20 nets; numbers of nets per boat continued to increase as boats increased in size, and in the 1860s a major advance was made with the introduction of nets of cotton: these were lighter and fished better, although they were less robust and more easily torn. This enabled boats to carry more nets, and it was calculated in 1878 that this substitution had allowed a boat to expand the area of netting put down in the sea by as much as five times without significant increase in the weight of gear (Buckland *et al.* 1878: xxii). By the

1890s the biggest boats were shooting as many as seventy nets or more, and a net train could extend to two miles or more compared with lengths of around quarter of a mile or less early in the century.

All hauling was manual originally, and the nets often simply tied together end to end, and stones had to be tied along the foot ropes of the nets to sink them into fishing position when they were shot. As longer net trains developed it was necessary to have a connecting rope, and in the 1850s the use of manual capstans began for the easier hauling of this rope. Towards the end of the century the heavier bush-rope started to be used for linking up the net train, and this passed under the nets when in the sea, and rendered unnecessary the use of the sinker stones.

From the data kept by the Fishery Board it has been possible to measure the increase in value of the boats and equipment in the period when improvement accelerated after the mid-19th century. The value of the equipment (boats and gear) per owning fishermen on the East Coast between 1854 and 1882 increased in real terms by 144% (Gray 1978: 98).

Although there was the general increase in boat size and the improvement in gear, expansion in the first half of the 19th century was primarily to be attributed to a big increase in the numbers of boats participating in the fishery. Once the corps of fishery officers was established around the coast, it became usual for most fishing boats to make some effort at the summer herring fishery; and with a general rise in population of fishing settlements at this time, there was a persistent increase in the size of the active fleet. There were already thousands of fishing boats on the Scottish coast at the start of the 19th century, and there is no clear measure of how many of them were involved in the herring fishery. The total strength of the fishing fleet in the major inquiry of 1855 was 6,744 (ARFB. 1855: 41), and by 1866 the number involved in herring fishing could be put at 9,047 (Buckland *et al.* 1878: xxii): this latter figure was very probably about the peak, as thereafter the expanding fishery became more capital-intensive and concentrated in fewer better boats, the number in 1876 being 7,345 (Buckland *et al.* 1878: xxii). However such data are incomplete in showing how the bigger boats were coming to dominate the catching power, and this is better shown by the figures for first class boats, which are those of 30 feet and over: in 1855 the total of first class boats had reached 2,743 (ARFB. 1856: 24) and in 1881 it had reached 5,101 (ARFBS. 1882: xv). By the latter date many boats were over 40 or even 50 feet and carried from 40 to 50 nets and could have crews of up to seven or eight men. The total area of netting in use by 1878 had risen to an estimated 230 million square yards, and the striking contrast that this made with the 1855 figure of 77 million square yards also emphasises the great enhancement of catching power that had come with the move to cotton nets in the 1860s. In fact the 1878 figure for netting represents a peak, as subsequent expansion was achieved by more intensive use of gear by fewer bigger boats. The trend in numbers of boats involved in the Peterhead district is a reflection of the general trend (fig. 7.2). After an increase in fishing effort in the early part of the century, there was some falling off from the late 1840s with the reduced market, but this was followed by a strong expansion to the 1870s which was achieved by an increase in the strength of the fleet. From the 1880s onwards expansion was achieved more by the concentration of catching power in bigger boats; and from 1900 onwards these were to include a rapidly increasing number of steam drifters (ch. 8).

Despite the various improvements, there was limited increase in the catches of the average crew until late in the century. Even in the early days a single lucky catch could exceed 50 crans, and by the 1880s 100 or even 150 crans was known in a single haul; but although the best boats might reach in a season the usual engagement limit of 200 crans, this was unusual until the 1860s. However by the 1880s the best boats might reach a seasonal total of over 400 crans, and by the early 1890s were known even to reach 1,000

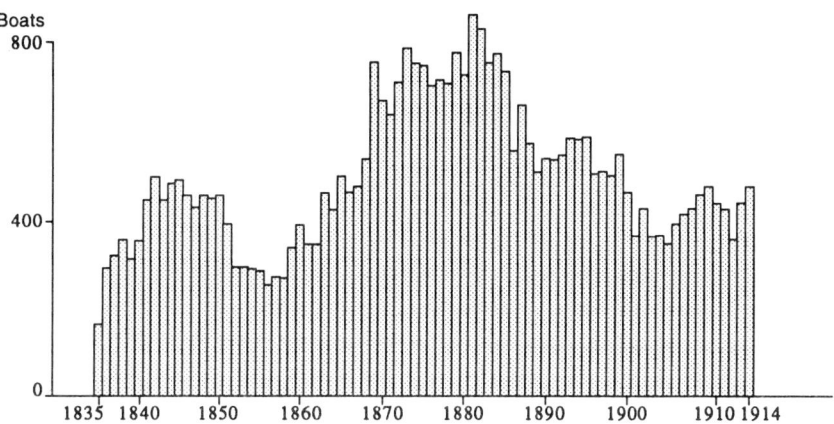

Fig. 7.2 Numbers of boats at herring fishery in
Peterhead district, 1830–1914.

crans. This meant that the best boats were establishing a long lead in the fishery, and this is shown in the earnings of both boats and individual fishermen. Seasonal earnings of the best boats by the 1880s were known to top £400, and for deck hands with no share in boat or gear might reach £30 (Coull 1986: 325): this was far above any average for even skilled workers at the time.

The great expansion in catching power was also accompanied by another important change in an expansion of the area fished. In the early 19th century the fishery was confined to within 10 or 15 miles of the coast, and this changed little in the first half of the century. Thereafter, however, a number of factors combined to widen the area exploited. Increasing numbers of boats and lengths of net trains were causing congestion on near-shore grounds, and the bigger boats could venture further off and fish with less competition, despite some increase in risk of not being able to get back with their catches in good condition from the greater distance. In addition there were increasing reports of herring being scarce on inshore grounds: this must to some extent be a reflection of more boats in competition for available fish, and of the increased catches that were needed by bigger and more expensive craft. It may well also have been due to some change in the migration patterns of the herring – a phenomenon which is well attested from time to time.

From mid-century there was a notable expansion in the area fished: and by 1867 in the Peterhead district, for example it was estimated that a full half of the catch came from beyond the former threshold of 15 miles (SRO. 34/4: 105), and boats were ranging as far as 30 to 40 miles off. This trend was to continue, and by the end of the century the area fished had extended to as much as 80 to 100 miles from land. By this stage there was a situation near to maximum efficiency, in that the herring came from as wide an area of sea as they had done with the Dutch, but the great advantage of onshore curing continued, with only a limited effort at curing aboard.

What curing aboard was done continued to be largely on the basis it had been in the previous century, with the busses mainly serving as depot ships for the boats which caught the herring. In addition to continuing to operate on the West Coast, this was also important in the early stages of the fishery at Orkney: here the busses were a main means of disposal for the boat fishermen, and often gave better prices that land-based curers. Busses too in the emerging main centre of Wick in the second decade of the 19th century are on record as being important in taking surplus production and preventing gluts (SRO. AF 36/12: Report for 1820).

CHANGES IN THE GEOGRAPHICAL DISTRIBUTION OF FISHING EFFORT

In the early 19th century, Caithness had already emerged as the main herring fishing area of Scotland, although there was also considerable activity in the traditional areas of the Forth and Clyde; and in the latter case, as well as the fishery on the Clyde itself, the Clyde towns like Greenock and Rothesay were also involved in sending busses to the Minch fishery and to Orkney.

The general trend in the first part of the 19th century was for there to be a continued tilting of the balance towards the East Coast, and indeed the whole development of the 'boat fishery' was largely determined by events on East Coast, although in time it spilled out beyond it. The Caithness fishery was already established and continued to expand; Orkney and Sutherland early became involved, and substantial fisheries were established elsewhere on the East Coast, especially in the Aberdeenshire ports of Peterhead and Fraserburgh. At each of the main centres, numbers of boats fishing in the season regularly ran into hundreds, and the overall trend was for their numbers to rise. Caithness regularly had a fleet of over 600 operating by the 1830s, of which upwards of a half were visitors; and by the early 1850s the total had risen to over 1,100.

The build-up of the Caithness fishery created great pressures on Wick harbour and town, and there was an early overspill to other places on the Caithness coast. At first this was mainly to places like Staxigoe and Ramsgoe in the immediate vicinity of Wick, but curers set up at Sarclet, Lybster, Whaligoe, Dunbeath and a number of other places to cater for the growing fleet. Wick was regularly the leading port of the area, but places like Lybster and Helmsdale might land as much as 20% or 30% of the total catch. Caithness was to reach its peak in fishing effort in the 1850s, by which time the number of boats at fishing stations in the county approached 1500, and Wick itself had as many as 1100. Inevitably this created a variety of strains, and congestion became a serious problem.

In the early stages Orkney played an important secondary role to Caithness: the main early centre of St. Margaret's Hope on South Ronaldsay on a sheltered bay was only 25 miles from Wick. There was a rapid increase in effort on Orkney from the establishment of the Fishery Office in 1816: in 1818 there were 160 boats engaged for the fishery (SRO. AF 29/79: 28) and in 1821 the number was 268 (SRO. AF 29/79: 79). At the latter date the cure on shore had risen to 19,135 barrels, but this was only part of the picture. It became an important part of the Orkney pattern that Clyde busses came in the season and based themselves in Widewall Bay on South Ronaldsay. As many as 30 or 40 of these vessels would come and as well as fishing with boats of their own, they bought herring from the local fishermen and dominated the market to the vexation of curers ashore. Stronsay, which in later years was to be the main base of the fishery in Orkney was also early involved mainly through the promotion of Samuel Laing, the local landlord. The fact that the stranger busses were to dominate the Orkney fishery for fully three decades could only inhibit the development of local shore-based curing, and this is an important factor in the archipelago falling somewhat behind as the Scottish fishery expanded. Even so, when expansion got under way again after the hiatus of the 1840s, Orkney was still capable of regularly putting into operation a fleet of well over 300 boats for the season in the later 1850s (AF 29/83: 220), although numbers thereafter did tail off as the general size of boat increased. In Orkney most of the herring fishermen continued to be crofters and farm labourers who participated seasonally in the fishery and did not have the commitment to invest in bigger boats.

As the fishery became established in response to the bounties for shore based curing, herring were landed all along the East Coast of Scotland. Fraserburgh (from the 1820s) and Peterhead (from the 1830s) emerged as rivals to Caithness. Curers in these ports could give terms which were as good as, or better than, those at Wick, and the town of Peterhead was also reported as taking measures to encourage curers as its former main activity of whaling declined (*John O'Groat Journal*: 24th Aug. 1849). Although the British Fisheries Society did undertake a major programme of harbour building at Wick to solve the problem of congestion, this was to prove an expensive failure (ch. 16). Harbour improvement was in fact more successful at Peterhead and Fraserburgh, and from the 1860s development at the Aberdeenshire ports outpaced that of Caithness: from the 1870s to the end of the century the general level of output at both Peterhead and Fraserburgh exceeded 200,000 barrels, and at the peak in the 1880s there were over 2,000 boats working from the Aberdeenshire coast. Here too there was an inevitable problem of congestion despite harbour improvements, and this was alleviated by curers setting up at smaller ports like Boddam, Port Erroll and Rosehearty.

Employment in the fishery rose to a peak in the 1880s; although there was to be subsequent expansion it was achieved by greater intensity in the use of both labour and capital. The available data refer to all personnel involved in the fisheries for which the catches went for curing; this means that a small proportion counted would have been engaged primarily in the curing of cod and ling, with little or no involvement in the herring fishery; and it also means that there were other workers in other fisheries who were unaccounted for. However the total personnel enumerated by the early 1880s approximated to 100,000: in 1882 almost half (48,296) were fishermen, while of the total of 51,100 on shore, the big majority (47,464) were gutters, packers and general labourers, the remainder being coopers and the curers themselves.

There were regional differences in the distribution of resident labour, although these were less than the regional differences in capital assets, as there were inter-regional flows of labour to help crew the boats and to work in the curing yards. The main concentration of capital and labour was in the North-East (i.e. the districts from Aberdeen to Findhorn), with a prominent outlier in Fife. The North-East had over 40% of the 5,101 Scottish boats of over 30 feet, and when the Fife fleet was added it was over one half. However the 13,368 fishermen enumerated in the North-East were 27% of all fishermen, while Fife had 7% of the total. In on shore workers the East Coast was more prominent, with a total of 36,748, or 72% of the overall total; and this included 20,939 (41%) in the North-East.

There were changes in the distribution of fishing effort which are much linked with the continuing development of mobility on the part of the curers and fishermen. This involved the growth of considerable fisheries linked to shore based curing from the 1840s in Lewis and from the 1880s in Shetland: these are discussed more fully in the chapter on mobility (ch. 13).

SEASONAL RHYTHM OF THE FISHERY

The engaging of boats by curers at particular places gave an essential stability to the pattern for any one season, and there was a regular rhythm in the build-up and decline of effort over the period.

This is illustrated by the trend in numbers of boats fishing at the main port of Wick for 1860 (fig. 7.3). At a port, there was always a locally based fleet, but the tendency was for this to be augmented at the start of the season as visiting (or 'stranger') boats arrived to participate in the fishery: and at main ports the visitors were regularly the main part of the fleet. They came in greatest numbers from the places along the Moray Firth and from Fife, where the home locations were less suitable for prosecuting the fishery than the peninsular locations of Caithness and Aberdeenshire. It was acknowledged too that the visitors were generally the best equipped, and their catch rates were distinctly above average. It might take as much as two or three weeks at the start of the season for the fleet to build up to full strength; and there were regularly herring sales at the start of the season outside the engagement system, when the still filling herring were generally less suitable for curing and landings were usually small. After all boats had arrived there was a period of stability in the strength of the fleet for over a month, although there were short-run variations in the numbers of boats putting to sea, mainly in response to the weather: in stormy conditions fewer would venture out. At the end of the season the active fleet dwindled: the quality and abundance the herring deteriorated then, and the visiting boats left, although there was a tendency for the home fleet to persist rather longer.

During the season the obvious goal of each crew was to reach its complement: in the early decades of the 19th century it was unusual for a boat to reach the normal 200-cran complement, but it became more and common for leading crews to do this as boats and gear improved, especially after about 1860, when the best boats fairly regularly reached totals of 300 to 400 crans or more; and they still tended to continue fishing till the end of the season, although over-complement fish might receive a reduced price.

FLUCTUATIONS IN THE YIELD OF THE FISHERY

Herring, along with many other pelagic species show great variations in abundance from year to year, and there can also be sizeable differences in the detailed direction of their spawning migrations. One of the consequences of this in the fishery often are high-amplitude variations in both the short and the long term. While the variations in output for Scotland as a whole are shown in the

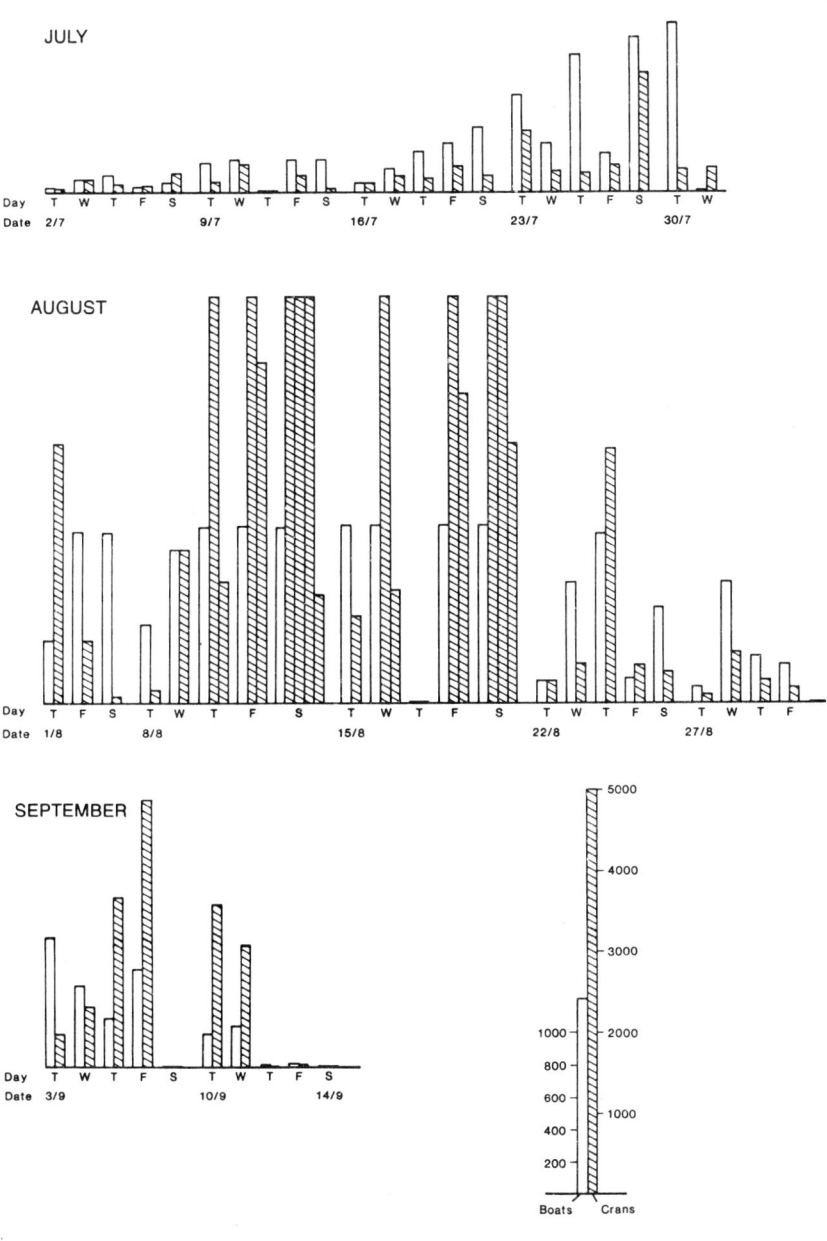

Fig. 7.3 Daily herring landings in Wick district
for 1860 season.

overall annual gross totals of production, this does mask the greater variations that individual ports frequently experienced.

In fact more striking are the short-run variations that occurred during the season, in which the effects of variations in weather (with gales and fogs) was superimposed on variations in location and abundance of the shoals. Not infrequently the great part of the landings of the season could be concentrated in a very few days, and this was especially so up to about 1860, when the great part of the fishing was within the restricted zone of 15 miles from the coast. In Caithness as early as 1838 it was known for a boat to land over 90 crans from a night's fishing, at a time when many boats had less than this for a whole season; and in the same year over 200 crans were landed by another boat in one week (*John O'Groat Journal*: 17th Aug. 1838). The degree of detail of daily landings itemised varies in different districts, but those from the then leading district of Wick about 1860 shows this day-to-day variation in graphic detail (fig. 7.3). As long as bad weather did not intervene there were five landings each week, as boats did not fish on Saturday and Sunday nights. In any week, the daily landings regularly vary by a factor of at least five, and in some weeks the factor runs into hundreds. There were also big variations in weekly totals, and it is striking that in an eleven-week season over 70% of the landings were made in the fortnight ending 17th August; and Saturday, 10th August alone accounted for 18% of the overall total. This example shows the great problem of gearing the work of curing to that of fishing: it must have been a massive strain on the fishermen and curers alike to cope with the alternation of gross overwork and enforced idleness.

MARKETING THE HERRING

It was official policy to expand both the yield of the fishery and the export of the cure, although after 1815 it was found expedient effectively to consolidate barrel and export bounties. From the earliest phases it was rare for less than half of the cure to be exported, although home consumption also regularly ran into hundreds of thousands of barrels. What is unclear for most of the century is the part of the catch which was consumed fresh, and this was not systematically recorded till the 1880s.

Although there was a rapid increase in exports to over 200,000 barrels annually (fig. 7.1), until the 1840s the great part of this was to the outlets of Ireland and the West Indian plantations, where a poorer quality of cure was acceptable. This market orientation was mainly responsible for a slackening in the expansion of the fishery in the 1840s, with the combined effect of the freeing of the slaves on the plantations and of the Irish potato famine. For over a decade at this stage the home market was the dominant sector, although it was showing a healthy expansion with rising average incomes and improving transport.

However there was from the start a market foothold in the main markets of the North European Plain in Germany and Poland, and in the second half of the century this was also to extend into the Eastern Baltic, to what is now Lithuania, Latvia, and Estonia as well as Russia. A comprehensive overview of import trends on the continental markets is scarcely possible, supplied as they were by Holland, Norway and other countries as well as Scotland. However for the main single importing port of Stettin, it is clear that by the 19th century the contribution of the Dutch was minor, and the essential competition was between Norway and Scotland. Norway like Scotland had problems in monitoring the quality of its cure (Östensjö 1963: 139, 140, 148, 149), and its herring were bigger although from a winter fishery in which fish are leaner; nevertheless there was a great surge of development in its fishery in the early 19th century and it dominated the market till about 1840. From mid century, however, the Scottish cure got a decisive advantage: the increase in the practice of sorting before packing, together with the inspection of the Fishery Board was now paying dividends, and Scottish exports forged ahead until in the latter part of the century they were regularly over a million barrels a year. The northern European market continued to be very much the main outlet, and here market expansion was fostered by the general rise in population and in living standards, and by improving communications, as the main arteries of the rivers were increasingly supplemented by the railway. Linked to this was a marked price rise for herring after the middle of the century, and for a period of about 25 years from 1855 the general price level at around 30/- per barrel was almost double that of the 1840s (Gray 1978: 58, 59). There were also important structural adjustments in the trade: in the early phases curers had links with agents on the continent and disposed of their herring on account. The importance of the trade to continental consumers became such as to prompt leading importers from the middle of the century to send their agents to the Scottish ports

(RBPFSJ. 1856: 12), and this allowed the curers to sell to them directly and eliminate one of the risks to which they were subject.

None the less it was a consistent concern of the Fishery Board to increase the sales in other parts of Europe, and it made unfavourable comments on various occasions about the disincentive to trade that was still present in the import tariffs that continued to be levied by the Russian and Austro-Hungarian empires for most of the century. Even so the Scottish product was, in fact, to dominate the market until the inter-war years.

THE CLYDE AREA: A SEPARATE PROVINCE

Before the big developments of the 19th century the Clyde area had come to the forefront as the main herring fishing area in the country, both in virtue of the fishery on the Firth of Clyde itself, and of the fitting out of busses by the burghs of the area for fishing in the Minches.

This continued to be an active area in the fishery, although on the national scene it became overshadowed by the East Coast. It did not have the emphasis of the rest of the country on curing for export: with the coming to the forefront of the Strathclyde area as the main industrial power-house and concentration of population in the country, fishermen and merchants in this area had from the early 19th century the substantial advantage of a bigger and more accessible market for fresh herring within reach than was possible anywhere else. In the 18th century, about two-thirds of the catch had been cured, and while curing continued, the considerable increase in production went to the fresh market. As the century progressed steamers increasingly linked with the catching boats at landing points, and the herring reached Glasgow in the best of condition.

In the herring fishery there was a distinct trend for fishermen who had been part-time and seasonal to become more fully committed to the fishery. This was primarily prosecuted in the early 19th century by the traditional drift net, and although there were improvements in equipment like those that had been introduced on the East Coast, there was less scope for the use of lengthened net trains, especially in the main area of the fishery in the constrained waters of Loch Fyne. A more significant development here was that of the gear that was at first termed the 'trawl', but can less confusingly be recognised under its later name of the ring-net: this was a type of gear that was particularly suited to use in the sheltered waters of this area. Essentially it was a long strip of netting which could be used to surround a herring shoal; in the early use of the gear this was then pulled into the shore, but in time the technique of closing it off below the surface to trap the shoal in the sea was mastered. Its effectiveness had been proved in the 1830s (Martin 1981: 6), but it led to a very serious confrontation with the users of the traditional drift-net. It was certainly very difficult for fishermen using the two gears to work in the same area, and there were also concerns that it might deplete the stocks. The ring-net was in fact cheaper as a catching instrument, and while its greater efficiency doubled fishermen's incomes, the greater supply to the market brought down prices (Gray 1978: 121–122). This conflict was to lead to its being outlawed in 1851; but despite the efforts of fishery officers, backed by the law to prevent its use, it continued to be employed on a clandestine basis, and a bitter dispute was prolonged over some thirty years. Eventually resistance to it did subside and the ban on it was repealed in 1867 (Martin 1981: 24–25) at a time when the general direction of national policy was to do away with restrictions in fishing. It was a case in the end it being justified by its results, and an improved method could not be indefinitely suppressed.

BUILD–UP TO CRISIS: THE 1880S

There were inevitable strains and adjustments during the expansion of the Scottish herring fishery to the position of world leadership, and there were always dangers that supply and demand would get out of balance. By the 1880s there had been a major increase in the numbers of people mainly dependent on it for their livelihood: in the Peterhead district, for example the numbers of fishermen, coopers, labourers, and gutters and packers had all more than doubled in the forty previous years of available statistics (Coull 1991: 136). When a major crisis did develop in the 1880s it struck many thousands of people. At this juncture and on top of a general recession in trade, there was a major increase in the supply to the market with the spectacular herring boom that occurred in the Shetland Islands in the 1880s (ch. 12). There were already serious warning signals for the trade with a glutted market in 1884, but in the absence of any consensus on restraint, production continued to rise in the following year: the consequent fall in prices created great difficulties, and indeed was the ruin of many curers and others. In the nadir year of 1887 there were about 900 herring boats left on the beach during the whole of the year, in a situation when crews could not get curers to give engagements before the start of the summer fishing, and prices were low during the season (ARFBS. 1887: xix). The trade continued in a depressed state for a run of years until there was renewed growth from 1893 (ch. 8). A consequence of this run of difficult years was that numbers of fishermen joined the national exodus which was occurring to overseas countries in search of new opportunities, and one result was that some of them established commercial fisheries at such places as Port Chalmers in New Zealand and Port Dover on Lake Ontario.

Even before the crisis of 1884s curers had mounting concern about the obligations they had under the engagement system. In the early stages of the fishery the general level of engagement prices had been of the order of 10/- per cran, and although there was some falling back in the 1840s with the decline of the plantation and Irish markets, the overall trend was upwards in a period of generally stable prices, and by the 1870s engagement rates were of the order of £1 per cran. At different stages when competition was keen in an expanding fishery bounties varied considerably, and in the enhanced competition of the 1880s they could reach anything up to £70 per crew, which was a considerable additional expense for the curers. There were thus big pressures and commitments on the curers, and the engagement system also effectively bound them to buy herring while these were still in the sea. In addition the agreement was made at a time when neither the quantity or the quality of the catch was known, while there could also be variability or uncertainty at the market end of the trade. Although the system did have elements of flexibility with the fixing of the seasonal price, and with the adjustments of bounties to crews, it was now proving too rigid, and the curers campaigned for a system that would allow the adjustment of prices during the season in the light of short-run developments. Eventually from 1885 day's prices were tried and this soon gave way at Peterhead to the auction system: in 1887 there were already 458 boats fishing for the auctions at the port, compared with 102 on engagements (ARFBS. 1887: xxxvii). The auction system spread to other main ports within a few years, although there was a time lag at smaller centres where there was less competition by buyers and it was more difficult to organise: it became an important centralising influence on the pattern of operation, and was especially attractive to the leading curers. While the fishermen were very generally opposed to sharing more of the risk of the trade, it was not all disadvantage for them, as on occasions when supply on the market was short prices were driven up to unprecedented levels. On the other hand it introduced the depressing experience of catches having to be dumped when the market was glutted.

REFERENCES

Annual Reports of the Fishery Board (ARFB.) 1810, 1829, 1855, 1856, 1858, 1868.

Annual Reports of the Fishery Board for Scotland (ARFBS.) 1882, 1887.

Bertram, J, G. (1865) *The Harvest of the Sea*, John Murray, London.

Buckland, F., Walpole, S., and Young, A. (1878) *Report on the Herring Fisheries of Scotland*, London.

Coull, J.R. (1991) 'The Development of the Herring Fishery in the Peterhead District of Scotland before World War I', *Sjöfartshistorisk Årbok 1990* (Norwegian Yearbook of Maritime History 1990), Sjöfartsmuseum, Bergen, 119-142.

John O' Groat Journal, Wick.

Gray, M. (1978) *The Fishing Industries of Scotland. A Study in Regional Adaptation*, Oxford U.P.

Martin, A. (1981) *The Ring-Net Fishermen*, John Donald, Edinburgh.

Östensjö, R. (1963) 'The Spring Herring Fishery and the Industrial Revolution in Western Norway', *Scand. Econ. Hist. Rev.* 31, 2. 135–155.

Report of Bonamy Price and Frederick St. John to the Commissioners of the Treasury on the Subject of the Fishery Board (RBFSJ.) (1856)

Scottish Record Office (SRO.) AF 29 Orkney District; AF 34 Peterhead District; AF 36 Wick District.

8

The Herring Fishery at its Peak
1893–1914

It was in the closing years of the 19th century and in the years before World War I in the 20th century that the herring fishery expanded to its peak: and at this stage it was by a long margin the leading fishery in the globe in volume of production. As well as including fishermen and curers, the organisation which had developed by this time also embraced fish selling companies; these in addition to conducting auctions were involved in keeping boats' books and indeed in investment and could have shares in boats. With the scale of the trade, there were various agents and intermediaries involved in marketing; and there were also specialist boat builders, sail makers, marine engineers, ships chandlers and coal merchants. Even so it is clear that the main capital in the industry was concentrated in the fleets. This was partly because the main activity of curing was essentially one of simple and limited capital requirements. An estimate of 1913 put the whole capital value of all fishing-related premises on shore at $c. £1,500,000$; and although herring curing premises were the biggest single fraction of this, the total value of these was put at $c. £400,000$. (ARFBS.1923:xli). Even so, at this time there were around 1,000 curing yards, employing a total of well over 40,000 people (ARFBS. 1910:xlii). By contrast the capital tied up in the herring fleet and its gear by 1913 was well over £2 million.

The emergence of more capital–intensive operation with steam boats was a fundamental technical change at this time, and it was accompanied by socio-economic changes, as in general only the most successful fishermen could command the resources for them; even so there came an innovation in that in a considerable number of cases shore owners acquired shares in boats. While it is evident that a bigger proportion of the fishermen continued to have a

126

capital stake through net ownership than had shares in boats, the overall result was that more of the fishermen became hired hands, depending only on their labour share. With the capital costs of steam boats there were perforce adjustments to the division of the proceeds among the owners and crew: it became the general practice that running expenses were first deducted for the gross returns, and that the remainder was divided into three equal parts between boat, gear and crew. It was also common practice that the cook and fireman got a weekly wage, but with no labour share (ARFBS. 1911:xvii). There had been adjustments too in the system for dividing up the proceeds on sail boats as the capital costs increased: here after the deduction of expenses from the gross receipts, the remainder was divided into 13 shares of which the crew got six, and the balance was divided between boat and gear (ARFBS. 1911:xix).

A number of factors inter-acted to drive the industry up to its peak. Basic was the recovery of the main cured market from the recession of the late 1880s, and this was supplemented by some expansion and diversification of the home market. Recovery was also strongly promoted by the eventual successful application of steam power aboard drift-net vessels from the end of the 1890s. Of great significance too was the extension of the operation of the curers to the major autumn fishery at East Anglia. There were also developments in the organisation of the fishery, among which the emergence of a small number of fish selling companies at the core of the industry is of outstanding importance: these companies in effect replaced the curers as the pivots of an expanded fishery. The fishery must also, ironically, have benefited indirectly from the economic impact of the arms race between Britain and Germany at this time. It is in addition at this stage that the published reports of the Fishery Board for Scotland are most detailed, and give much statistical and other data relating to the fisheries. The more complete data available show that at the peak in the years before World War I the total landings of herring in Scotland fluctuated around 250,000 tons, and in the top year of 1907 reached 315,700 tons.

After cured production had stagnated in the mid-1880s at between 1,300,000 and 1,400,000 barrels annually, there was a hiatus before growth resumed from 1893, after which output climbed to a peak of over two million barrels (fig. 7.1); and if the major contribution from the East Anglian fishery is considered, which was then dominated by the Scottish fleet, the peak total approached three million barrels. These trends were determined primarily by events on the continental market. After the export to that market had stagnated at around 900,000 barrels in the mid-1880s (fig. 7.1), there was a hiatus before growth resumed in the mid-1890s, but it was then on a steeply rising trajectory and in the peak years before World War I it exceeded 1,500,000 barrels; indeed when the contribution from East Anglia is considered, it was well over 2,000,000 barrels.

As well as the main salt curing sector, there was also expansion and diversification of other parts of the herring market. The home fresh market expanded to a peak which could take up to 40,000 tons annually in the years around 1900, and this was essentially due to the facility of rail transport, supplemented to an extent by steamer connections on the Clyde and to the Western Isles. It was also helped by the increasing practice of distributing the herring in ice, and by rising purchasing power in the cities and elsewhere. The fresh trade also encouraged a regular winter fishery from mainland ports and from Stornoway: although herring were generally scarcer at this period, prices tended to be higher, and with leaner winter herring and lower temperatures the spoilage problem was reduced. However as the 20th century unfolded this market sector went into decline, partly through the rise of kippering, but mainly because other types of fish which were more easily kept came more within the popular price range. It was improved transport in Britain which also was to encourage the expansion of the kippering sector, which involved light smoking: this acted as a short-term preservative, and gave a distinctive flavour to a product for the home market. This trade sector was established from the late 19th century at the leading mainland ports of Peterhead, Fraserburgh and Wick; Eyemouth became an important centre, thanks to the winter fishery on the Forth, and Glasgow was a also a main centre, supplied from the Clyde fishery. In addition, it was possible for island ports like Lerwick, Scalloway and Stornoway also to conduct kippering and for the product to reach the mainland markets in good condition. By World War I, production of kippers was running at upwards of 200,000 barrels (*c.*20,000 tons) annually.

From the 1890s developed the 'klondyking' sector, in which herring were bought in Scotland, and transported lightly salted in steamers to continental destinations to supply their processors: there was a regular trade especially to Hamburg and Altona in Germany, and it also extended to Norway. This gave a level in the market with a price level above that for curing, and as well as being significant in summer, it could also become a major outlet for the lower herring landings in winter.

With these developments in the trade, a prominent result was elements of regional specialisation. While diversification applied to an extent everywhere, there were variations in emphasis related both to the season of the fisheries and to market opportunities. The main centres on the East Coast of Peterhead, Fraserburgh and, Wick along with Shetland were very much dominated by the main summer fishery. This catered mainly to the basic cured trade sector, and in these centres around 80% of production continued to go to pickle curing. In its peak year of 1905 in the leading district of Shetland production actually touched the million barrels in the one district. In contrast, the East Neuk of Fife, where landings were mainly from the biggest winter fishery on the East Coast, were overwhelmingly for the higher value fresh market. On the West Coast as a whole, only a minority share went to curing, and on the Clyde there was virtually no pickle curing: here steamer connection allowed the herring not only to reach the main market of Glasgow, but also to be dispatched onwards to various cities in England as well as Scotland. Steamer connection also allowed Stornoway an important stake in the fresh market from the winter fishery in the Minch. On the other hand the Shetland fishery was overwhelmingly in the summer, and the fat fish spoiled more easily, while the distance to mainland ports was greater: Shetland never had more than a minor share in the fresh market, although it was able to be an important centre for kippering.

Although the crofting communities of the Highlands and Islands continued to have an important interest in the herring fishery in supplying an essential part of the hired male labour for crewing the boats and women for gutting and packing herring, the role remained an auxiliary one. The full reasons for this are complex, although lack of capital and the remotenesss of the region have been recognised as important factors (Gray 1972: 113, 114). The history-making Crofters Commission which reported in 1884 had seen fishery development as a main way to betterment for these communities, and suggested that crofters be given special loan facilities to help them acquire better boats (Coull 1986: 193). The Fishery Board were directed to provide low-interest loans for the purchase of fishing boats between 1886 and 1891: this resulted in a total of 131 loans with an average value of £98 being provided for fishermen in the area (ARFBS. 1899: 171): but this could only make minor inroads to the problem. In fact, this initiative could scarcely have been worse timed in the light of the wider events which plunged the herring trade into crisis from 1884 onwards. The crofters of the Western Isles had generally found the bigger sail boats beyond their means; more so was the much more expensive steam drifter which came on the scene at the turn of the century. Only in Shetland and Caithness did a minority of crofters manage to acquire a handful of steam drifters. Although in Shetland especially crofter-fishermen retained an interest in ownership of herring boats, they continued primarily to use sail boats until the installation of motors began.

THE CONTRIBUTION OF STEAM POWER

The 19th century growth of the herring fishery took place against an expanding use of steam power in an increasing range of fields. Although from the early part of the century there was the development of the marine steam engine, in the first half century of its development it had no direct effect on fishing, and as a very general rule the increase in costs involved made its use impossible. Even when it started to be installed in trawlers from around 1880, it still appeared doubtful if it could be justified for drift-net fisheries.

The first impact of the marine steam engine on the herring fishery was made by 1882, from when there was the occasional use of paddle-tugs to tow sailboats to sea in calm weather (ARFBS. 1882: xxiv). At about the same time steam freighters started to displace the sailing transports which had conveyed the barrels of cured herring to the continent, and by 1893 the steam freighters were dominant in the trade (ARFBS. 1893: 153). There were also sporadic efforts to use steam powered vessels as herring drifters from as early as the 1870s; but although they had an improved mobility, being much less independent on wind and tide, they gave no advantage in the fishing operation itself, in contrast to what occurred with the trawlers; and the great increase in capital and operating expenses at first rendered their use prohibitive.

Aboard the regular drifters, the first use of steam power was not to propel the boats, but to ease the heaviest tasks on board. The most important of these was the hauling of the 'bush' (main) rope, to which the nets were attached when in the sea, and which by the late 19th century was regularly about two miles long. Also very important was the raising of the sail and boom on the main mast:

on the biggest sail boats, between boom and sail the total weight was between two and three tons, and they had to be lowered not only in port, but also when the boat was riding at her nets, to prevent excessive rolling. To raise them with block and tackle was a very arduous task, especially when at sea. The installation of the steam capstan was the answer to both the hauling of the bush rope and the hoisting of boom and sail, and it was from about 1896 that the most progressive crews started to have small steam engines costing £80 to £90 installed to power the capstan (ARFBS. 1896: 164).

From 1898 there were not wanting bold spirits who were prepared to face the extra cost of investing in steam-powered boats. The initial reaction to the advent of steam drifters in the herring fishery was a serious questioning of whether in the herring fishery they warranted the great increase in costs, but the year 1902 was to show matters in a new light. There was a great increase in performance of steam drifters over sail-boats, and their advantage was especially marked when the fishing was at some distance offshore: in 1902 the best hauls were often made at 50 to 60 miles range, and at Peterhead, it was stated that the steam drifters might make three or four landings in a week, while sail boats could only make one or two (ARFBS. 1902: 193). Two years later the fishing ranged up to 100 miles off, and in that year at Peterhead sail boats had a good average at 399 crans, but the steam drifters averaged 836 crans (ARFBS. 1904: 205). From 1902 onwards there was a rapid transition, and by the outbreak of World War I the steam drifter dominated the fishery. It was clear that steam boats could more easily move to where herring shoals were reported, and had the great advantage of a more rapid and sure return to port; in addition now that auction markets were established in all the main ports, with their earlier return in the day the steamers generally got better prices. It was soon evident that average catches of steam drifters were well over double that of sailboats. However it was also evident from the start that steam drifters meant a change in dimension of costs. The cost of the biggest sail-boats in the 1890s had increased to the £500 to £700 range, and the installation of a small steam engine and capstan added another £80 to £90. Ready for sea with spars and sails the cost of such craft had risen towards what had been considered the formidable price of £1000. However the custom-built steam drifters cost £3000 to £4000, and they also ran up considerable coal bills. They were by 1905 earning average grosses fully double on average those of the big sailboats, but on the other hand increased expenses were swallowing up upwards of half the gross. It was this which made adjustments in the methods of dividing up the proceeds inevitable, with the amount of capital now tied up in the boat (see above).

Steam drifters were originally built at a limited number of yards in places like Aberdeen and Montrose which could cope with the construction of steel boats, but once they had proved successful there was also the building of the rather cheaper wooden drifters at a range of places, especially in the North-East: these vessels were priced at £2200 to £2400, or about one third less than the steel drifters, and also cost less to run (ARFBS. 1911:202).

Although there were some attempts to set up companies to own and operate steam drifters, the great part of the Scottish fleet was owned by fishermen, with some help from 'shore' owners who put capital into them. However to buy them the fishermen had greatly stretched their credit, and many were still deep in debt when World War I broke out; and in the changed and difficult conditions of the inter-war period, this debt was to be a continuing millstone round the necks of the fishermen.

The coming of the steam drifter also accentuated another long-term trend, that of extending the effective working year. The East Anglian autumn fishery was to reach a new level of importance (ch. 13) and was to become of near the status of the home summer fishery. There was also an expansion of winter herring fishing, and although catches were limited at this season they might reach high prices for the home market. Drifters also regularly sought to increase earnings by engaging in the great line fishery in the spring; and some developed a new form of versatility in the early months of the year by going with both nets and great lines to the Minch, and switching between herring and line fishing according to catching opportunities and markets. In all the coming of the steam drifter expanded catching power, but it also generated a new level of commitment for the fishermen.

The coming on the steam drifter brought into finer focus a long-term trend which had been apparent in the herring fleet since the middle of the century, and indeed earlier: this was the concentration of more and more of the catching power in progressively bigger and more expensive craft. The numbers of first class boats (i.e. those of over 30 feet) had been growing at a brisk rate from the mid century: the number recorded in the detailed investigation of 1855 was

2,743 (ARFB. 1856: 24), and by 1881 it had risen to 5,101 (ARFBS. 1882: xv). With the crisis of the later 1880s the number of boats started to fall quite sharply, but after a brief hiatus catching power continued to increase with the continued concentration on bigger craft. After this there was the separate recording of boats in excess of 45 feet, and by 1900 the size of the herring fleet was recorded as 3,590: and although numbers had halved in the space of a quarter of a century, now the fleet included 2,115 boats of over 45 feet (ARFBS. 1900: 14, 15, 126). The trend was accelerated with the building of steam drifters: the great majority of them were built in the years between 1900 and 1914, and the total strength of the Scottish herring fleet in 1913 was 2,576 of which 884 were steam drifters (ARFBS. 1913: xiii–xiv). This meant an acceleration in capital accumulation. While the value of the Scottish fleet was published from 1896 onwards, it is only in the period from 1902 that details are available giving the break-down between the different fleet sectors. Even then steam drifters are regularly bracketed with steam liners: this was because the drifters fairly regularly spent part of the year in lining as a subsidiary activity, especially in the spring; however it does fail to distinguish the fleet of full-time steam liners that operated, mainly from the leading white fish port of Aberdeen, and numbering around 50. Even so it is evident that the bulk of the 1896 figure of £1,873,870 for the total value of boats and gear derives from the relatively small number of boats in the trawler sector, although from 1900 onwards the steam drifter fleet was the main sector of capital formation. In 1902 the total value of Scottish fishing vessels was £2,177,500 and of this the value of the non-trawler fleet at £1,151,000 was over one half; and included were 100 steam drifters and liners, valued at £205,620.

The pace of development accelerated markedly in the period up to World War I, and was brought to a climax as the profits of good years in 1912 and 1913 were reinvested. The total value was £4,100,000 for the fleet and gear in 1914, and now the steam drifters and liners with their gear at £2,475,000 was over 50% above that of the trawler fleet. On the other hand, in the period between 1902 and 1914 the value of sail–boats and gear declined by almost half from £1,460,000 to £763,993.

These trends also meant considerable changes in the regional balance of the fleet. The West Coast had been more had more left behind during the 19th century, and although there was a measure of success along virtually the whole East Coast, the most successful herring fishermen were concentrated in the North-East between Peterhead and Nairn, and in the East Neuk of Fife. These had always been to the fore in the moves to bigger boats, and indeed other parts of the coasts of Scotland often progressed by buying boats which were being discarded in these leading areas. The bigger sail-boats were mainly in these areas, and this is illustrated by the regional break-down of the fleet when data first become available in 1894: the West Coast had only 94 (5%) out of the total of 1,727 boats of over 45 feet on the register, and the Clyde area had none. Despite this the Clyde area was doing better than the remainder of the West Coast, essentially because it was concentrating on the ring-net fishery. For this fishery smaller boats, as well as being considerably cheaper, were better suited for manoeuvring in waters which could be narrow and sometimes shallow. There was rapid installation of motors in the Clyde skiffs at this time, and also adjustment in the share system, although it was a smaller scale change than was needed with the transition from sail boats to steam drifters. On the Clyde the system for two sail boats operating the ring net was that there were a total of 12 shares: with four-man crews, each man, each boat and each net all got one share. When motors were installed, an extra half share was added for each engine: this meant that two motor boats had between them 13 shares, while a motor boat partnered with a sail boat had 12 1/2 shares to distribute (ARFBS. 1911: xix).

The bigger boats were represented in all East Coast districts, and they were most concentrated on the south shore of the Moray Firth and in Fife: in 1894 these areas had 1004 between them and accounted for 58% of the total. When capital-intensive operation took a big and decisive step with the move to steam drifters, change in the structure of the fleet accelerated and concentration of ownership in the leading areas became more emphatic and only about 60 boats of the total of 981 steam drifters and liners were owned outside the North-East in 1914. While boats of less than 30 feet continued to be relatively numerous for a variety of small fisheries and other purposes, the class of boat that now had least place was the boat of 30-45 feet: in 1913 their numbers had dwindled to 410 against a figure of 2,646 boats of over 45 feet.

It was at this peak period that the North-East emerged decisively as the power-house of the Scottish fishing industry. By 1914, the ownership of over 85% of the total value of Scottish fishing boats was concentrated on about 120 miles of coast between Aberdeen and Nairn: and although the district of Aberdeen led because of its trawler fleet, the value of the steam drifters in the area was about 50% above that of the trawlers; and the one district of Buckie had almost one-third of the Scottish steam drifter fleet.

Inevitably the investment in steam drifters brought other pressures, and most prominent here was an added impetus to extend the season of operation to help recoup the expenditure. There were already increasing pressures to extend the season from about the 1870s with the bigger sail boats, and with the steam drifters these became insistent. A consequence was the extension of mobility to different bases in Britain at different seasons, and this is discussed in the chapter on mobility (ch. 13).

Landings of herring reached their ceiling in the years prior to World War I, and fig. 8.1 shows their distribution by districts in Scotland for 1910. It also gives an indication of the fishing effort with the numbers of boats and fishermen; and of the scale of the curing organisation in different districts with the numbers of other workers. There were inevitable complications in compiling statistics by this stage, as boats could go freely between the main ports where the engagement system had been abandoned. Also no one year can fully typify the fishery, and although 1910 was overall something of an average year, the fishing was rather poor on the West Coast.

There was further impetus to the long-term trend of centralisation of operation on main ports, and curing almost deserted other ports, apart from in Shetland: although Lerwick became very much dominant here, there was some survival of 'country stations' in other parts of the islands: access to these continued to be easier from many grounds, especially for sail boats. None the less, even the former major centre of Balta Sound went into steep decline. However although the main cause of change in Shetland was the pull of the one auction market at the all-weather and all-tide market of Lerwick, there was a great controversy on the effect of the whaling from centres established in the north-west of the islands from 1903: this occurred at the time that the early summer herring fishery prosecuted to the west and north of the islands went into precipitate decline, and pollution caused by whaling was strongly suspected of being the cause of this (Coull 1994: 55–68).

The Fishery Board regularly enumerated numbers at the peak week in each district, so that it is possible that boats and personnel might be counted more than once. Even so, the landings figures give a basis for comparison between districts, and the numbers of

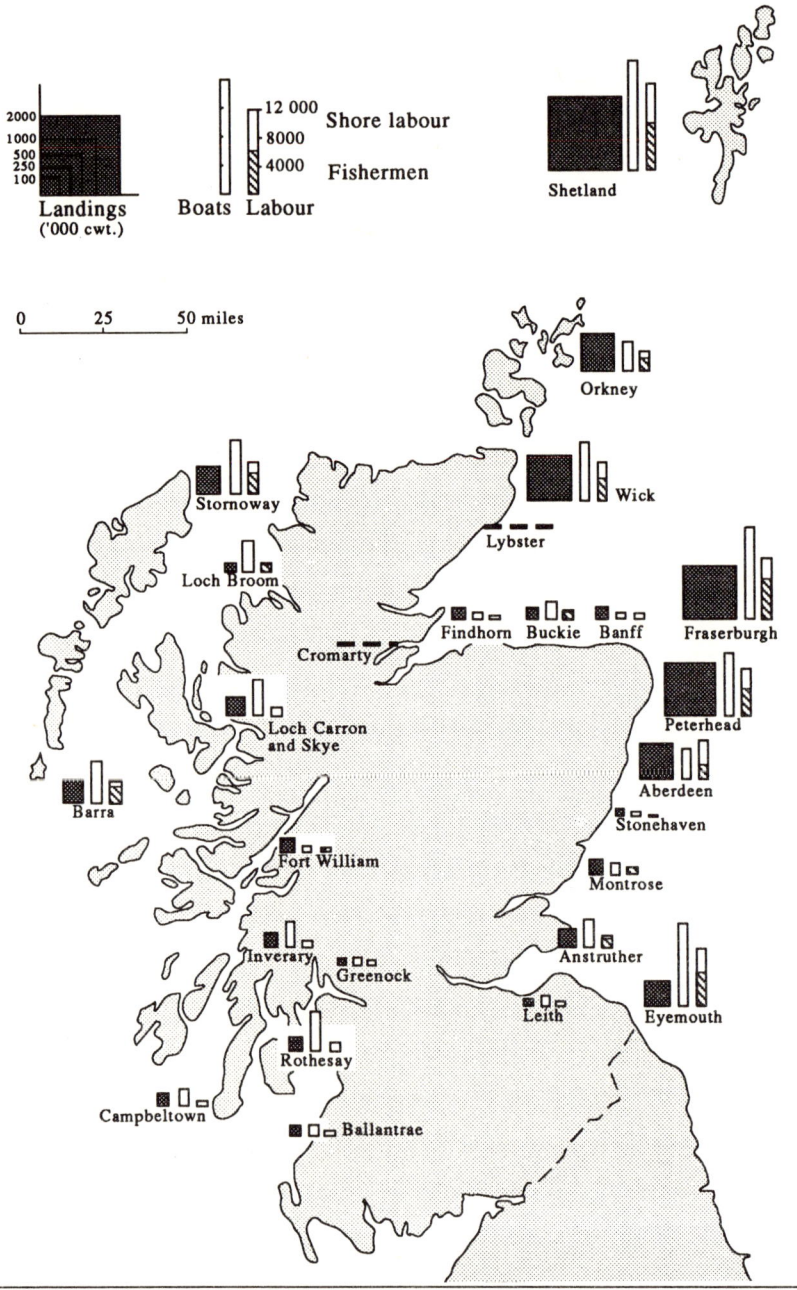

Fig. 8.1 Herring landings in Fisheries districts, along with boats and personnel at peak season, 1910.

personnel are in broad accord. At the peak season at all main centres the numbers of boats ran into hundreds, and the personnel involved, both ashore and afloat, ran into thousands. The big majority of Scottish fishermen had some involvement in the fishery, which means that the total was over 30,000. Shore labour was not enumerated in the same detail as fishermen, but in total probably approximated in numerical strength to the number of fishermen; and in distribution it was certainly of most importance in the main seats of curing, which was much more labour intensive than other means of disposal. The main curing districts were in the North-East, Wick and Shetland.

The leading single district was that of Shetland with upwards of 1,000 boats, around 7,000 fishermen and about 8,000 shore workers. The Aberdeenshire districts of Peterhead, Fraserburgh and Aberdeen were a coherent group which together exceeded Shetland by a fair margin: there was a fleet of over 1,400 boats, with over 10,000 fishermen and around 10,000 shore staff. Wick had upwards of 300 boats, over 2,000 fishermen and upwards of 2,000 shore workers. Stornoway and Barra were ahead of Wick in importance in numbers of boats, due to their importance in the early summer and winter fisheries; between them they had over 800 boats, around 6,000 fishermen and around 2,000 workers on shore. The districts of Anstruther and Eyemouth were the location of the main East Coast winter fishery and Eyemouth was also significant for summer landings; between them these districts had over 500 boats, with around 3,500 fishermen and around 1,500 shore workers.

The Clyde area was the most distinctive in the whole country, with a fishery geared to the restricted waters of the firth and its lochs; it was dominated by smaller boats using the ring-net and landings dominated by the fresh market. The total numbers of boats evidently approached the 1,000 mark, but with four and five man crews. The leading individual districts like Campbeltown and Rothesay might have around 300 boats at peak season and fishermen's numbers above the thousand, but shore workers were never more than a few hundred. On the rest of the West Coast, the sizes of boats owned by native fishermen also continued to be relatively small, and in effect the only districts of major landings were those of Stornoway and Barra to which the bigger East Coast boats resorted seasonally in strength.

There were main shifts in the pattern which occurred during the year (ch. 13): these were associated with the participation of a substantial part of the fleet in the early summer fishery at Shetland and to some extent in the Minch from May to early July before going to other East Coast ports, generally in Aberdeenshire for the main season; and with important winter fisheries from January to March in the Stornoway and Anstruther districts. There was some tendency for particular groups of boats to make a practice of working from particular bases within a general pattern that allowed and promoted mobility. English drifters generally spent the whole summer at Shetland; Lossiemouth drifters in the early season tended to work from Castlebay; and another sector of the extensive Moray Firth fleet worked much of the season from Stronsay in Orkney. In the winter fishery most of the big Moray Firth fleet worked form Stornoway, while the fishery on the Forth, as well as employing the local fleet attracted boats from as far north as Peterhead and Fraserburgh.

REFERENCES

Annual Report of the Fishery Board (ARFB.) 1856.

Annual Reports of the Fishery Board for Scotland (ARFBS.) 1882, 1893, 1896, 1899, 1900, 1902, 1904, 1910, 1911, 1913

Coull, J. R. (1986) 'The Herring Fishery in Peterhead at the Turn of the Century. Revolution by Steam Power' *Aberdeen University Review* 175, 323–332.

Coull, J. R. (1986) 'The Importance of Fishing in the Napier Commission Report of 1884' in Ritchie, W., Stone, J. C., and Mather, A. S. (1986) *Essays for Professor R. E. H. Mellor*, Dept. of Geography, University of Aberdeen, 190–195.

Coull, J. R. (1994) 'The Whaling Controversy in Shetland and the Hebrides in the Early Twentieth Century', *Northern Scotland* 14, 55–68.

Gray, M. (1972) 'Crofting and fishing in the north-west Highlands', 1890–1914', *Northern Scotland* 1, 89–114.

Gray, M. (1978) *The Fishing Industries of Scotland 1790-1914. A Study In Regional Adaptation*, Oxford U.P.

9

The Advent of Trawling:
White Fishing for Industrial Markets

The modern age has witnessed basic changes in fishing, as in so many other things: and the modernisation of fishing on the world scale owes not a little to the development of trawling in Britain. Trawling, by pulling a bag net over the sea bed was an innovation in catching white fish *en masse* rather than on individual hooks. While the basic innovation was made with sailing trawlers in the southern North Sea in the early decades of the 19th century, trawling from about 1880 was also the first mode of fishing to prove economic with power-driven vessels. Such was the rise in productivity from steam trawling that it led to a rush of investment, and in subsequent times the use of trawls by power-driven vessels has become the most important method of fishing in the globe. However, trawling could not easily work on the same grounds as the traditional long-lining, and conflict was often acute: in fact, steam trawling generated shock waves all around the North Sea.

In Britain, the development of trawling inter-acted with the extension of the rail network in making fresh sea fish available in quantity in inland locations for the first time ever. Moreover, the price of the fish was sufficiently low to give it a major cost advantage over meat for several decades, and it became a major protein food of especially the poorer classes in the cities during the Industrial Revolution.

THE DEVELOPMENT OF TRAWLING IN SCOTLAND

Scotland was at first isolated by distance from the development of trawling, which originally concentrated in supplying markets in London and other English cities. However the success of trawling caused it to expand, and there was a continued search for new grounds to exploit. This brought English sailing trawlers into Scottish waters in the 1860s, and started to give concern for fish stocks off the Scottish coast; and for three years in the 1860s the Fishery Board banned trawling over the 'Fluke Hole' in the Firth of Forth to allay the concern of Scottish fishermen in the area, although it removed the ban when it found no definite evidence of the adverse effects of trawling.

Trawling continued to increase in the waters off the east coast of southern Scotland, which was within the range of English trawlers. With the thousands of line fishermen there were on especially the East Coast at this time, concern became acute, not only because of the fish the trawlers were catching, but also because of repeated gear conflicts and damage done by trawls to lines shot in the sea. The trawlers, like the line boats, in general preferred the inshore grounds, both because of the usual greater abundance of fish, and for the great difficulty of hauling the trawl by a man-powered windlass on sailing trawlers: this was an exceedingly heavy task, and restricted their operations to depths of at most 50 fathoms. The response of government was to mount official inquiries, but initially these came out clearly in favour of trawling, on the grounds that it led to a great increase of the supply of fish on the market, and there was no real evidence of damage done to fish stocks.

The advent of steam trawling in the 1880s could only exacerbate the conflicts, especially as the greater manoeuvrability of the powered vessels meant that they could also work with safety on small patches of ground inshore that sailing trawlers had had to avoid. In addition, steam trawling in Scotland was originally mainly an intrusion from England, which made it additionally resented. However there was fairly rapid investment into steam trawlers in Scotland; this was especially in the port of Aberdeen, although there were also trawlers at Granton, and to a lesser extent at ports like Dundee, Peterhead and Fraserburgh. They did have to depend initially to a considerable extent on experienced skippers and other personnel from England.

Fig. 9.1 shows the rapid growth of the Scottish trawler fleet which occurred between 1892 (the first year of complete records) and World War I. Although there were still more English than Scottish trawlers landing in Scotland in 1892, they were overtaken by the Scottish fleet in the subsequent year, and there was a brisk increase in the Scottish fleet thereafter to a figure of around 350;

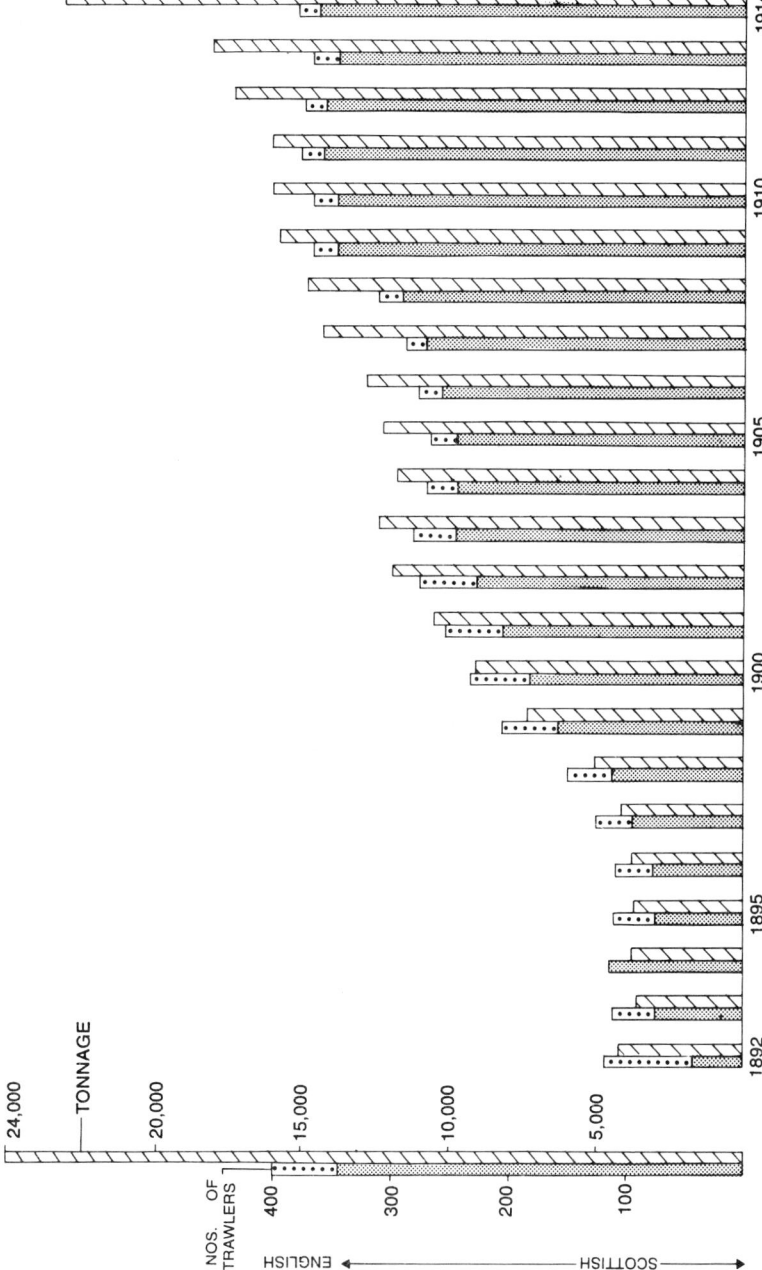

Fig. 9.1   Numbers and tonnage of steam trawlers landing in Scotland, 1892–1914.

and the tonnage increased more than in proportion with a trend towards bigger vessels so that by 1914 the tonnage of the average trawler was in excess of 60 tons.

While the revolution in fish supplies was wrought mainly by the steam trawler, there was also a significant development of the development of the steam liner: this was generally a vessel comparable to the trawler in size, but using great lines on a new scale to catch mainly the more valuable fish species like halibut, cod and plaice. It also operated a power-driven line hauler, although there was still considerable labour-intensive work in line-baiting.

The development of trawling also brought in capital from outside fishing, and this was associated with a new level of investment and a business organisation in the form of company ownership which in large degree separated it off from the rest of the Scottish fishing industry. While trawling was profitable, the cost of steam trawlers was anything from £3000 to £8000 each, which represented an outlay in the range of ten times or more what even the most prosperous traditional fishermen were paying for boats at the time. Labour for the trawlers was recruited partly from traditional fishermen whose livelihood was being undermined, and who generally found they could earn more on trawlers, even if the working conditions were often more arduous than those to which they had previously been accustomed; in Aberdeen fishermen were pulled in from villages to north and south as well as recruited from the fishing communities of Fittie and Torry within the city; and men were also pulled in from the general population. There was also a significant immigration from English ports like Hull and North Shields in which trawling had been longer established: this appears to have been particularly the case for key personnel like skippers and engineers. While the trawler labour force was not entirely isolated from that of other fisheries, it became largely so and had something of a definite hierarchy in it, with the superior position and level of payment of the skippers and mates. The deck hands on the other hand were often recruited on a casual basis and trawlers seldom had the stability of crews of traditional boats. They were paid mainly on a wage basis, with only a minor addition from the proceeds of the voyage, although the skippers and mates had the incentive of payment as a percentage of the value of the catch. By the years prior to World War I some successful trawler skippers were buying or building their own vessels, although the big majority of the fleet continued under company ownership (ARFBS. 1911: xvii). In fact, the trawling sector was the part of the Scottish fisheries that showed the characteristics of an Industrial Revolution structure.

In Aberdeen especially part of the fishing effort was conducted by steam liners which were often company-owned, although individual skippers too could become owners. The usual method of division of the proceeds on liners was that of the 'half catch' which meant that after the deducting of expenses half went to the boat and gear and half to the crew. In detail the division was often more involved, as some of the crew got a share from lines while others might have a weekly wage, like the trawler men (ARFBS. 1911: xviii).

As well as trawling companies owning their own fleets, there was the necessary growth of special services like firms dealing in gear and equipment and the provision of ice. Auction selling of fish replaced earlier arrangements between fishermen and fish mongers, and at the centre of the system emerged a small number of fish selling companies, which sold fish on commission and had links with all sectors of the trade (Gray 1978: 168–169). There was too the proliferation of small fish merchants engaged in the curing and the marketing of the fish. The development of this inter-dependent business organisation confirmed and enhanced the position of Aberdeen, as it developed the capacity and infrastructure to service the fleet as well as to market its catch. One result was that much of the port land use of Aberdeen became devoted to the fish trade. The city, realising that a major new component had come into the urban economy, built a new fish market in 1889, which was twice extended subsequently. The area to the west of the fish market basin, which was adjacent to the railway sidings, acquired many premises of fish merchants, and these also extended into the Torry part of the city on the south side of the River Dee.

The development of trawling had a remarkable effect in the extent to which it centralised at main ports white fish landings in Scotland and indeed in Britain (Coull 1972: 158–164). Previous to the rise of trawling white fish landings were distributed around much of the coast, and auction selling of fish was rare. Trawlers obviously needed bigger harbours and various other facilities, and competition of many buyers at the auctions raised average prices. Aberdeen quickly became the main white fish market in Scotland, and also one of half a dozen main white fish markets in Britain. In addition to wide distribution within Scotland, there were also fish trains to various centres in England, of which London was of course most important.

TRENDS IN TRAWL LANDINGS AND RANGE OF GROUNDS WORKED

Another important development before World War I was that boats from other ports and other countries were attracted by good prices to the Aberdeen market; as well as boats from various Scottish ports, English trawlers working in the northern North Sea continued to land some of their catches; boats from continental countries were also attracted, and landings from German trawlers fishing at Iceland became a prominent feature.

Fig.9.2 shows that trawl landings rapidly came to dominate Scottish demersal catches. From a figure of 12,626 tons in 1889 the trawl catch at 53,658 tons in 1900 was a full half of total demersal landings, and by 1913 at 127,097 tons it was over 75%; it had multiplied by more than a factor of ten in 14 years. Linked to this was the centralisation of demersal landings to an unprecedented degree, especially on the port of Aberdeen, and in the years before World War I accounted for over 70% of the Scottish total. Aberdeen in fact became one of the six main trawl ports of Britain: although never of the status of the main Humber trawl ports of Hull and Grimsby, it was next in rank. Granton on the Forth developed as another important trawl port, and Dundee on a smaller scale also became a regular trawl port. Trawl landings were made at times elsewhere, including ports like Peterhead and Fraserburgh although they were mainly committed to the herring fishery. Trawlers might work at times on the West Coast, and from grounds off south-west Scotland might land at ports like Oban and Ayr.

Trawling in Scotland was at first concentrated on inshore grounds on the East Coast and the Firth of Clyde, especially the former. This was certainly more economic in fuel use, although it aggravated the inevitable friction with traditional line fishermen; and this was to lead to the banning of trawling within three miles of the coast and the closing off of the Moray Firth and the Firth of Clyde (see below). It also led to the depletion of inshore grounds, especially those within reach of the main port of Aberdeen, and this became a serious problem by the early years of the 20th century.

Above: 1. Cairnbulg, Aberdeenshire. A fishing village on record from *c*. 1600. The site is by a shingle beach, on which boats could be pulled up, and fish split, salted and spread to dry. Despite the exposed site the houses are close by the beach, so that fish and equipment did not have to be carried further than necessary between the houses and boats (*J. R. Coull*).

Right: 2. Crovie, Banffshire. This village is at the foot of a steep brae above the beach on a very constrained site. Such were the advantages of having the houses as close to the boats and the beach as possible that even sites such as this might be used (*J. R. Coull*).

Right: 3. Sandend, Banffshire. The name of the village is significant, as it is built not on the sand but by a cove where the rock coast starts. As well as the rock being better for house foundations, a cove gave depth of water for loading and unloading boats, and split and salted fish could be spread on the rocks to cure (*Aberdeen Journals*).

Below: 4. Whinnyfold, Aberdeenshire, on a cliff top above a creek. Where there was not space for building beside the beach on stretches of coast like this, the houses were regularly built on the cliff top, and the boats had to be reached by paths down the cliff (*J. Livingston*).

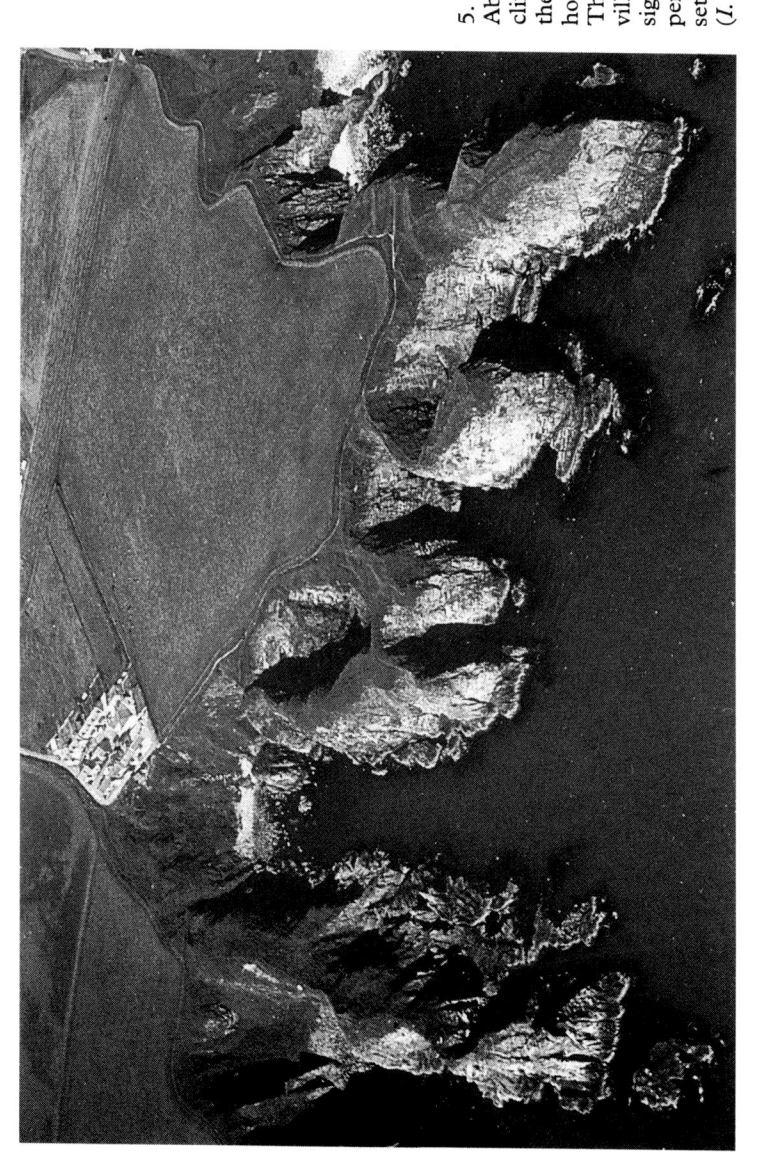

5. Bullers of Buchan village, Aberdeenshire. This village on a cliff-top site takes its name from the spectacular enlarged blow-hole in the centre of the picture. This is one of the few fishing villages which did not expand significantly in the modern period, and illustrates the size of settlement that was once typical (*I. B. M. Ralston*).

6. Baiting lines with mussels, Cruden Bay, Aberdeenshire. The traditional method of inshore fishing for white fish was essentially a family effort in which the women spent long hours in collecting and shelling mussels to bait the lines before the men put to sea (*North East of Scotland Library Service*).

7. A Shetland sixern being rowed. Although these boats did have masts and sails, oars were much used, not only when manoeuvring the boat at lines, but in making way against the wind. Such boats might fish as much as 40 miles offshore (*Shetland Museum*).

8. The steps at Whaligoe, Caithness. There are various places on the Scottish coast where access to coves used for fishing has been a considerable problem. Outstanding is the case of Whaligoe, a creek in a stretch of bold cliff coast in Caithness cliff, where about 1800 David Brodie, tenant to Sir John Sinclair of Ulbster, built this staircase of 330 stone steps from the cliff top to the quay below (*J. Livingston*).

9. At sea aboard a line boat from Gourdon, Kincardineshire. The lines are coiled up in the line sculls. Line fishing persisted longer here than anywhere else on the East Coast, partly because the hard ground offshore was more difficult for trawlers to work (*Scottish Fisheries Museum*).

10. 'Haaf' station at Stenness, Northmavine, Shetland, 1880s. The sixerns used for the fishing are hauled up on the beach. Also visible are tubs for washing the fish and the vats in which they were salted before being spread on the beach. The tripods were used for weighing the fish. Behind to the left are the lodges which were the fishermen's accommodation for the season (*Shetland Museum*).

11. Fisherrow fishwives. Selling the fish was formerly mainly by the fishwife, a well-known figure in traditional Scottish life. At Fisherrow they were fortunate in having a big market relatively close in Edinburgh, but fishwives elsewhere might in a single day's trip cover thirty of forty miles as they walked around farms and crofts in the landward area selling fish or exchanging them for such produce as eggs and meal (*Scottish Fisheries Museum*).

12. The harbour at Keiss, Caithness. This was one of the many harbours built with the help of funds from the Fishery Board to help promote the fisheries. Its greatest importance was in the 19th century, when the demand for harbourage for the herring fishery in this area was intense (*J. Livingston*).

13. Burghead, Morayshire. This settlement was replanned and rebuilt from the early 19th century, and it shows the typical grid lay-out of the period. The harbour has been built in the lee of the headland, and since the approach is from the north-west on the sheltered inner Moray Firth, it is particularly safe (*Economic Development and Planning Departrment, Grampian Regional Council*).

14. Peterhead harbour and town from the air, 1960s. Because of the two small protecting offshore islands, visible in the foreground, this was one of the best sites for developing a harbour on the whole exposed East Coast. The harbours were built in a series of phases, mainly from the 18th century; their earlier development was dominated by trade and whaling, but from the middle of the 19th century harbour development was primarily for the herring fishery. On the island of Keithinch in the left foreground was one of the main areas of herring curing yards. With its position on the North-East shoulder it was one of the best located harbours for the herring fishery, and later from the 1960s it emerged as the main white fish port in Scotland and Britain (*Aberdeen Journals*).

15. A good herring catch, Wick. Open boats were used in the fishery well into the era of the camera, and from the earliest days the possibility of securing such bumper catches was always one of the main incentives to engage in the fishery (*George Washington Wilson Archive, Aberdeen University Library*).

16. Fifies being rowed out to catch the wind, late 19th century, Wick. Close manoeuvring in harbours often involved this even with big boats. These were herring boats, and in the stern can be seen the 'iron man' winch which was used to haul in the main rope, and the nets are made up forward of the rear mast (*Wick Society*).

17. The sailing herring fleet with their dipping lug sails putting to sea, probably off Fraserburgh. All the old herring towns have such photographs: it was normal for hundreds of boats to congregate at each main harbour for the season, and when the fleet left port there were scores of sails across the horizon (*Scottish Fisheries Museum*).

18. Boats pulled up on the beach, Burnhaven, Aberdeenshire, late 19th century. The bigger boats on the upper part of the beach are the decked fifies, 40 to 50 feet long, which were the herring boats at this time. They were regularly pulled up between seasons, although this was a major effort, which might involve all the people of the village (*North-East of Scotland Library Service*).

19. Loch Fyne skiff under sail, Campbeltown Loch. These boats were relatively small, but for working in the Firth of Clyde where the waters are often confined and sometimes shallow, they proved the most efficient craft in the early days of the ring net herring fishery in the 19th century (*Argyll Studios, Campbeltown, per I. McGeachy*).

20. Steam drifter *Golden Rod*, Peterhead, probably inter-war. These vessels lifted the herring fishery to its all time peak in the years before World War I, but in the difficult inter-war period their running expenses became a great problem. The nets are made up a forward of the wheelhouse, and the steam capstan used for hauling the 'bush' rope can be seen forward of the foremast (*Scottish Fisheries Museum*).

21. Steam drifters, racing to catch the market, entering harbour, Great Yarmouth (inter-war). There was a great concentration of boats in the autumn herring season at the two ports of Yarmouth and Lowestoft. The Yarmouth harbour extends for nearly two miles inside the mouth of the River Yare and there was great congestion with a combined fleet of English and Scottish boats that at the peak approached a thousand (*Yarmouth Mercury*).

22. Hauling drift herring nets at Shetland in the 1960s. The nets were shot at night and hauled in the early hours of the morning. Most of the crew were involved in the manual hauling of the nets, and the herring were shaken from them as they came aboard. With a good shot, the hauling alone might take ten hours or more (*J. H. Goodlad, Hamnavoe, Shetland*).

23. 'Redding up' herring nets in port on a steam drifter (probably inter-war). This was frequently necessary to have them arranged so that they could be thrown overboard in turn in an orderly manner when the time came to shoot again. As the nets were shot they were tied to the 'bush' (main) rope and 'buoys' (visible on the right) attached to each net in turn to hold them vertical in the sea for fishing (*Scottish Fisheries Museum*).

24. 'Barking' herring nets, Peterhead, 1960s (also known as 'cutching'). Nets of natural fibres rotted easily with repeated dousing in sea water, and frequent dipping in a solution of hot tan bark was usual to minimise deterioration (*J. R. Coull*).

25. Women mending nets, St Monance (probably inter-war). Herring nets were made of relatively thin cord, especially after cotton came in for their manufacture in the 1860s, and they got a great deal of wear and tear in use. The necessary mending of nets was done mainly by the women folk in mending lofts in the winter. Here on a fine day the net has been brought out of doors to be mended (*Scottish Fisheries Museum*).

26. Curing scene at Wick in the early 1870s. The boats are still open, and nets and furled sails can be seen aboard many of them. The women gutters are gathered around the 'farlanes', and are putting the gutted herring into baskets, from which they will be transferred to the barrels. The filled barrels ready for shipping are seen on the right (*Illustrated London News*).

27. A big fleet of herring sail-boats at Castlebay, Barra. Castlebay has one of the best natural harbours in the Hebrides, the Island of Vatersay effectively enclosing and protecting the bay. It was, along with Stornoway at the other end of the Long Island chain, one of the two main bases for the herring fishery on the West Coast, and was especially important for the early summer season (*George Washington Wilson Archive, Aberdeen University Library, no.C7406*).

28. Concentration of curing yards at the north end of Lerwick (inter-war). Here the curing yards were built along the shore and had their own jetties at which boats landed herring directly after they had been bought. This was to become by the early 20th century the greatest single concentration of curing yards in Scotland. At this time the fleet was dominated by steam drifters, but sail boats and motor boats were also in use. Piles of barrels awaiting shipment can be seen, and in the bottom right is one of the accommodation huts in which the gutting women were accommodated (*Shetland Museum*).

29. Cooper putting hoops on a barrel, Lerwick, 1950s. Barrel making was a skilled craft as the staves had to taper away from the middle and still be watertight. The coopers made many barrels in between seasons, and generally supervised the curing during the season. In the 19th century, when a main function of the fishery officers was to inspect cured herring, they were regularly recruited from the coopers (*Shetland Museum*).

30. Aberdeen steam trawler *Lord Learney*. The trawl winch can be seen forward of the wheelhouse, and the two horse-shoe shaped gallows at the side from which the trawl net was slung to shoot it (*Scottish Fisheries Museum*).

31. Steam trawlers in the fish market basin, Aberdeen. The fish market was originally built in 1889 to cater for the booming trawling industry, and it was twice enlarged afterwards. At the peak between the wars there were over 300 steam trawlers and over 60 steam liners operating from Aberdeen (*City of Aberdeen, Art Gallery & Museums Collections*).

32. Buckie seine-netter *Carinthia* in Caledonian Canal locks at Fort Augustus. For decades the main use of the canal was by fishing boats going between East and West Coasts. The seine net is visible in the stern and the ropes on the fore deck. Hundreds of boats of this type were built with the help of government grant and loan schemes in the decades after World War II. Many of them were dual-purpose, alternating between drift-netting for herring and seine-netting for white fish at different seasons (*J. R. Coull*).

33. Peterhead Seine-netter *Fidelia*. From the late 1960s these boats were increasingly built of steel rather than wood, had rope reels for coiling the ropes and deck shelters to protect the crew at work (*J. R. Coull*).

34. Shetland purse-seiner *Charisma*. A major innovation from the 1960s was the building of boats like this which vastly increased the catching power in pelagic fisheries. The net is several acres in extent and is operated by surrounding the shoals (*Shetland Development Department*).

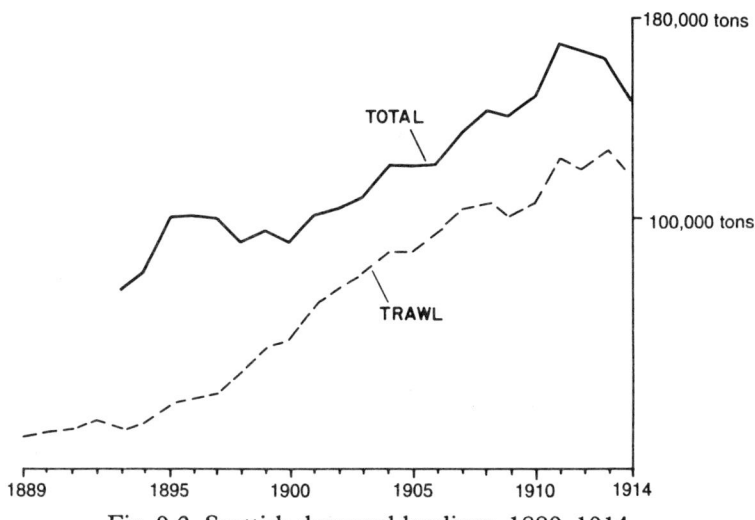

Fig. 9.2 Scottish demersal landings, 1889–1914,
including trawl landings.

However with the continued growth of trawling, and with the building of bigger trawlers, the area worked expanded in a spectacular manner.

By the end of the 19th century the intensified exploitation of the trawlers from Britain and other countries had extended over the whole North Sea, and also to the limits of the continental shelf elsewhere around the British Isles; and they were able to range everywhere apart from places where ground that was too rough and rocky would damage the trawl. In the early years of the 20th century, Scottish trawlers joined those from England in extending their operations to Faroe and Iceland, although landings from these grounds at Aberdeen were to come mainly from German trawlers. While working such grounds lengthened the time lost in making the passage to and from port, this was more than compensated for by increased catch rates; and market prices were sufficiently competitive to attract them. Although the North Sea still provided the main bulk (61.5%) of the Scottish white fish catch of 165,000 tons in 1913, Iceland and Faroe between them accounted for 22.5%, while 12.9% came from the west side of the country. The most important single species now was cod at 61,700 tons, largely due to the big landings form Iceland and Faroe. Haddock accounted for 36,500 tons, and of this three-quarters was from the North Sea (ARFBS. 1913: xxvii).

Fig. 9.3 shows the distribution of catches for 1912 by statistical rectangles (1° lat. x 2° long.) for the trawler fleet landing at Aberdeen. While the main weight of catches came from waters off the north-east and north of Scotland within about 200 miles of the port, there was also significant fishing effort across the whole of the north part of the North Sea; Faroese grounds at between 300 and 400 miles were important, while the catches from Iceland at between 700 and 800 miles are the most prominent single feature of the distribution. In addition the steam liners extended the areas worked by conducting part of their operations on grounds which were too rough or too deep for trawling, and they might range as far as Greenland as well as Iceland. With so much fishing at long range, there were necessary changes in the methods of caring for the catch. On long trips, spoilage of fish was an obvious danger, and gutting at sea and the carrying of ice became essential. Trawlers also had holds or fish rooms in which the catch could be shelved.

Although trawlers were as a general rule bigger craft than had previously been employed in fishing, it was not automatic that they increased safety. They fished summer and winter, and often enough faced difficult weather conditions. Trawler losses in fact were relatively frequent and there was considerable accompanying loss of life. One of the more moving modern books on the Aberdeen trawlers has catalogued these losses and has the title of 'The Real Price of Fish' (Ritchie 1991).

DISPOSAL OF TRAWL CATCHES

While from the start trawl catches increased the supply to the fresh market in and around the ports of Aberdeen and Granton, a great part of the catch was cured by salting or smoking. The main function of trawling was in fact to increase the supply of smoked haddock in Scotland. However, trawling also resulted in increases in the landings of various other demersal fish, and advantage of this was taken from 1889 onwards by sending flat fish (especially soles) to the London market; although rail freight increased the price significantly, the fact that most of these fish were landed from short trips gave them a quality advantage over supplies from the Humber ports; this rendered the trade viable, and special fish trains from Aberdeen to London were to be a regular feature till the 1970s.

On a wider view, a useful and available mirror of the rapidly

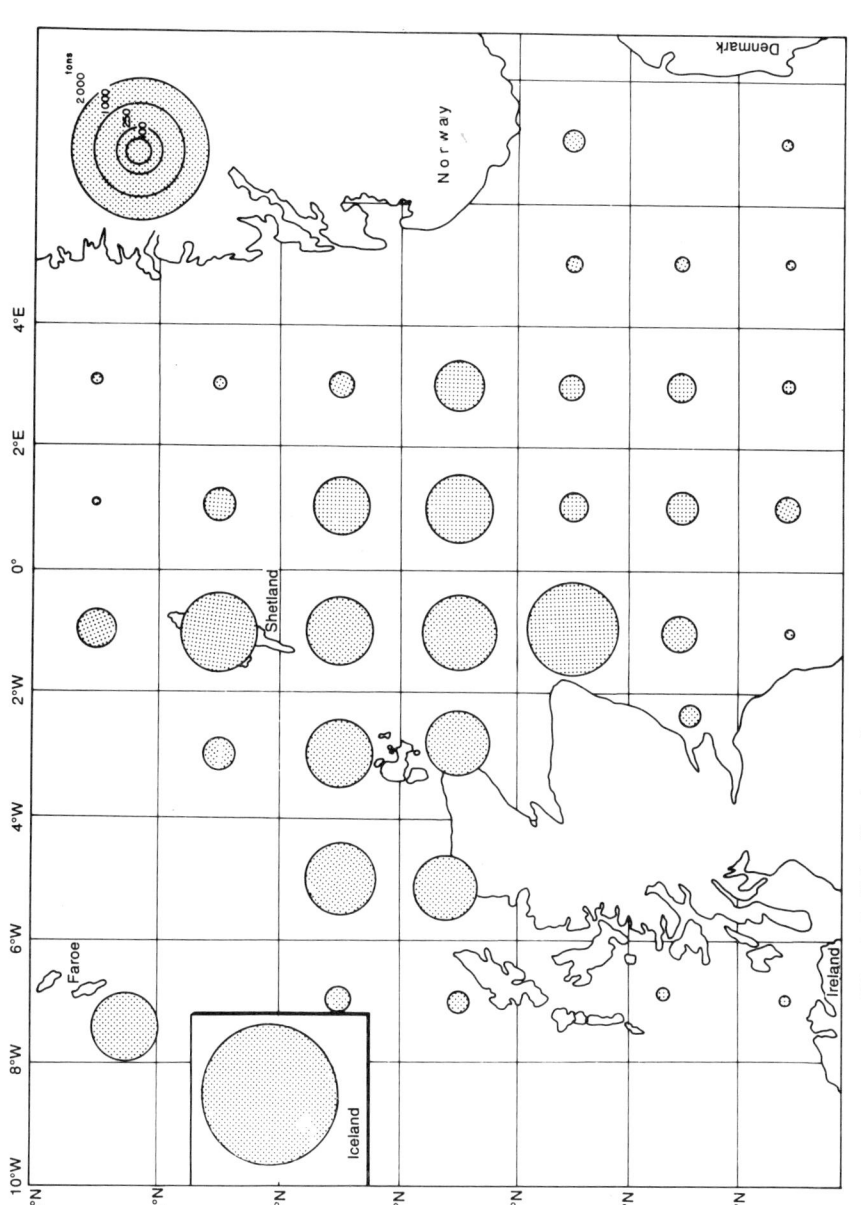

Fig. 9.3 Distribution of trawl catches landed at Aberdeen, 1912.

changing situation in white fishing is provided by the comprehensive recording of the fish carried by rail in Scotland from 1897; however this is at a relatively late stage of the rapid transition, and tends to show a new pattern already established as opposed to one emerging. The total tonnage carried rose from 105,001 tons in 1897 to 164,288 tons in 1909, a rise of 56% in eight years; and by the latter date the rise of overland motor transport must have started to complicate the pattern. The main function of the railway for the fishing industry was the transport of fresh white fish, although other sectors of the industry also benefited, including especially the trade in fresh herring. While a great number of fishing settlements benefited to a minor extent from having access to a rail link, what is already clear in 1897 is the great and increasing concentration of landings at main ports: Aberdeen here was clearly far ahead of all rivals, and in 1897 shipped 31,873 tons out of the Scottish total of 105,000 tons; and by 1909 the Aberdeen share had risen to almost half the Scottish total of 164,288 tons. The general trend at other places was a very modest rate of increase, and even Leith also showed a rising but much more modest total than Aberdeen. The main trend for many places was actually a decrease, as line fishing declined and the greater buying competition at trawl ports attracted more landings and overland consignments. Peterhead, Fraserburgh and Mallaig showed increased shipments because of their importance as herring ports.

Linked to the great change in transport of fish from the coast, there were great developments in the pattern of distribution to the consumer. Inland wholesale markets, like those in Glasgow and Manchester as well as Billingsgate in London, became nodes in the system, and acted as secondary centres of distribution in their own regions. For retailing there was the development of specialised fishmongers and fish friers. Although sea fish had now been brought within the price range of most of the population, the perishability of the product and the irregularity of supply did limit the efficiency of the system, and it was common for the price to double between first sales at coastal auction and retailing.

THE IMPACT OF TRAWLING ON TRADITIONAL LINE FISHING

The counterpart of the coming to the fore of the trawl as the dominant method in white fishing was the rapid decline in traditional line fishing from the late 19th century. As with many traditional activities, the first effect of the accelerating economy of the Industrial Age was to stimulate lining for the expanding markets and this was the story of the great part of the 19th century; but the longer term effect of the industrial mode of production in trawling was to give a sustained shock to traditional line fishing, and ultimately to lead to its eclipse. However, this only occurred after a contest with trawling that lasted for decades (Coull 1994: 107–122). There was prolonged agitation against trawling in inshore waters from the 1860s to the end of the century. The instinctive fear of the line men was that the accelerated exploitation of trawling was destroying fish stocks, although the earliest government inquiry of 1864–66 found no evidence of this and came down decisively in favour of trawling as 'one of the most copious and regular sources of supply' of fish (RCBTR. 1885:x) for the national market. Continued unrest among inshore fishermen led to another inquiry in 1878 in England and Wales, where trawling had a longer history, and this did recognise that there was 'considerable injury' – by the trawlers to gear of line and drift-net fishermen (RCBTR. 1885:xi). While this arose partly through trawlers at times working in the dark and through poor marking of gear by the other fishermen, it was evident that trawlers were often negligent of the other fishermen's gear and might tow right through it. While government was slow to act, it was clear that the problem would not go away, and ultimately a Royal Commission on Beam Trawling was appointed which covered the whole of Britain in its report of 1885. By this time several continental countries had reacted to the threats from British and other trawlers by banning trawling on inshore grounds, and the new act passed in 1885 gave the Fishery Board for Scotland powers to close inshore waters to trawling. The main bays and firths of the East Coast which suffered most from trawling were closed off by 1888 (fig. 9.4) and a general ban on trawling within three miles of the shore was imposed from 1889. By 1892 there were in addition selected bays protected in all parts of the country, and the main embayments of the Moray Firth and the Firth of Clyde had also been put out of bounds for trawlers (fig. 9.4). As well as restricting Scottish based trawlers, these measures also worked against the many English trawlers which were now operating in more northerly waters after depleting the grounds nearer to their home bases (ARFBS. 1892:xiii).

These measures did help the small line men, and it was also

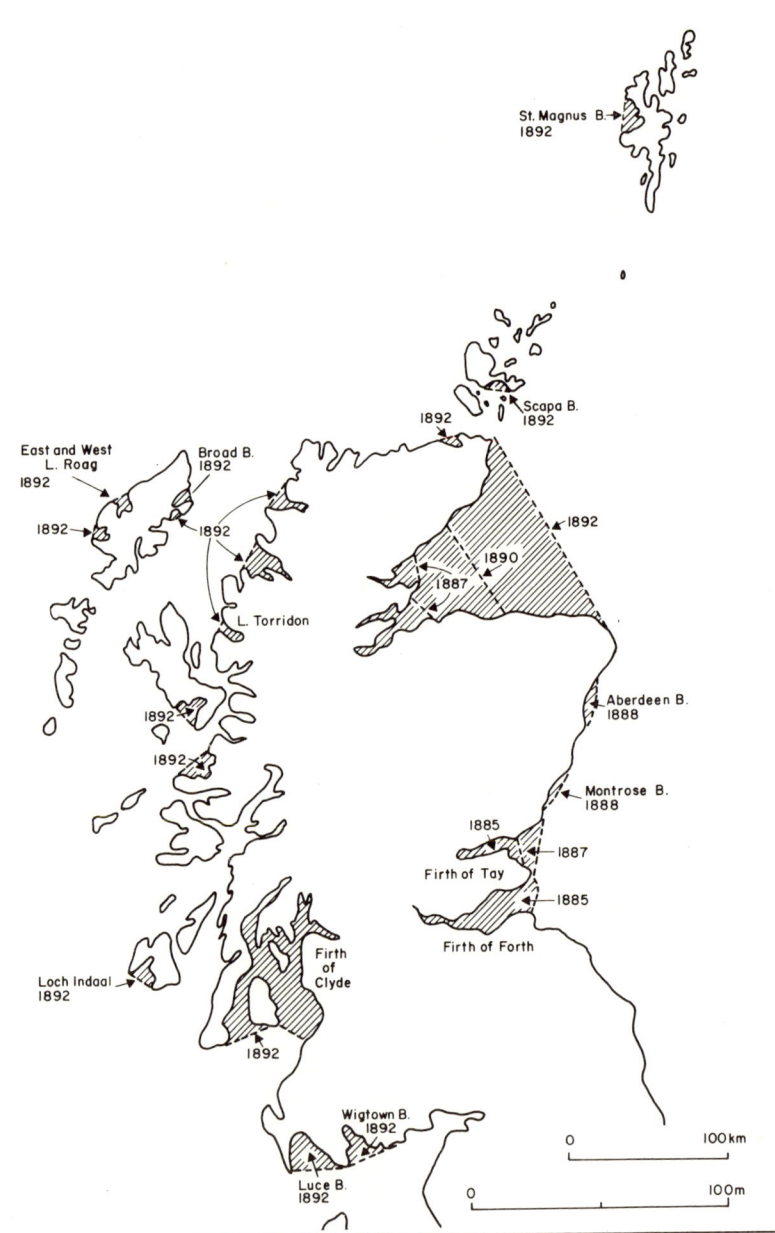

Fig. 9.4 Closing of Scottish Coastal Waters
to trawling 1885–1892

arranged that fishery officers could investigate complaints of damage by trawlers to the gear of other fishermen, and recommend an appropriate level of compensation. There was however now a new task in fishery protection, and there was to be an on-going problem of trawlers 'poaching' in forbidden areas, despite the fact that the master of a trawler convicted of this offence might face a fine of up to £100 or imprisonment of up to 60 days. The chances of being detected by one of the fleet of five fishery cruisers available were in effect limited, as these cruisers also had the task of superintending the other fisheries, and the leading herring fishery itself involved thousands of boats. Although other evidence could legally be the basis for a conviction against illegal trawling, the production of such evidence which would stand up in law was infrequent. There was also another anomaly in the arrangements in that several continental countries declined to recognise Britain's right to exercise fishery protection beyond three miles. British trawling interests took some advantage of this loophole by registering trawlers in Scandinavian countries, and 'bona fide' trawlers from continental countries might also fish in waters banned to the home fleet. Such trawling in the Moray Firth especially became a major problem. Although in the celebrated 'Mortensen v Peters' case of 1906 the prohibition of trawling in protected areas beyond the three-mile limit was confirmed by the High Court in Edinburgh as also applying to continental flag vessels, the prosecution was never closed, and continental trawlers continued to fish waters officially closed to home boats. It became a stock complaint in Scotland that the real reason for this was that by this time English trawlers were fishing on a big scale off the coasts of Iceland and north Norway, and that there was a powerful behind-the scenes lobby which wished to retain the British right to fish close to the coasts of other countries.

The decline of lining was to an extent slowed down in the early 20th century with the installation of motors in craft formerly propelled by sail and oar: this was a cheap way to provide power on board, and one in its early phases better suited to small and middle sized boats (ch. 15). After a tentative start, from the end of the first decade of the 20th century the installation of motors was remarkably rapid: the number recorded rose from 75 in 1909 to 694 in 1914 (ARFBS. 19141: xv).

The Fishery Board kept a series of statistics for line fishing boats during the peak period of transition, and the numbers of traditional line boats fell from 5,715 in 1898 to 3,478 in 1911, a 39%

reduction in 13 years. It is noteworthy that in the leading fishing districts, whether dominated by the trawl or the herring fishery, the scale of reduction over the period was considerably greater. At Buckie the reduction was from 375 to 158 (58%), at Peterhead 149 to 76 (49%) and at Aberdeen 156 to 47 (70%). In many districts (such as Montrose and Eyemouth) there was a run of several years of reduction, followed by effective stabilisation, as boats could be bought relatively cheaply from leading districts where they were being disposed of, and as motors began to be installed. Numbers in fact remained substantially steady in several districts where the ban on trawling in the firths gave the line fishermen more scope: such were Cromarty, Helmsdale and Lybster, although in these districts numbers were in any case modest and each of them always had fewer than 100 boats. The installation of motors also revived lining for more than a generation in several places, notably in the Anstruther and Montrose districts, and in the village of Whitehills in Banffshire. It was shown in the Anstruther and Banff districts in 1913 that line catch rates with motor boats had fully doubled, although inevitably expenses were rather higher (ARFBS. 1913:xx). On the northern periphery, numbers of boats in Orkney actually rose from 180 to 242 and those in Shetland remained substantially stable throughout at around 500. In Orkney this was due more to the lobster than the line fisheries and in Shetland it was related to the modernisation of the islands' fleet with the herring boom, and the need to continue to engage in line fishing outside the summer herring season.

REFERENCES

Annual Reports of the Fishery Board for Scotland (ARFBS.) 1892, 1911, 1912, 1913.

Coull, J.R. (1972) *The Fisheries of Europe. An Economic Geography*, G. Bell and Sons, London.

Coull, J.R. (1994) 'The Trawling Controversy in Scotland in the Late Nineteenth and Early Twentieth Centuries', *Internat. Journ. Maritime Hist.* VI,1, 107–122.

Gray, M. (1978) *The Fishing Industries of Scotland 1790 –1914. A Study in Regional Adaptation*, Oxford U.P.

Ritchie, G.F. (1991) *The Real Price of Fish. Aberdeen Steam Trawler Losses 1887–1961*, Hutton Press Ltd., Beverley, Humberside.

Royal Commission on Beam Trawling Report (RCBTR.) (1885), HMSO., London.

10

The Herring Fishery Between the Wars: Crisis and Readjustment

World War I has been accepted as a major watershed in world events and in the development of the global economy. The readjustments which followed in British industry are legion; and the fisheries, like other industries had to undergo major readjustments. The disorganisation which came in the herring fisheries were particularly severe, and on more than one occasion evoked responses of special government action; and while the white and other fisheries were less severely affected, they also had formidable problems to face. Part of the background was that, although the fisheries in Britain by the early 20th century had become the most productive in the world, their costs were relatively high and their dominant position had been reached essentially through greater volume of production and turnover. Other countries had begun to catch up even before World War I; and although the inter-war years were disorganised and at times chaotic for the global economy, competing countries continued to undermine Britain's previously leading position.

Within Scotland there was some regional variation in the position: inevitably it was the fishing settlements most committed to catching herring for curing for the main continental market suffered most. This included the great part of the Banffshire coast, Wick, the Peterhead and Fraserburgh area of Aberdeenshire, and Fife. On the other hand, the problems were less severe in the trawling ports of Aberdeen and Granton, from which the great part of the white fish went to the home market. There were also some other parts of the coast which escaped relatively lightly. The section of the East Coast between Arbroath and Stonehaven had been less committed to the herring fishery, and was able to operate and survive mainly on the basis of inshore lining with motor boats. In Orkney as the herring fishery reached its maximum from the late 19th century there had been limited investment in the bigger sailboats and none in the expensive steam drifters. Orkney men had largely withdrawn from herring fishing, even if curers and visiting boats came in big numbers to Stronsay: the Orkney men found employment in lobster fishing, often in association with crofting or farming; and lobster fishing was relatively little affected by the interwar problems. For the ring-net fishermen on the Clyde there was the comparative security of the market for fresh and kippered herring to the Glasgow area and beyond, and the cured market represented effectively an outlet for surpluses. Even within the areas that had been most committed to catching herring for salt curing there was some variety of response. At Lossiemouth, and to a more limited extent in Buckie, there was the scaling down of herring fishing and the diversion of a significant part of the effort into seine net fishing for white fish. In the Shetland Islands, where numbers of steam drifters were small, and where there were still many crofter-fishermen, there was a somewhat better accommodation to the difficult situation by continuing in the home summer herring fishing with the relatively cheaply operated motor boats; and at the same time in winter haddock lining from small motor boats also generated significant income. The situation in the Western Isles was worse than in Shetland, as employment opportunities were more directly tied to the main East Coast centres of the fishery: here a main employment for men was going as part crews on the boats, while many women also found employment in curing yards; and both of these opportunities were seriously cut back. One of the effects of the difficulties of the period was to prompt another pulse of migration of Scottish fishermen to Commonwealth countries like Canada and New Zealand to ply their trade or find other work there: this included both men from the main fishing centres on the East Coast, and crofter-fishermen from the Western Isles, the West Coast and Shetland.

THE HERRING FISHERY DURING WORLD WAR I

The herring fisheries were badly disrupted at the outbreak of war in 1914 for a combination of reasons. There was the obvious risk to fishing boats of naval operations, especially on the North Sea side of the country which faced Germany. Thousands of the fishermen volunteered for naval service or were called up, and many steam fishing boats were requisitioned for naval service, in which their main function was to be patrol boats: 829 steam drifters were among the 1,264 vessels commandeered by the navy for war service (ARFBS. 1919: vi). Not least was the fact that, unlike the white fisheries, the herring fisheries depended primarily on export markets: and the great irony was that Britain had gone to war with the country that was the main export market. The herring trade was inevitably badly dislocated, but even in war time the fishery was sufficiently important in contributing to food supplies to continue at a lower level, and with the emphasis on production switching partly to the West Coast.

The dislocation in the fishery was at its worst in the first year of the war, and in the nadir year of 1915 landings were down to just over 35,000 tons, or about 15% of the prewar average. Thereafter there was the working out of an accommodation to the restricted situation, and for the remainder of the war, production hovered around 100,000 tons annually. In the absence of many of the steam drifters this was achieved partly through the speeding up of conversion of sail-boats to motor power.

READJUSTMENT PROBLEMS IN THE EUROPEAN CONTEXT

While it was appreciated that after the Armistice at the end of World War I the fishery could not immediately return to prewar patterns and levels of operation, the initial reaction was that it was only a matter of time until the trade resumed where it had left off in 1914. It was to take some years before it was appreciated that the trade had not only been interrupted but also crippled in a way that was to prove permanent.

Essentially, the long lead that Britain had enjoyed from the late 19th century in catering to the markets in cured herring in Europe was no more. Other countries were developing and modernising their fisheries, and the struggle for survival in the difficult market conditions indeed was a stimulus to do so. Norway, Holland, Iceland and Germany all advanced to the first rank among European herring producers, and had in general lower production costs than the fishermen in Scotland and the rest of Britain. By the 1930s the total production of cured herring in Europe was running at levels between 2.2 and 3 million barrels annually, which was of comparable dimensions to the scale of the trade before World War I. Although

Britain with a production fluctuating between 700,000 and one million barrels was still the leading single producer throughout the interwar period, its dominant position had been irretrievably lost; and by the end of the 1930s, Germany was vying with Britain for the leadership, and had taken the leading position in supplying its own market (ARFBS. 1938: 27). To add to the problems, even during the stress of the interwar period there was a slow rise in British living standards, and one of the consequences of this was that the home market in herring was in serious decline.

Production of herring was inevitably curtailed while costs rose considerably faster than prices, and what was now an over-sized and ageing herring fleet struggled grimly to survive. A variety of measures were taken to ameliorate the situation, both within the industry itself and by government and government agencies: but so deep-seated was the maladjustment that there could be no real solution to the industry's problems, and the measures taken appear with hindsight as palliatives. It was to take the new economic climate of the post-1945 years to set Scottish fisheries once again on a path of development and growth.

THE AFTERMATH OF WORLD WAR I

The initial post-war years saw the industry try to re-establish itself with government help, which took the form of market guarantees for two years and the provision of boats to replace those lost while commandeered for war service. However the industry was thrown back on its own resources for most of the 1920s and a main objective became that of matching production with a reduced demand. In the peak years before 1914, landings generally had been around 250,000 tons, but in the 1920s only in three years did they exceed 200,000 tons, while in the still more difficult decade of the 1930s they fluctuated around 125,000 tons (fig. 10.1). Even more significant are the figures relating to exports to the main continental markets: this often surpassed one and a half million barrels at the peak in the early years of the century, but only two years in the 1920s reached a million barrels, and in the 1930s it was unusual to reach half a million (fig. 7.1). By the 1930s there were serious reservations at both international and national levels on the laissez-faire philosophy of economics, and in 1935 government intervention to help an industry now at crisis point took the form of the setting up of the Herring Industry Board (HIB) to help regulate and

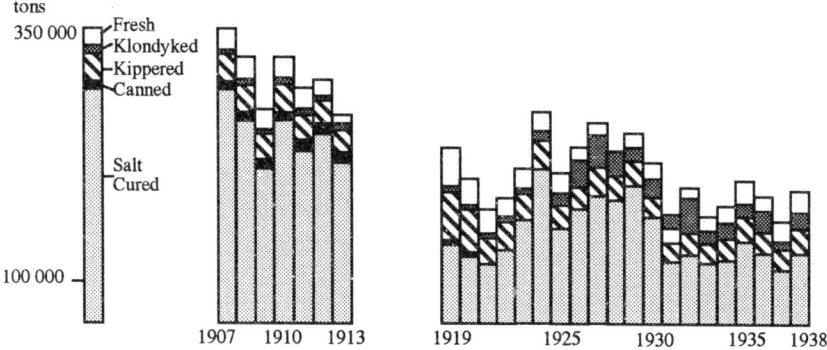

Fig. 10.1 Scottish herring landings and disposal, 1907–1938.

rehabilitate the industry, although its efforts were to be cut short by the renewed outbreak in hostilities in Europe in 1939.

While the inter-war years were difficult for all the countries involved in the herring fisheries, their aggregate effect was most severe in Scotland as opposed to Britain. The major part of production was in the hands of Scottish fishermen, and the fact that the Scottish drifter fleet was mainly owned by share fishermen and did not have the backing of company capital now became a source of weakness. Also curing for the continental market, the worst hit sector of the trade, was a near-monopoly of Scottish curers. While there was a considerable recovery in the market for cured herring by the late 1920s, by that time it was becoming clear that British dominance was being subjected to a mounting challenge, especially from Holland and Germany. The German policy of expanding its own catching power (in the face of exchange and balance of payment problems) had by the end of the 1930s raised the dimensions of the German cure to that of the U.K., while that of Holland was running at 60% to 85% of the U.K. (ARFBS. 1938: 27). The British market share was thus increasingly constrained.

In the long-term fortunes of the herring fishery, there was to be a structural change that was to be even more important than the inter-war market upheavals and the increasing competition from fishermen of other nations. Even in the difficult period between 1918 and 1939, living standards were edging upward, and one of the main consequences of this was changing consumer preferences, with the move away from salt-cured herring to other fish and other sources of food protein. In the 1950s it was to become evident that the once dominant cured sector of the trade had shrunk to be a mere shadow of its size half a century before, and the essential basis of what had been the world's biggest single fishery at the start of the 20th century had been removed. The other market sectors of freshing, kippering, klondyking, and canning did fare better, although this had the effect of a limited amelioration rather than a solution. Partly by market promotion it even proved possible to effect some expansion in the canning sector through increasing exports; but this did little to conceal the overall decline.

LONG-TERM STRUCTURAL CHANGES

The fishermen were distinctive within the structure in being very generally independent operators who owned their own boats on a share system, although there was an important distinction between the share fishermen and the hired hands who made up part of the crew. The other sectors of the industry were organised in companies, although curing especially was conducted at a range of scales (SFC. 1934: 26). There were of course different categories of employees in the firms, and the dominant position of the curing sector is reflected in the regular enumeration in the official statistics of coopers (male), gutters and packers (female) and labourers (male): the first two of these categories were very generally employed by the curers, but labourers could include carters and others employed by other types of firms.

While there was of course a common interest among the different sectors in the industry's general prosperity, there were always inevitable tensions between the various groups, and these became more acute in the constrained situation of the inter-war years. There was little overlap between the groups, so that interests generally divided along well defined lines. Within the web of relationships that bound the different groups together, it is clear that it was the exporters who exerted most power: with their key middleman position they were able to bring strong pressure to bear on other groups – especially on the curers from whom the brought directly (SFC. 1934: 26–27). The difficult situation of the inter-war years did lead to efforts to co-ordinate the activities of the different sectors within the industry: this was done mainly by the setting up local committees at all the main ports with the encouragement of the Fishery Board (ARFBS. 1920: 14), although there were also periodic meetings and discussions at national level.

THE FISHERMEN, THEIR BOATS AND GEAR.

The fishermen were the most numerous group dependent on the industry. While separate statistics for Scottish herring fishermen were not systematically kept, the number of boats primarily engaged in herring fishing indicate that crew manpower was *c*.18,000 immediately after World War I, and declined by about half by 1938. Although in the main the boats were owned by fishermen, they did not present a coherent body of interest, as they included those who had an ownership share in a boat and gear, and hired hands who had none. Crews were characteristically from eight to ten men, of whom two to four would be hired hands. There was a slight upward trend in average crew size over the period, as a bigger proportion of the sail- and motor-boats (which had smaller crews) withdrew from the fishery. While there was some variation in arrangements, it was still usual for running expenses to be deducted first and for the proceeds to be divided into three equal parts – one each for the boat, the gear and the labour share; and those members with a capital stake got proportionately more, according to the size of that stake. In the most difficult years it could actually be a liability to own a share in boat or gear, as depreciation costs might not be covered; but inevitably the greater hardship usually came on the hired hands whose only source of income came from the labour share. The constrained situation inevitably gave much scope for argument about the division of the proceeds, and a major complication came when the national scheme of compulsory unemployment insurance came into force in 1933. This rendered it advantageous for hired men to have part of this remuneration as a regular wage, rather than depending completely on their allocation from the labour share: in addition to guaranteeing a minimum income this would allow then to collect unemployment benefit between fishing seasons. This resulted in a campaign for a minimum wage, and the force behind it can be illustrated with the results from the poorer seasons. Thus in the difficult year of 1926 the average labour share for the two-month Scottish season was less than £8, and when food costs were deducted this was reduced to a net total of £2, or 5/- per week (ARFBS. 1926: 22); and the average labour share at East Anglia in the next year was at the same low level, and can be compared with the £2.10/- per week which was unsuccessfully demanded for the East Anglian season in 1923 (ARFBS. 1923: 7). The clearest statement of individual annual earnings comes from the report of the Sea-Fish Commission in 1934: there was non-earning time between seasons which had the effect of bringing down annual incomes, although share fishermen at these times were involved in vessel and gear maintenance. Thus in the relatively good year of 1929 the average annual labour share was £51, and in the disastrous year of 1934 it was only £12 (SFC. 1934:11).

There was a partial change to wage payment from 1934, and the wage element was later increased; in 1936 a general level of £1.10/- was reached through agreement between hired hands and individual skippers (ARFBS. 1936:6). Even this, however must be seen as near a starvation wage.

Fishermen were also involved in making representations to government, Members of Parliament, and government agencies for amelioration of their problems. Here the main voices were those of share owners, and they could be joined with those of other sectors of the trade to generate more impetus. Independent action by fishermen was actually most prominent in the early post-war stages of the crisis. It was the fishermen who, in the early part of the 1919 season when prices were unprofitably low, stopped operations to bring pressure on the government to fund the guarantee price scheme which operated for that year (ARFBS. 1919:viii). In the following year, when the government at first refused to implement another price guarantee but offered a scheme of export credits instead, the fishermen again successfully brought pressure to bear to suspending operations, and secured a price guarantee scheme for a second year (ARFBS. 1920:13).

While the fishermen made direct representations to Parliament at various other times during the crisis, in the main they were part of a lobby that campaigned for the whole trade, although they did secure a special scheme of loans for the replacement of the big numbers of nets which were ruined in the disastrous 'Armistice Gale' of 11th November, 1929 at East Anglia (ARFBS. 1929:7).

Fishermen also played a reluctant part in schemes which were put into force to curtail production, but such means as limiting numbers of nets in boats' net trains, and at times limiting the number of fishing days in the week as well as the length of fishing seasons. Such schemes generally operated by voluntary consensus, and there were issues on which it could be difficult or impossible to secure unanimity. The most difficult issue was setting a starting date for the main summer season, which had always begun earlier in Shetland and the Western Isles where herring became available as early as May, as opposed to early July on the East Coast. However, early season herring were less suitable for the main curing outlet, and there were frequent arguments by Shetland fishermen, supported by various habitual visitors in favour of an early start; but there was generally a compromise date fixed by which the season in Shetland, and also usually the Western Isles, started earlier than on the East Coast.

The fleet used in the fishery was dominated by the steam drifters constructed in the boom period between 1900 and 1914, and it continued to be dominantly fishermen-owned; in all about 75% of the ownership was with the share-fishermen and 25% with other owners (ARFBS. 1928:8). A smaller component of the fleet also consisted of motor-boats, which were for the great part sail-boats of the era before steam drifters but now with engines installed; but they also included from the end of the 1920s a small but growing number of custom-built diesel boats. It was universally agreed that the steam drifter was technically the best boat for the work, but that in inter-war circumstances there was a big saving in costs with motor operation. However most motor boats actually had a small auxiliary steam engine for driving the winch which was required for hauling the main rope to which the nets were attached as efforts to devise an adequate motor winch had been unavailing. As a very general rule the steam drifters continued to have the biggest gross earnings, although there was an exception in 1926 when coal was in short supply during the General Strike. During that dispute coal was rationed to 25% of normal consumption; and even when it could be obtained prices could multiply two or three times and reach levels of £3.10/- or even £4.10/- per ton (ARF.S. 1926: 6, 19).

Some of the steam drifters had been lost after being requisitioned for government service during World War I, but these were replaced by government built 'standard' boats, and by 1920 pre-war catching power had been restored (ARFBS. 1920: 68), and there was a fleet of *c.* 850 steam and *c.*1,100 motor boats engaged in the fishery. The great part of this fleet concentrated on drift-net fishing in the open sea, but a substantial minority of the motor boats (perhaps between 300 and 400 of them) were ring-netters which operated mainly on the sheltered waters of the Firth of Clyde, and a large part of their catches went to the more reliable Glasgow market.

The steam drifter fishermen had strained their resources to buy their boats, and even in the prosperous conditions before World

War I the majority were heavily mortgaged (NSFI. 1914: 220). This added to concern especially on the Moray Firth: this was the area in which the fishermen had complete ownership and control of their boats, but this had been achieved by fishermen raising mortgages on the value of their houses (ARFBS. 1911: xvii). In the inter-war situation the ageing and now oversized fleet got increasingly into the bonds of debt. It was observed in 1930 that costs had risen 60% to 70% from 1913, but gross earnings were about the same, as any increase in price was compensated for by lower landings (CEAC. Report 1932:57). By 1933, a survey showed that the average steam drifter was over £580 in debt to fish salesmen alone, and in one port the average total debt was £1,800 per boat (SFC. 1934: 12). The fishermen were locked into a system with very limited opportunity for advance or retreat. There was virtually no opportunity to sell out without heavy loss, and in any case other employment was very scarce during the whole inter-war period. Some effort could be diverted into catching white fish by long-lining or by the seine-net method borrowed from Denmark, but while this occurred on a significant scale, the white fish sector continued to be much dominated by steam trawlers. The seine-net fishery was the main activity into which herring fishermen diverted part (or in some cases all) of their effort, but even in 1938 seine-net landings were only 12% of all Scottish white fish landings by value.

The herring fishery was failing to generate significant capital for re-investment even in the relatively good years of the late 1920s. In 1928, one of the best inter-war years, the fishermen's earnings represented a living wage without significant surplus for capital replacement (ARFBS. 1928: 7). In 1929, running expenses alone took up half the average drifter's returns, and with the subequent deterioration of the early 1930s the proportion had risen to three-quarters in 1933 (SFC. 1934: 10). Even if the position improved slightly after this, the Herring Industry Board found that in 1937 the average drifter was making only *c.* £1,500 from herring fishing, against a minimum figure of £2,200 needed for a 'reasonable profit' (HIBR. 1937: 7). In the next year the position was put succinctly: it was 'absolutely impossible' for a drifter to pay if it fished only during the curing seasons, and the market restricted employment outside these seasons to only a few boats (HIBR. 1938: 16).

In the 1920s there was some decline in the motor section of the fleet, some of the converted sail boats being of extreme age; and although the numbers of steam drifters were fairly constant their capital value dropped steadily. In the 1930s the drifters perforce began to be scrapped, and this was accelerated by the Herring Industry Board's scheme after 1936 when steam drifters were bought for scrapping; by 1938 130 had been disposed of in this way (HIBR. 1937: 20: HIBR. 1938: 16). This helped accelerate the run-down of the drifter fleet and by 1938 it was reduced to half the size of the 1920s.

The Herring Industry Board also instituted a system of loans to help fishermen recondition their boats, but this had only a minor effect, the men being reluctant to increase their debt burden; and by 1938 the total value of the steam drifter fleet was down to only 12% of its 1920 level (fig. 10.2), and 85% of the fleet was over 20 years old (ARFBS. 1938: 10) and in extreme need of renewal.

While the problem of inadequate boats was the main operational problems of the fishermen, gear (mainly nets and ropes) was also an expensive item the capital value of which was about half that of the boats; and nets especially suffered considerable wear and tear and normally required frequent renewal. For proper operation, a steam drifter had to replace nets at the rate of about forty per year (SFC. 1934: 12), the boat at any one time using about seventy or eighty. The downward graph in aggregate value of nets was less steady than that of boats because of three different government schemes to aid gear renewal: these were in 1924–25 after the early post-war losses, the second in 1930 after the heavy losses in the 'Armistice Gale' of 11th November, 1929, and the third under the Herring Industry Board from 1935 onwards after the nadir of the crisis in the early 1930s (HIBR. 1937: 40–41; HIBR. 1938: 22, 23).

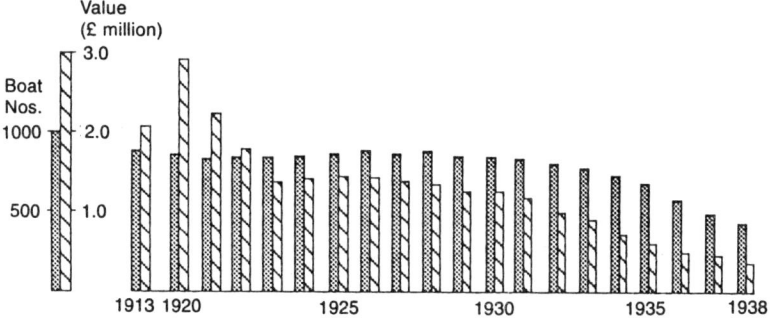

Fig. 10.2 Numbers and values of Scottish steam drifters and liners, 1913–1938.

Under these schemes a total of over 1,000 loans were made to a total value of over £30,000. While they did help, their effect was less than intended because of the reluctance of fishermen to get ever deeper in debt, and they tended to struggle on with worn-out gear. By 1938 the total area of netting in use was only about one-third that of 1920, and its value had gone down to c.15% (fig. 10.3).

An innovation of the later part of the inter-war period was the engaging in herring trawling by part of the Aberdeen trawler fleet in the later summer. By that stage of the year the herring having spawned were less active and could be taken with a bottom trawl. The main part of the herring stock was returning towards the wintering grounds on the edge of the Norway Deep, and trawlers might catch them at around 100 miles form the Scottish coast. As this was the time of year when the white fish market tended to be over-supplied, trawlers found this a useful alternative activity.

THE CURERS AND THEIR EMPLOYEES

Curing continued to represent the dominant sector in the shore-based part of the industry in terms of number of firms and total employment. The number of curing firms probably approached 300 in 1920, but by 1933 had been reduced to c.200 (SFC. 1934: 26). It is difficult to estimate accurately the number of individuals in the labour force, as it continued to be counted for the peak week in each district, which could mean that many who moved during the season might be double-counted; however such moves were relatively rare, and hence the total can be estimated as slightly below the aggregate of district totals. On this basis if appears that the total number of women gutters and packers in 1920 was c.9000, and that this had about halved by 1938; and the coopers went down by rather less than half from a 1920 figure of c.1500. Numbers of general labourers are less clear but appear to have been approximately equal to those of coopers.

In the inter-war period the position of the curers was to be eroded, not only by the declining cured market but also by the increased market share of other outlets; nevertheless Scottish curers continued to exercise a leading position in curing not only in Scotland, but also at herring ports in England and Ireland, where the fleets operated for part of the year. However curing had always been to a greater or lesser extent a speculative enterprise, and it

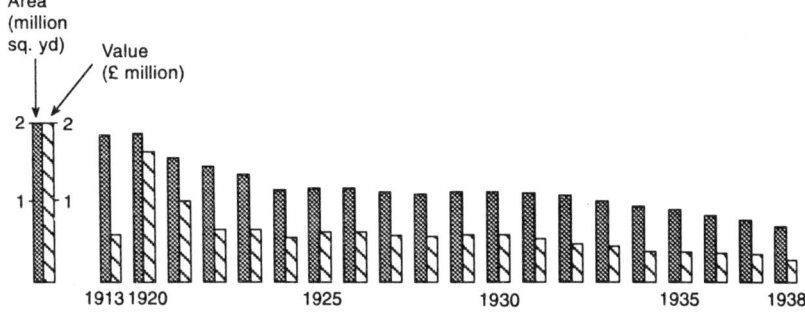

Fig. 10.3 Area and value of herring netting in
Scotland, 1913–1938.

was not uncommon for especially the smaller firms to fail even in prosperous periods. There were many instances of failure in the early post-war years, when numbers of new small firms tried to take advantage of the resumption in the trade, and there was to be continuing attrition in later years.

The sizeable curing lobby was of course involved in Parliamentary lobbying, sometimes on its own account but more usually with other trade sectors; and curers also took direct initiatives in attempting to re-establish the market. They had their most prominent role in the early post-war period, when the cured sector retained its dominance. They were much involved in securing the government price guarantees of 1919 and 1920: they were guaranteed 90% of the cost price of the herring in 1919 (ARFBS. 1920: 14). In addition a deputation of curers eased a difficult market situation in 1919 by securing a contract with the German government for the supply of 250,000 barrels (ARFBS. 1919: x). Later, in 1925 the curers joined with continental importers to form a 'combine' which attempted with some success to clear a serious glut on the market by taking measures to restrict production (ARFBS. 1925: 6).

In fact the main influence exerted by the curers was to be in restraining production: as the dominant single outlet, they had a strong voice on the various port committees. The main measures taken related to the length of the seasons, but it could involve suspension of operation and other restrictions in mid-season. A primary concern was that the summer fishery should not begin too early, as the herring in the early season were immature and produced a poor quality cure that could spoil the market. Curers actually petitioned the Fishery Board unsuccessfully in 1928 to secure legal powers to fix the starting date of the summer season (ARFBS. 1927: 7). In the light of the fact that fishing had always started earlier in Shetland and the Western Isles where herring became available as early as May, it was no simple matter to fix starting dates for curing, but it was agreed that in the Western Isles curing could begin about one month ahead of the East Coast, and at Shetland one or two weeks ahead. The dates at Shetland tended to be especially contentious as there were more local boats and curers than in the Western Isles, and they always wanted to take advantage of the early season market.

Curers might also suspend operation in mid-season for short periods if they judged the herring quality poor, and they were also primarily responsible for closing the season. In the later season a greater proportion of spent herring appeared in the catches, and the closure date was generally set on a basis that combined an assessment of deteriorating herring quality with market prospects. In the autumn season at East Anglian poor weather conditions for the fishermen could be the factor that ended a season, as happened in 1919 (ARFBS. 1919: xi) and 1926 (ARFBS. 1926: 22); however the length of the East Anglian season was also influenced by the amount which had already been cured during the Scottish summer season, and in 1928 (ARFBS. 1928: 21) and 1937 (ARFBS. 1937: 19) for example, there was a premature closure because the market was already fully supplied. After 1935 the Herring Industry Board had the responsibility of setting the length of the seasons, but the curers continued to have strong voice in the background.

Another bone of contention between curers and fishermen were the latters' tendency to use small-mesh nets: while these did catch more herring the increase was due to the taking of smaller fish which were less suitable for curing, and this was especially a problem in the early inter-war years although it faded into the background thereafter. It was not unknown for the curers to have differences with other trade sectors, and the dispute between them and the fish salesmen in 1933 was sufficiently acute to delay the start of the summer fishing (ARFBS. 1933: 8). It could also be difficult for the curers to get concerted action among themselves. They attempted, for example, to agree a maximum price of £2.7/- per cran to be paid when landings were low in July of 1920, but this 'broke down at one port after another' and they often bought at prices of £2.10/- or £3 per cran (ARFBS. 1920: 15).

Although in 1934 the ten largest curing firms accounted for *c.*25%

of all production, the big number of small firms with little bargaining power weakened the whole trade sector. A sample of returns from the largest firms showed a net trading profit as low as *c.*3% in 1929, and by 1933 this had become a loss of nearly 10% (SFC. 1934: 29).

The relationships of the curers with their employees were inevitably strained, and at different times disputes broke out. While in the immediate post-war situation in 1919 it was recorded that there was a shortage of all types of skilled labour (ARFBS. 1919: x), the position thereafter was that there was an excess of personnel looking for work. The situation was complicated by the fact that those recruited tended to be under-employed, despite curers' attempts to cut costs; this was related partly to the unpredictable level of the landings, and also to the general freedom that boats enjoyed to land at any port, which aggravated the uncertainties for the curers. The difficulties in gauging capacity for the curers are illustrated by the fact that at the 1913 peak of the industry a three-women curing crew had an average output of c. 575 barrels in the Scottish summer season and *c.*625 barrels at East Anglia, while the corresponding figures for the 1933 nadir were *c.*275 and *c.*335 barrels (SFC. 1934: 25).

In the long term, the curers' interests were most seriously compromised by both the shrinking market and a shrinking market share. By the 1930s consumer demand on the continent as well as in Britain was shifting towards fish species that did not require the heavy salting that herring needed for preservation, and also to other protein foods. Also the big market advantage that Britain had enjoyed in the early 20th century through the economics of scale of a much bigger industry were decreasing as the competition from other countries built up. This competition came from Norway, Iceland and Holland, and from the main consuming countries of Germany, Poland and Russia themselves. The rise in production in the main single market of Germany is especially noteworthy. In the early years of the 20th century Britain alone produced between two and three million barrels of cured herring annually, and in the 1930s the estimated total for the whole of Europe was running at this same level; but now the British share of this fluctuated around 900,000 barrels, a level of production which Germany herself had reached by 1937 (ARFBS. 1937: 25).

OTHER TRADE SECTORS: FISH SALESMEN, KIPPERERS, FRESHERS AND CANNERS

The other trade sectors were all considerably smaller, and information relating to them in official and other publications is much less detailed, although their position did get some specific recognition in various of the inter-war reports. It is clear that over-capacity persisted virtually throughout the industry, and while all sectors shared the problems of a far-flung industry with unpredictable supply, the different sectors were slow to seek the advantages that could have accrued through better organisation and co-ordination within them.

In the case of fish salesmen's firms, there was again a considerable variation in scale: in 1934 there were *c.* 44 firms and *c.* 40% of the business was concentrated in the hands of three of them (SFC. 1934: 22). Many of the firms had extra expenses incurred by maintaining branch offices in different ports in serving the trade; and they 'largely acted as a financial buffer' in a series of years when credit to the fishermen was refused by the banks (SFC. 1934: 22). These firms had brought themselves to a very low ebb by shouldering a disproportionate part of the debt burden.

The kipperers produced a distinctive market product which normally justified higher prices to the fishermen than the curers paid: they could generally secure supplies even when herring was scarce. Also by the 1930s they were leading the way in mechanising the splitting and gutting of the herring. There was a kipperers' association which in 1934 consisted of about 60 firms which had a total labour force of between 1,200 and 1,500 (SFC. 1934: 29) and included big numbers of both men and women. If kippering suffered a lesser degree of contraction than curing, the overall trend in a fluctuating market was certainly downward (fig. 10.1), and this trend was aggravated by problems of maintaining quality as competition from other food products increased. By the end of the 1930s output had decreased by about one half, and while it was recognised that there was excess capacity in kippering, it proved difficult to see how it could be rationalised and costs reduced (HIBR. 1938: 19).

The 'freshers' were merchants who distributed herring in a fresh state and like the kipperers catered for a market of which the geographical extent had been greatly expanded with the growth of the railway system. Much of the trade here was in small parcels that had to withstand relatively high transport costs (SFC. 1934: 33, 34), and the merchants involved were poorly organised for securing lower freight rates. On the other hand, it was recognised that improving living standards did mean that there was greater potential in this sector, and one of the first initiatives of the Herring Industry Board was to mount a sales promotion campaign that featured fresh and canned herring (HIBR. 1937: 14). This was to result in a significant market recovery in this sector in the late 1930s, with the result that this part of the trade did achieve a slight overall increase in the inter-war period. (fig. 10.1).

Canning was a sector of significance from the late 19th century, but it was always a minor component of the industry and involved only a few firms; it was also a subsidiary activity in these firms which were mainly involved in canning other foods. There was little urgency in getting the canned product to the consumer, and a considerable sector of the market was overseas. In a review of the industry in 1938, this was recognised as the one sector of the herring trade without substantial problems (HIBR. 1938: 19, 20).

The exporters consisted of a limited number of firms: in 1934 there were only about two dozen, evenly split between British and continental firms, while in addition there were about a dozen continental importers who bought direct in Britain (SFC. 1934: 26). With their key middleman position these firms played a leading role in setting prices, and this often involved forcing curers and others to accept slim margins or even losses. Despite their strong bargaining position, the exporters could not be immune from the general decline in the trade, but there was some compensation for the continental importers in the 1920s with the relatively strong increase in disposal by klondyking (fig. 10.1). While this had some effect in maintaining prices to fishermen, it had adverse results on shore employment as it effectively transferred processing to continental ports. Moreover, the increase in klondyking was not sustained in the 1930s, as supplies to the continental countries from their own fleets increased. The exporters too were left with excess capacity which continued and was still recognised in 1938 (SFC. 1934: 19).

THE ROLE OF GOVERNMENT AND GOVERNMENT
AGENCIES

While government agencies (especially the Fishery Board for Scotland) were involved with the industry throughout the period, the most prominent direct actions of government were in the immediate post-war period, and in the latter 1930s after the nadir of the Great Depression. The main response of government in the first inter-war years was to underpin the industry in the hope that trade would soon return to normal; but by the 1930s it was clear that the industry was involved in big-scale permanent readjustment, and the response was the creation of the Herring Industry Board in 1935 with the objective of regulating and rehabilitating the industry. Government also became involved in direct negotiations with governments at the market end on the continent in the attempt to surmount trade difficulties, and the governments included now not only those of Germany and Russia, but also of the newly independent countries of Poland, Lithuania, Latvia and Estonia. In addition government also monitored the winter imports of herring from Norway which supplied much of the home market at that season; an annual quota of 25,000 tons and an import tariff of 10% were imposed (SFC. 1934: 19, 20).

The role of the Fishery Board continued in administering regulations and collecting data, and this included the long-standing practice of branding barrels of cured herring as a guarantee of quality if curers wished. There was some expansion of the functions of the Board, and they were instrumental in 1920 in the removal of the restrictions on the supply of barrel staves and salt that the curers needed (ARFBS. 1920: 6, 7). The Board's fishery officers at the ports were also involved in the area committees that came into being with Board encouragement from 1920 (ARFBS. 1920:14). and it was usual for senior inspectors and executives to visit the continent to aid the marketing effort.

It was in the first two post-war years (1919 and 1920) that the government agreed, under considerable pressure, to operate market guarantees. With chaotic markets in 1919 the guarantee was for the purchase of up to one million barrels: these were to include 400,000 for the Scottish summer fishery, and 600,000 from the East Anglian autumn fishery – the other main component of the industry; and curers were to pay a minimum price of £1.15/- per cran to the fishermen (ARFBS. 1919: vii–xi). In view of the fact that the industry regularly produced and sold between two and three million barrels in the years before World War I, this guarantee might hardly appear to be an excessive commitment: but such it was to prove. In 1920 the government acceded to further strong pressure with a guarantee scheme which had a £3 million ceiling

(£1.8 million for the Scottish summer season and £1.2 million for the East Anglian autumn season), with minimum prices of £2.5/- per cran for fishermen and £3.2.6. per barrel for curers (ARFBS. 1920: 13). This led to a situation where by the end of 1920 there were 250,000 barrels unsold in Britain along with great unsold stocks on the continental markets (ARFBS. 1921: 5, 19); but in face of a deteriorating national situation the government refused further aid, and the industry was thrown back on its own resources.

However, government was active in another way: it had to discharge its obligations to replace to repair vessels which had been lost or damaged while commandeered for war service. It was stated by 1920 that catching-power had been re-established (ARFBS. 1920: 68) and by 1922 70 ex-naval boats had been disposed of to fishermen (ARFBS. 1922: 32).

Despite the definitive position taken by the government at the end of 1920, it continued to be petitioned for further aid, and substantial difficulties were again encountered by the end of 1923 – a year when the problem was abundance of good quality herring, when the prices to the fishermen on a glutted market came down to an average of £1 per cran for the Scottish summer season (ARFBS. 1923: 6), which was only half the level of 1919 and 1920. This led to unsuccessful representations for a government guaranteed wage of £2.10/- per crew man for the East Anglian autumn fishery in 1923 (ARFBS. 1923: 6); but there was a government scheme in 1924 and 1925 which did make loans available for gear replacement (ARFBS. 1937: 40, 41), as worn-out gear had become the single greatest problem in fishery operation. A similar scheme was put forward after the widespread destruction to gear in the disastrous 'Armistice Gale' of 11th November, 1929 at East Anglia (ARFBS. 1929: 7), although the later 1920s were the period of a partial recovery in the trade.

However this was to be but the prelude to the industry being plunged into a deeper trough in the disruption to international trade that followed the Wall Street crash of 1929, and this was to lead to a new level of Parliamentary concern as it became clear that the symptoms of distress in the fishing ports were among the worst in Britain. The first response of government here was to call for a special report from the Economic Advisory Council, and this was published early in 1932. While this report noted that problems of the different sectors of the trade, including the big fall in profit margins to the fishermen (CEAC. 1932: 57), its one substantial recommendation was for the creation of a board to regulate curing and export (CEAC. 1932: 153).

Continued deterioration in conditions in the fisheries in the early 1930s led to the creation of a Sea-Fish Commission in 1933, and in 1934 the government took the exceptional step of guaranteeing the fitting out expenses for herring boats provided their numbers did not exceed 1,000 at the summer fishing (ARFBS. 1934: 7). The Sea-Fish Commission in their first report in 1934 addressed a situation that for Scottish herring fishermen had become disastrous. It recommended the creation of a special board with powers to license all sectors of the industry from catching through to exporting: the objective was to be to streamline the industry and curtail capacity, but it was also to help new initiatives in the development of more economic motor vessels and of new markets. The proposals of the Commission were accepted by Parliament with minor amendments; the Herring Industry Board (HIB.) was created under an act of 1935, and it was given enhanced powers under a fresh act of 1938. While the greater part of the dealings of the Herring Industry Board were with the herring fishermen and the herring trade in Scotland, it was also responsible for these in England also, and most of its published data do not distinguish the Scottish and English components.

In the few years during which it was able to operate before World War II, the Herring Industry Board (probably inevitably) did more to ease the pain of downward readjustment than to bring in positive new developments. At the outset its main efforts were in giving loans for the reconditioning of boats, the replacement of gear, and in the buying of boats for the purpose of scrapping them. It is significant that it was the last of these measures that made the greatest impact: in the year 1936–37 116 drifters were brought and scrapped (HIBR. 1937: 20) although the numbers thereafter tailed off substantially. There was limited response by the fishermen to the schemes for reconditioning boats and even for the purchase of gear, because of their reluctance of getting even further into debt. By 1938 there had only been 25 loans made for reconditioning boats, and loan repayments were seriously in arrears (HIBR. 1938: 23).

The programme of licensing all sectors of the trade made limited impact. It did put the Board in a position of control of such matters as the opening and closing of seasons, and overall the regulation of the trade was better coordinated. On the other hand a very limited amount was achieved in cutting excess capacity in most sectors of the trade; the scrapping of redundant vessels was the main effort in this direction.

The Board engaged in direct market negotiation with representatives of continental governments, but were not able to make any noticeable improvement to sales by producers in Britain. On the other hand, some success was reported in their sales promotion in the much smaller but higher value fresh and canned sectors of the trade (HIBR. 1937: 14).

It was under the new act of 1938 that the Board made what was to prove, in the long term, its most important initiative. A scheme was put forward to allocate £250,000 to provide fishermen with grants of up to a third of the cost, as well as loans to invest, in the more economic diesel boats. In the short term, this scheme was stillborn, as it was to run for the four-year period from 1940 to 1944; but government grant and loan schemes were to be the instruments by which a new and more prosperous era for the fishermen were ushered in during the changed economic climate of development and expansion which came in after World War II.

The onset of World War I had broken the patterns created by a century of growth and resulted in cut-backs, readjustment and hardship that lasted through the inter-war period. It was not until after World War II, in a changed economic and political climate, that Scottish fisheries were again set on a path of development and growth.

REFERENCES

Annual Reports of the Fishery Board for Scotland (ARFBS) 1911, 1919, 1920, 1921, 1922, 1923, 1924, 1925, 1926, 1928, 1929, 1933, 1934, 1935, 1936, 1937, 1938.

Committee of the Economic Advisory Council (CEAC) (1932) *Report on the Fishing Industry.*

Herring Industry Board (HIB) *Reports*, 1937, 1938.

Scottish Department Committee (1914) *North Sea Fishing Industry* (NSFI), Vol. I, Appendix 18.

Sea-Fish Commission for the United Kingdom (SFC) (1934) *The Herring Industry.*

11

The White Fisheries in the Interwar Period

While the difficult trading conditions of the inter-war period inevitably caused problems for the white fisheries, these were less daunting than those facing the herring fishery. This was mainly because the catches continued to be primarily for the home market and did not have to face to the same extent the disorganised situation in international trade. In addition white fish continued to be for a big part of the population a cheap source of food protein, and there was not the same market contraction that confronted even the fresh and kippered sectors of the home herring market. The production of white fish continued at the same general level which it had reached prior to World War I; and although still dominated by trawl landings there was some revival of both small and great lining with power-driven boats, and by the end of the period the seine net was starting to emerge as the main rival to the trawl (fig. 11.1).

None the less there were readjustments in the white fisheries, especially in the immediate post-war period when a number of factors combined to create formidable problems. The dominant trawling sector was considerably less profitable than in the boom days before World War I, mainly through costs increasing more than prices. Capital formation was much slowed down, and fleet renewal was much restricted; in addition there was some excess catching capacity, and it became a fairly regular practice for trawling companies to lay up about 10% of their fleets in summer when there was the tendency for prices to drop as landings increased with better fishing weather. As well as cutting operating costs, these seasonal lay-ups served to keep supply more in step with demand.

The main weight of the fishing effort continued to be in the

North Sea, especially in the early post-war period, with average lengths of trips between four and five days. While this reflected in part the location of the main resource base, it was also to some extent dictated by the needs for fuel economy which resulted in more effort on grounds at shorter distances from the trawl ports. There was some working of West Coast grounds, although this was partly by part of the Granton fleet basing itself at Ayr and Oban for part of the year. With the rather easier situation from the late 1920s there was more working at longer range, and by 1933 it was stated that slightly over half the trips were to distant waters in the North Sea and off the Butt of Lewis (ARFBS. 1933: 33). The bigger trawlers also extended their operations to Faroe, where extra costs could be compensated by higher catch rates. The full-time steam liners were less constrained by the issue of fuel economy from their mode of fishing and also as their fish (such as cod, halibut and skate) were of higher unit value, and could best be fished in the deeper water at the edge of the continental shelf. These liners operated to the west of the Hebrides and of Orkney and Shetland; and to a considerable extent they went the longer distances to Faroe, Iceland and Greenland.

In the immediate post-war situation the first impulse was to resume developments where they had left off in 1914, but this fairly rapidly led to a situation of overcapacity. There were a number of other problems also, and post-war readjustment became complicated and involved a total of 302 trawlers. The big part of the fleet had been requisitioned for war duty (ARFBS. 1919: vi), and these were released to resume fishing. To these were added vessels which had been built for mine-sweeping and patrol duties and now converted to trawling, and ports like Dundee and Peterhead increased their trawler fleets significantly, while in Aberdeen the trawler fleet had by 1920 reached the unprecedented figure of 260 (ARFBS. 1920: 66,67). This led to congestion in Aberdeen, and with the limited purchasing power on the depressed post-war market prices were low. In 1920 also the miners' dispute in the autumn caused coal to be rationed to 50% of normal consumption and drove up coal prices; and with fishing material prices up three times from 1914 levels, some of the bigger trawlers were laid up for three months. Steam liners, with lower rates of fuel consumption, did better: and this fleet sector was expanded in 1920 by 30 trawlers and numbers of steam drifters converted to lining. At one time 120 steam liners were fishing grounds spread as far as the west of

Ireland, Rockall and Faroe while landing at Aberdeen (ARFBS. 1920: 68).

Even when the immediate post-war difficulties were beginning to subside by 1922, it only became clearer that catching capacity in white fish had outgrown the available market; and the market was itself depressed by the national problem of unemployment. This problem was seriously aggravated by the resumption of German trawler landings in quantity, and this contributed to the serious fall of 27% in average white fish prices between 1921 and 1922 (ARFBS. 1922: 5). The German trawlers had previously concentrated their fishing at Iceland and supplied material mainly for curing which had a limited effect on the main fresh fish market. However that outlet had become constrained by shortage of skilled labour and limited market opportunities. When the German trawlers switched their main effort to North Sea and other grounds and their landings reached *c.* 18,000 tons in 1922, the home trawler men saw fishermen from the country with which they had recently been at war having a serious impact on the market, and this resulted in the laying up of the entire Aberdeen white fish fleet for three months in protest (ARFBS. 1922: 6; 1923: 8). After extended negotiations the Germans agreed to limit their landings, although in 1925 they reached a peak of over 20,000 tons; however they did subsequently begin to land more at home in Germany and at other British ports, which did defuse the situation. The problem was briefly to resurface in the early 1930s when with the German problems of foreign exchange the trawlers found that they could get better prices in Britain. While there were foreign landings by boats from several nations throughout the inter-war period, the attraction of the British market for these boats was slipping away with deterioration of Britain's relative economic position. There were landings of groups like Belgian trawlers and Swedish seiners, but these had very limited effects on price levels.

Trawling was also beset throughout the interwar period by a problem of small-sized fish. There were repeated complaints especially about the amount of small haddock being landed. This was in fact only part of the situation, as many more were shovelled overboard at sea. While this was not a new problem in the trawl fishery, it was aggravated in the early post-war years by the fact that exploitation levels had been reduced for several years during the war and there was a superabundance of young fish on the grounds. However the Fishery Board were able by 1924 to report that the situation in trawling was improving and that profitability was being improved partly through the use of improved gear: prominent here were better designed otter boards which were reducing fuel consumption (ARFBS. 1924: 8). There were inevitable problems during the general strike of 1926 when coal was rationed to 25% of normal consumption, and for nearly all vessels increased expenses swallowed up earnings and left no profit; but it was a mark of a new resilience in the industry that white fish landings decreased by only 2.5% from the previous year (ARFBS. 1926: 6). By 1927 it was stated that steam trawling and lining were at the point of overcoming their difficulties, and that increased efficiency of operation was contributing to this. In 1930 it was reported that trawling was now much more regular again throughout the year, and that trawl companies were adding to the fleet with several new and second-hand boats (ARFBS. 1930: 34). A significant improvement from around 1930 was the beginning of the use of radio receivers by the trawlers, which for the first time allowed them to keep up with weather forecasts: this had an obvious use as a safety measure.

A development of this time was that fishermen whose previous work had been on herring drifters sought berths, often with some reluctance, on trawlers. This was particularly the case on the Banffshire, which had had the greatest concentration of man power committed to the herring fishery, and where unemployment among fishermen had become a particularly severe problem. This was to set a pattern whereby for the rest of its history, men from Banffshire played an important part as crew in the Aberdeen trawl fishery.

Behind the relative success of the trawler fleet was the fact that it catered mainly to the home market, and it continued to provide one of the cheapest sources of food protein. Unlike that for herring, the trend of the price of white fish did keep somewhat ahead of that of the national price index, and this made for a more viable catching sector. Although the movement of the main single species of haddock barely kept up with the trend of the price index, the prices for most other white fish like cod and lemon sole kept at least 50% ahead of it; and in the case of skate the increase was over 200%. Even so the real price margins in trawling must have deteriorated with the post-war increase in costs.

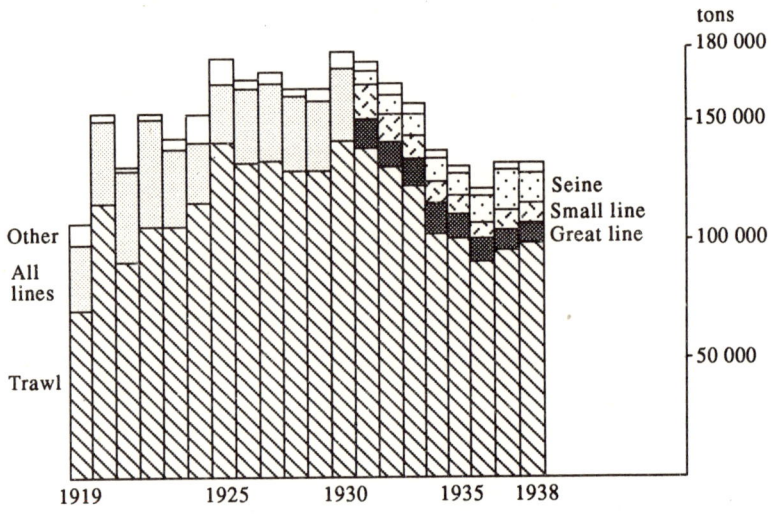

Fig. 11.1 White fish landings by method
of capture, 1919–1938.

LINE FISHING

While trawling continued to dominate the white fish market, it had
to contend with some competition from other catching methods.
Fig. 11.1 shows that line-caught fish in the immediate post-war
period continued to supply up to 30% of the white fish, although
this declined by the late 1930s to between 10% and 15%. In the
immediate post-war period there was something of a boom in
great-lining, which was less demanding on fuel than trawling, and
the great-line fleet was supplemented by steam drifters because of
the major market difficulties encountered by the herring fishery.
Now that the inshore fleet was generally motorised, inshore lining
was also more economic and there was some revival of it; and it still
in general provided the highest quality of white fish. Such fish were
landed daily, whereas much trawl-caught fish could be a week old
or more at first sales; and in addition line-caught fish never suffered
compression in the trawl cod-end. In the 1930s small lining was of
the same order of importance as great lining (fig. 11.1). However line
fishing was still considerably more labour-intensive than trawling
because of the time-consuming work of baiting that still generally
fell on the women folk of the fishing communities. Its long-term
decline was inevitable, especially when an alternative arose which
dispensed with baiting: this was the seine net.

THE DEVELOPMENT OF SEINE NETTING

What was to become a more important competitor to the trawl was the seine net: the main method of working this gear, which is a bag net with ropes, was borrowed from Denmark and required less engine power in a boat than did the trawl. However the way to this development had been partly prepared by previous events at the end of the war and in the immediate post-war years. Under war-time exigencies in 1917 when inshore lining was proving unremunerative, inshore grounds between the Red Head in Angus and the mouth of the River Ythan in Aberdeenshire were opened to seining on the pattern that was allowed in the Forth and Clyde (ARFBS. 1920: 67). Various crews in this area bought seine nets and had a profitable fishery which was dominated by flat fish like plaice and sole. There was subsequent pressure to extend the area sanctioned for inshore seining south to Fife Ness and north to Rattray Head, and this was acceded to.

The Danish way of working this gear had developed from the middle of the 19th century in the Limfjord, and inshore waters in the Kattegat, the Belts and the North Sea; and with the motorisation of the fleet in the early years of the 20th century its use was extended offshore (Thomson 1981: 1–10). Danish boats, which were bigger than the Scottish inshore craft, using this method of fishing and attracted by the big British market, were landing at ports in England soon after the end of World War I. Their landings gained enhanced importance during the miners' strike of 1920 when trawl landings were reduced; and they proved competitive in subsequent years, their success being largely due to ability to produce at much lower costs than trawlers. A number of boats in Scotland were stimulated to experiment with the technique from 1921 onwards; and the port of Lossiemouth was to come to the fore as the main port involved in the innovation.

The Danish method of operating this gear was to put the ropes out in a long arc or circle, with the net being dropped at its mid point (fig. 6.1). The boat dropped its anchor after shooting the gear was complete; it then slowly winched in the ropes in a direction against the flow of the tide, and heaved up the net when it came to the boat. This method was well suited to a fishery for species like plaice and sole which were particularly important on the extensive shallow sandy grounds on the Danish side of the North Sea, and the early use of the gear in Scotland was mainly from steam drifters near to the East Coast, although by 1923 they were also trying the

gear with a measure of success as far afield as the Dogger and
Fisher Banks in the North Sea (ARFBS. 1923: 8). It was seen from
the start that the method was considerably more economic on fuel
than was trawling, as a boat at anchor merely needed enough power
to operate the winch. While plaice was also the most important
species in the early days of the fishery in Scotland, here round fish
were more important than the flat fish on which the Danes concentrated. The round fish swam more quickly than flatfish and
tended to escape the net when it was worked by the Danish
method; and while the basic method was adopted, in Scotland it
became the practice for the boat not to anchor after shooting the
net, but to maintain forward motion and winch in the net more
quickly: this made it easier to catch the round fish. Even so, early
seine netting in Scotland caught a high proportion of flat fish,
including especially plaice and lemon sole; and when separate
statistics for weight and value of seine net catches started to be
published in the 1930s, it was notable that the unit value of landings was distinctly above those of both trawl and line caught fish.

Seine netting was still at first restricted by the legislation that had
been passed in the late 19th century to restrict trawling in inshore
waters and in the firths.

With the limited exception of flounder nets on the Forth and
Clyde, all drag-netting had been prohibited, and there was also
controversy on the legal standing of any net having a pouch or cod
end. Once seining had begun off the East Coast, there was pressure
to have it allowed within the Moray Firth, but line fishermen were
resolutely opposed to any relaxation on drag nets outside three
miles. While there were some prosecutions against the early use of
the seine net, the law was to be clarified and amended: it became
an accepted method everywhere outside territorial waters which
were limited by three miles from the low water mark: this meant
that it could be deployed in major firths like the Moray Firth and
the Firth of Clyde. It became sanctioned for smaller boats (under
40 feet) for the great part of the East Coast south of Rattray Head
with the main exception of the inner part of the Firth of Forth, and
it was also permitted within the three-mile limit within the Firth of
Clyde. At Shetland seining was sanctioned for boats of under 40 feet
off part of the east coast of the islands from 1926 (ARFBS.
1926: 9). Later, seining was to be allowed within all territorial waters
at Shetland by boats up to 50 feet in length (Ritchie 1960: 4, 5).

With the great contemporary difficulties in the herring fishery, the authorities were in general sympathetic to a development like seining which gave an alternative employment to at least some fishermen. Even so, any wholesale conversion to seine netting at this time was far out of the question with the available national market already mainly supplied by trawling. In fact there were some attempts to convert steam drifters to trawling, although these in general came to little and were handicapped by inadequate engine power in the boats which made the attempts. Steam drifters had often gone long-lining in spring, a season when herring are particularly scarce, and there was also some increase in their long-lining effort at other seasons, but this came into competition in the market sector already being supplied by full-time steam liners. In all there was no general solution to the problems of the herring fleet in the white fisheries.

The first efforts at seine netting were mainly by steam boats during the early months of the year, which were always a low period in herring fisheries even in more prosperous times. They worked mainly in the Moray Firth where cod spawn at this season, and where the cod (as well as haddock and flat fish like plaice and sole) were available in quantity at no great distance offshore. Landings were made at convenient ports, which included Buckie and Macduff as well as Lossiemouth; and boats fishing further east landed mainly at the main market of Aberdeen. Some steam drifters reverted to the herring fishery in the summer, but most persisted with seining. The big part of them operated in summer on banks further east in the North Sea and landed at the main market of Aberdeen; but some at this season also tried seine netting on the West Coast.

From 1924 for a run of five years the situation in the herring fishery became somewhat less adverse, and steam drifters which had experimented with seining generally reverted to what they still saw as their main livelihood of drift netting for herring. While seine netting was more economic on fuel than trawling, steam drifters were really too big and cumbersome vessels for this method of fishing, while their large crews of nine or ten men added to the cost of the operation (Thomson 1981: 15). The development of a viable seine net fishery was essentially associated with the evolution of a suitable design of motor boat for the task: this was to be smaller than the steam drifter and was much cheaper to run. It was possible for motor boats to operate the gear and winch without the disadvantage they suffered in the drift-net herring fishery.

In 1924, when the future of the seine net fishery appeared in serious doubt, as many as 45 boats in the Arbroath district took up the method: these were nearly all motor boats, and their crews were attracted by the use of a method which obviated the time-consuming line baiting. More significant, however, was the custom building of new boats for the purpose by men who had originally gained experience of the fishery in steam drifters. The initiative here of John Campbell of Lossiemouth has been recognised. He in 1926 built the 50-foot 'Marigold' on a modified fifie design, with a Gardner semi-diesel engine, a forward cabin, and the winch immediately forward of the wheelhouse (Thomson 1981: 15, 16): it immediately proved profitable and acted as a spur to a series of other skippers. These new boats were in the 40 to 50 feet range: they had ample carrying capacity for a fishery based on daily landings and their motor engines were vastly more economic than those of the steam drifters. From the mid-1920s the main effort passed to motor boats, and by the end of the 1920s Lossiemouth had become the main seine net port in Scotland. Seining was also taken up as a main activity on the Moray Firth at Whitehills, Macduff, Lybster and Helmsdale, in which its main attraction was the saving of the trouble of line-baiting: it was also taken up after a time lag at Wick, which had lost its position as a main herring port, and which needed an activity to make this good (Ritchie 1960: 7). When the 60-foot 'Olive Leaf' was built in 1934 it was in effect to be the prototype of seine net boats for more than a generation (Thomson 1981; 18, 19). Her winch was immediately aft of the foremast, and the cabin was in the stern.

Approximately three-quarters of all Scottish seine net landings were from Moray Firth grounds (Ritchie 1960: 12); and the growth of the fishery was aided by the continued prohibition of trawling in the firths. The taking over of the main seine net effort by motor boats is plain in the official statistics: of the 762 tons landed in 1922, almost 94% came from steam boats: in 1925; of 2,701 tons in 1925 it was 58%; but by 1930 out of 5,411 tons only 17% were contributed by steam boats. While catches thereafter continued growing and reached 17,350 tons by 1938 (fig. 11.1), the catch of the steamers in this had dwindled to insignificance.

There were other results of the establishment of seine netting in that it gave a new lease of life to numbers of small ports that had declined, both with the centralisation of herring fishing at major ports and with the later decline of inshore line fishing at smaller settlements. Seine netting also came at the time when road was beginning to complement rail as the main transport medium for marketing, and it allowed in some cases places which had no direct rail link to participate in the fishery.

MARKETING DEVELOPMENTS

There were significant changes in the structure of the white fish market during the inter-war period. In addition to the continuing predominance of the fresh sector, there was a considerable rise in the proportion that was filleted before going to the consumer, and by 1937 this was recognised as a main demand sector (ARFBS. 1937: 37). While there was some tendency for the proportion of the catch that was cured to drop, this was still a major sector throughout the period, taking around 30% of the landings. The big part of this was short-term preservation by smoking of haddock, and to some extent species like cod and whiting; and by the early 1930s there was a rising demand for smoked haddock fillets (ARFBS. 1933: 39).

The trade outlets with inland wholesale markets, and the wide distribution of fish shops and fish friers were well maintained during the inter-war period, and the main change that was affected came with road motor transport becoming a rival to rail, especially over the shorter distances. It was recognised in 1936 that competition between road and rail within Scotland had become keen (ARFBS. 1936: 40); and by 1933 there were increasing numbers of mobile fish shops that were bringing the product to the consumer's door in town suburbs as well as in country districts (ARFBS. 1933: 43).

The cured sector that did fall off more rapidly was that of dry curing, which concentrated mainly on cod. This product was mainly for export to South America and the Mediterranean, and was more affected by the disorganisation of international trade. The market for this had come to be supplied mainly with cod form Iceland by German trawlers, and this led inevitably to some dissension in the port of Aberdeen in the early 1920s, when the local merchants engaged in the trade wished for the landings of these trawlers to continue at a time when other interests in the port wished them to be banned to help maintain market price levels.

REFERENCES

Annual Reports of the Fishery Board for Scotland (ARFBS.) 1919, 1920, 1922, 1924, 1926, 1927, 1930, 1933, 1936, 1937.

Ritchie, A. (1960) *The Scottish Seine Net Fishery 1921–1957* D.A.F.S. Marine Research 1960, no.3., HMSO, Edinburgh.

Sutherland, I. (1985) *From Herring to Seine Net on the East Coast of Scotland*, Camps Bookshop, Wick.

Thomson, D. B. (1981) *Seine Fishing. Bottom fishing with rope warps and wing trawls*, Fishing News Books, Farnham.

12

From World War II to the
European Community:
Modernisation and Diversification

During World War II fisheries activity was again inevitably much
disrupted by hostilities as it had been twenty years previously. The
great part of the trawler and steam drifter fleets were again requisitioned for war service, along with many of the bigger motor boats:
in all a total of 671 boats were called up for war service. The majority of the fisheries manpower were called up, and 10,000 of the
17,000 Scottish fishermen did service in the navy and the merchant
marine (RFS. 1938–1948: 5, 6). Although boats were operated
mainly by older men and boys, fishing none the less made a significant contribution to food supplies, and the remoteness of Scotland
from most national markets was largely countered for the period by
the government applying a flat freight rate to market: this was done
by the transport cost equalisation scheme, which was financed by a
levy paid by the buyer of the fish at first sales (RFS. 1938–48: 67).
There was less disruption to inshore than to offshore operation,
and partly associated with this was the main war-time change of
the continued growth of seine netting for white fish. This was taken
up as an important fishery in the important fisheries districts of
Shetland, Peterhead, Fraserburgh, Anstruther and Eyemouth:
despite the inter-war problems these centres had persisted with the
herring fishery, and previously seine netting had been on a minor
scale. East Coast men made more frequent visits to the West Coast
and Mallaig became an important landing point. Seining was also
extended to the Clyde, mainly by East Coast men coming seasonally
to land at Ayr and Campbeltown. In Shetland it became imperative
in wartime to find a fishery to replace the main herring fishery. Up

to this time seining had made little headway in these islands, but by 1946 there was a fleet of 57 boats, landings were over 2,000 tons (Ritchie 1960: 15, 16), and a foundation had been laid for post-war expansion. By the end of the war in 1945 the seine net catch at nearly 30,000 tons was well over double what it had been in 1938, and over 90% of it was landed at East Coast ports (Ritchie 1960: 8).

The operation of the trawler fleet was much constrained in the North Sea: here fishing was confined to a narrow band along the coast, and there was more emphasis on West Coast grounds, including those to the west of Orkney and Shetland (RFS. 1938–48: 11). The Granton trawler fleet found it better to operate from the ports of Ayr and Girvan on the Clyde for most of the war; and Aberdeen trawlers conducted some of their fishing from the English port of Fleetwood.

In the herring fishery, war-time operation on the East Coast was almost confined to the Moray Firth until 1943, but the West Coast was mainly unrestricted and a considerable part of the available catching power was deployed there. There was control of prices for herring, as for white fish during the war: maximum prices were fixed by Ministry of Food order and were for different years in the £4 to £5 per cran range; and port committees also determined minimum prices at the start of the war, while later the Ministry of Food applied a floor price of 30/- for herring unsold on the market. For the herring fishermen able to remain active the war proved a blessing in disguise, with prices and consumption both going up. A system of controlled prices was maintained for several years after the war, which obviated any real risk of the sort of dislocation that followed World War I.

NEW POST-WAR GROWTH

Following World War II fisheries in Scotland witnessed a period of development and prosperity that was a marked contrast to the struggles for survival of the inter-war years. This was a period of world-wide expansion in fisheries in which developed countries in Europe along with Japan played a leading role in modernisation of boats, gear and other equipment; and in Scotland as in many other places it was to see considerable diversification in the fisheries. In the established fisheries white fish came to be the most important sector as herring fishing declined in importance. Although the new developments in time were to generate by the 1970s pressures on resources that were to cause re-appraisal on fishing rights at the top international level, there was a generation in the post-World War II world that enjoyed the continued freedom of fishing on the high seas, and whose essential concerns were largely limited to catching fish at a price which could pay. With renewed economic progress at the national and international levels at the period, the main official concern was to promote efficiency in an industry that was in danger of falling behind in profitability and income generation.

TRENDS IN THE LABOUR FORCE

In developed countries fishing, like agriculture, has seen long-term contraction in its labour force over a period of over a century (fig. 12.1). This certainly can be seen in Scotland. At the peak in the 1880s the number of fishermen had approached 50,000. Most of these would have been full-time, but the total also included a big number of crofter-fishermen in the Highlands and Islands: although not separately counted at the time, they must have been at least 15,000 in number. It is clear that the numbers of fishermen were falling rapidly at the end of the 19th century with the rise to dominance of the more capital-intensive fisheries with steam trawlers and steam drifters. Although the numbers steadied at about

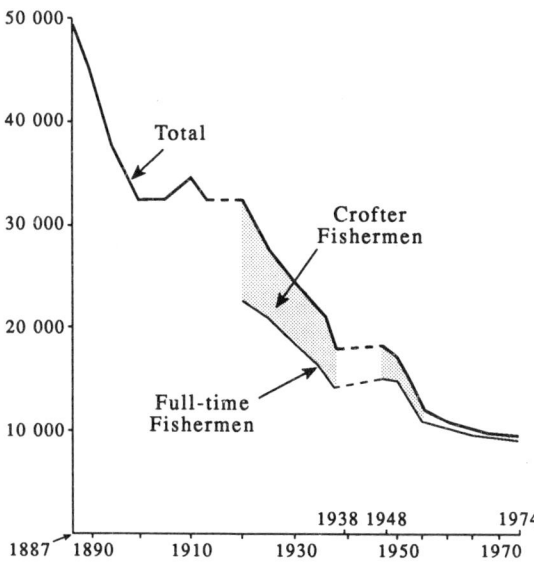

Fig. 12.1 Numbers of fishermen 1887–1974.

32,000 in the early 20th century when these fleet sectors were at their peak, the difficult inter-war period saw continued contraction in the labour force, despite the difficulty of finding alternative employment, and by 1938 the total was below 18,000. Crofter-fishermen were first enumerated after World War I, and at 9,829 in 1920 accounted for 30% of all fishermen. Even in the inter-war period the long-term trend towards full-time, as opposed to part-time employment continued and by 1938 the crofter-fishermen were down to 23% of the total. There was a temporary increase in the fishing labour force with good fish prices and better opportunities immediately after World War II, but thereafter the long-term run-down resumed, although from the mid-1950s it has decelerated. Since 1965 the total has been below 10,000. The enumeration of part-time fishermen was reorganised in 1960: this made it clear that other part-time men now outnumbered the crofter-fishermen; by the early 1970s the part-time category now accounted for under 15% of the allover total, and the other part-timers now were in a ratio of 5 to 1 over the crofter-fishermen. The dominant trend over the period was towards whole-year operation of boats to justify the enhanced capital investment, while the decline in drift-netting for herring cut down on the demand for seasonal employment in herring crews. The general decrease in working hours at the same time gave new opportunities to men whose main employment was in other fields to engage in such activities as lobster-fishing as a side-line.

While less detail is known of numbers employed in fishing-related activities on shore, this has also been a major employment sector that has witnessed great contraction and adjustment. Earlier data available applied very much to the fisheries for curing - particularly herring, but also of cod and ling. These took no account of such important categories as the numbers of women employed in line baiting for the white fisheries, or in net mending in the herring fisheries. What is clear is that the numbers employed in shore curing were of the same order as the numbers crewing the boats, and in the peak years around 1880 were around 50,000, and there was little reduction before World War I. Following World War I numbers of fish workers were at around 40,000 during the 1920s, although there was some subsequent contraction from the trough of the inter-war depression onwards. There has been a restructuring of this sector of the labour force which started in the inter-war period and gathered momentum after World War II: then the curing sector dwindled rapidly to near extinction, but there was an expansion of employment in the premises of white fish and shell fish processors and merchants. It is clear that for more than a century the employment generated on shore by fishing has been significantly in excess of that on boats, as in addition to the work in processing there has also been considerable employment in such activities as ship building and maintenance, and in transport.

There has been the continuation of a sizeable on-shore labour force, which is mainly now involved with the handling and processing of white fish and shell fish. The main category here is still processing workers, which are represented on some scale at almost all important landing places, and there is a continuing main concentration in Aberdeen. Figures collected in the mid–1960s showed a total of around 18,000 shore-based workers in fish merchants' premises, about 63% of them being in the North–East and 39% in Aberdeen alone. Outside the North–East the main concentrations were in the Leith and Ayr districts. At the former there were 1,800 related mainly to dealing with trawl landings, while in the Ayr district a total of over 2,000 were employed in dealing with a variety of species including shell fish, white fish and herring. More recently there has been some growth of employment in fish plants outside the main centres of fishing effort, including several districts in peripheral locations where such new employment was particularly welcome. This until the later 1970s was mainly in the freezing of shell fish, and in the case of the Stornoway and Eyemouth districts actually led to more than doubling the numbers employed in fish plants in the decade to the end of the 1970s (Steel 1980: 24); the numbers in the Stornoway district reached over 280 (Steel 1980: 22). In Shetland the main effort in the development of fish plants was in freezing white fish for export markets; and although there was an overall decline in numbers employed in fish processing in Shetland in the 1970s with the closure of the once supreme herring fishery, the figure of around 600 now meant mainly all-year work, compared with employment in the summer season only. With well over 200 also employed in other ancillary trades (Coull, Goodlad and Sheves 1979: 8), the proportionate importance of fishing-related employment in Shetland has been the greatest in the country. In addition, in the West Highlands, Western and Northern Isles the boom in salmon farming generated more employment in fish plants from the 1970s.

The data for workers in fish merchants premises give only a

partial picture, as there is employment in a number of other ancillary trades, and with the range and sophistication of the equipment now used by the fleet, some of these ancillary trades such as electronics have significantly expanded, while such activities as gear manufacture, ice-making, and transport continue to be important. Although the numbers in fish merchants' premises had gone down to *c.* 4,500 in Aberdeen by 1972, in this main port the total number of other ancillary workers was put at a minimum of 3,700 (Aberdeen Fishery Office: pers. comm). With the facilities the modern fleet needs there is employment in virtually all districts in providing and maintaining equipment and a variety of services.

FLEET REBUILDING

In Scotland this period saw the continued decline of the once-dominant drift net herring fishery, along with the rapid scrapping of the high-expense steam drifters and their replacement by more economic diesel-powered boats. The modern developments in types of boats and their equipment are more fully discussed in ch. 15. There was overall a run-down in the size of the fleet, although with modernisation this meant that catching power was maintained or enhanced; and although there was a contraction in crew man power, labour availability was approximately in balance with demand for it. An important trend was to have as much of the work as possible done by machine (or winch) rather than by men, so that fishing as a general rule has become less labour-intensive and more capital-intensive. This included a general decline in line fishing, with its time-consuming baiting; it also saw the run-down of the once supreme drift-net fishery, which still essentially involved manual hauling of nets. Even so, in the white fisheries a main part of the work has continued to be hand gutting of fish to minimise deterioration. The declining herring drift net fishery was effectively replaced by the expansion of seine netting to the point that this catching method became the leading source of white fish (fig. 12.2). In seine netting there was a general tendency for the size of boat employed to increase, and for trips to extend beyond the former one-day length and to range over longer distances. This was taken up for part or the whole of the year by men whose traditional livelihood had been the drift net; and others of them also developed as a main activity the trawling for nephrops (ch. 14) to provide for a new market sector which had opened with rising national living standards and purchasing power.

The rapid development which incorporated this transition was much helped by government aid. Part of the post-war philosophy was increased help and security for food producers. Although farmers were of course the main recipients of such aid, in fishing there were schemes of capital injection to help modernise the fleet, and there were also for a time operating subsidies to help profitability. Under the government 'grant and loan' schemes it was generally possible to acquire a new boat by putting down 15% of the cost, while 30% to 35% was paid by a grant and the remainder covered by a loan. There was a distinct trend for the average size of boat and engine to increase as it became more difficult for smaller boats to compete, and ultimately from the late 1960s there was the beginning of a transition to the stronger steel-built boats for the main part of the fleet.

The earnings of the white fish and herring fleets were effectively underwritten during the 1950s and 1960s by government operating subsidies. For the smaller boats these were paid according to the weight of the catch. For the smaller fleet sector such subsidies could amount to up to 20% of a boat's gross earnings and even for the bigger craft they often contributed around 10% of the gross. The trawlers were also paid subsidies, at what was effectively a rather lower rate on a daily basis. For the smaller trawlers this could exceed 10% of the gross returns, although for the bigger ones it was generally of the order of 5%.

While trawling continued to provide a big part of the supply of white fish, it was slower to modernise and in the early post-war years continued to operate largely with an ageing steam-powered fleet. This was eventually replaced in the years around 1960 with a more economic diesel-powered fleet under a government 'scrap and build' scheme; however in the process the numbers of the fleet were reduced by more than half. Even so with catch rates that improved by almost 50% over pre-war rates and a healthier market situation, the total trawl catch in the mid–1960s was only about 10% down from 1938 (Coull 1968: 13).

In the renewal of the fleet, there was also some replacement of steam long-liners by more economic diesel-powered boats, although the liner fleet was in long-term decline. Line trips often extended to three weeks or even more, and continued to involve the extra work of line-baiting: for these reasons the work became increasingly unpopular with crews. This sector of the fleet was supplemented by a number of diesel-powered boats, mainly from Fife, which continued to fish with long lines in middle and distant waters, although this eventually came to an end with the general extension of international fishing limits to 200 miles in the 1970s.

The regional distribution of catching power in the country continued much as before with the North-East dominant, and a major outlier in Fife. There was investment in dual-purpose diesel herring and seine net boats in Shetland, where continued interest in fishing had become largely concentrated in the two islands of Burra and Whalsay. The severe regional problems of the Western Isles became recognised from 1960 in the special training scheme for fishermen resident there; and this was later reinforced by extra financial help in the acquiring of new and second-hand boats from the Highland Board from 1965. There had been continuation of the fishing tradition in Scalpay (Harris), Eriskay, and (for shell fishing) Grimsay (North Uist), and the fleet additions were largely concentrated in Stornoway and Barra. The West Coast boats took increasingly to nephrops trawling as their best opportunity at this time: this fishery could be pursued by boats of moderate size and running expenses were also moderate for men starting in fishing. The extension of the financial aid from the Highland Board to Shetland and Orkney also helped development in those islands.

SEINE NETTING

The greatest development of the period was undoubtedly the rise of seine netting to be the dominant activity among Scottish fishermen. This was the continuation of a trend that had begun in the inter-war period (ch. 11), and which also was able to maintain momentum during World War II when it helped provide for national food supplies, and was aided at the time by a policy of flat freight rates which facilitated marketing.

With the government-aided rebuilding of the fleet and the continued decline of the herring market, more and more boats took up seine-netting. While a considerable number of the new boats continued to spend some time in herring fishing, especially in the East Coast summer season, more and more of them took to full-time seining: this avoided the expenses of gear duplication and white fish catches were more dependable than herring catches. Although catch rates were in general lower with the seine net than the trawl, it allowed boats to sweep a wider area of ground than did the trawl, and it could be worked with a less powerful engine. An important part of the developing pattern was the extension of the method of fishing to the West Coast, although for the great part operated by East Coast men. They were drawn there originally by higher catch rates on less exploited grounds, and a series of West Coast ports developed, including Lochinver, Kinlochbervie and Oban: from these boats operated for part or all of the year, while crews became involved in weekly commuting, returning to their homes at the week-end.

Boats having seine-netting as their major enterprise were first separately enumerated in 1963, when the total was 602, when they outnumbered the traditional trawlers by about four times. By this time the total seine net catch had edged ahead of that of the trawl; however the number of seine netters was now actually in decline, as with the installation of more powerful engines in the boats, numbers of them began to convert to the light trawl. Subsequently the trawl became again the main method of capture as boats outside the main trawl port of Aberdeen also became capable of using heavier trawls (fig. 12.2).

An important accompaniment to the growth of seine-netting was the reduction in relative importance of Aberdeen as the main white fish port. Although Aberdeen in the post-war period soon became the most important single port for seine-net landings, there were also landings at a variety of ports on both East and West Coasts (Coull 1968: 17). However there was also a considerable growth of the practice of consigning fish to Aberdeen by road for first sales, especially from the West Coast, where there were few buyers and where prices at the quayside markets were consistently lower.

With many of the strongest-going seine-netters from a range of East Coast districts landing at Aberdeen there was the development of congestion in the fish dock and at the market. Even so, landings dues were higher at Aberdeen than elsewhere. At this time Aberdeen was a port in the Dock Labour Scheme, and the fish market porters had a monopoly of landing fish from boats. At the start of the 1960s, because of costs and labour problems it was decided to discontinue the former practice of having Saturday auctions at Aberdeen, in addition to those through the week. It was at this juncture that a Saturday auction was set up at Peterhead, 33 miles away, a port not in the Dock Labour Scheme, which had lower pier

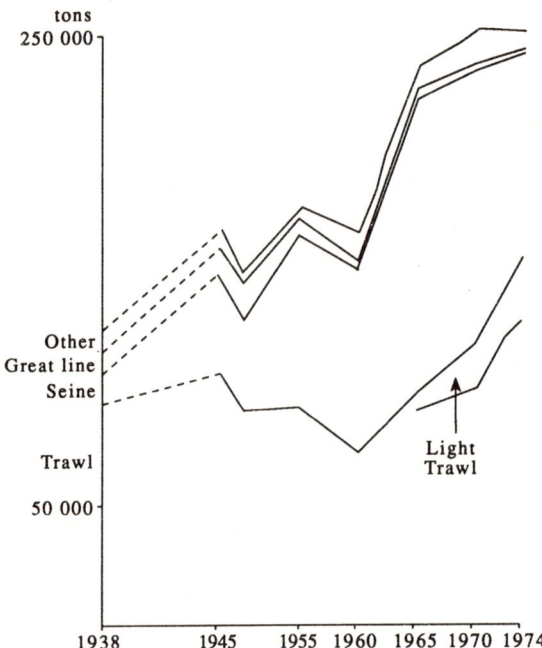

Fig. 12.2 White fish landings by method
of capture, 1938–1974.

dues and the fishermen could land their own fish. In effect, the buying firms transferred their operations to Peterhead on Saturdays, and took the fish to their Aberdeen premises by truck. The situation was further complicated when a berthing scheme was brought in to give the Aberdeen trawler fleet preference for berthing at the fish market over seine net boats. The dissatisfaction caused by this to the seine net men soon made them start landing at Peterhead through the week as well as on Saturdays; and the overall result was that with the dominance of the seine net over the trawl, Peterhead by the 1970s became the leading white fish port in Scotland. The Aberdeen trawler fleet continued to decline in a situation in which its former advantage in greater catching power was largely gone, and the new situation of extended international fishery limits also severely restricted the operations of its bigger vessels. This produced the modern situation in which Aberdeen continues to dominate the processing and marketing of white fish, with Peterhead as the main white fish port and landings at the Aberdeen market of secondary importance.

THE PELAGIC FISHERIES

The dominance of the herring in the interests of Scottish fishermen was now at an end, although the landed tonnage continued to be substantial (fig. 12.3). While there was a continuing effort by a declining number of drift net boats until the 1970s, more and more of the boats converted seasonally or permanently to seine netting for white fish. Drift netting in its latter years was itself improved by the use of rot-proof nylon nets which in any case were more efficient than the former cotton nets; and the use of the echosounder in finding shoals helped improve average catch rates. Even so, despite these improvements the drift net was to be superseded by more productive and less labour-intensive methods.

There were to be major innovations in fishing methods. The Danes and Swedes developed mid-water trawling for herring used for reduction in the eastern and central North Sea, and in doing so installed more powerful engines to pull the trawl at a sufficient speed. This was a major labour-saving method of catching, as it eliminated the manual shaking of herring from drift nets, an operation which could last for eight or ten hours with a big catch. Attempts in the 1950s and early 1960s to copy the Danes and Swedes by Scottish boats in the west side of the North Sea had poor success. On the west side of the North Sea the herring were in a more active and mobile phase of their cycle, and at this stage the boats did not have sufficiently powerful engines to catch them. However in the early 1960s the mid-water pair trawl was proved successful in fishing herring off the West Coast, and was mainly used in the winter when the herring were less active and mobile. This was later extended to the North Sea, as more powerful engines were installed on boats, but the great innovation in the North Sea fishery was that of the purse-net. This was copied from Norway where it had had great success in the Norwegian winter herring fishery: it involves the encircling principle, but on a much bigger scale than with the Scottish ring-net, and it was now possible to use it in the open sea because of the ability to haul it mechanically with the power block. It was also possible to make nets bigger from the stronger modern synthetic fibres, and purse nets spread out flat may be the size of five or more football pitches. This gear made a great impact on the herring fisheries in the later 1960s: and although at this stage the Norwegians much dominated the purse-net fishery, there was the emergence of a significant Scottish fleet of purse-net

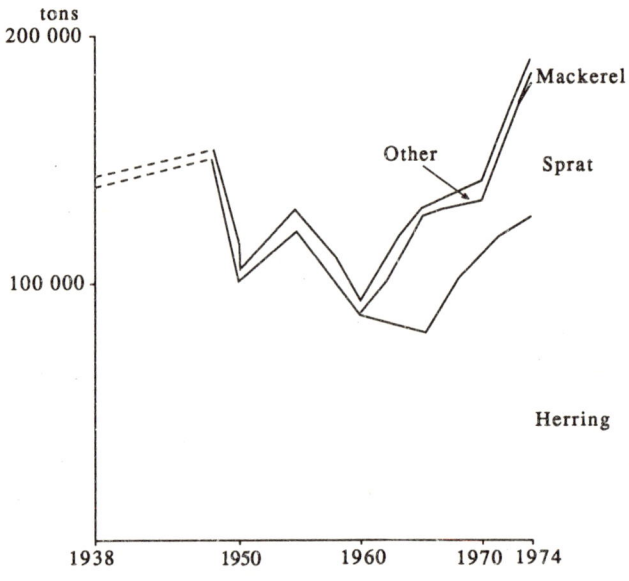

Fig. 12.3 Pelagic landings, 1938–1974.

boats that soon came to dominate the pelagic fisheries in the country. Purse-net boats came to be built bigger and bigger and to be concentrated in fewer hands: they commonly came to exceed 100 feet and later 150 feet; and some are now over 200 feet. This meant a real revolution in catching and carrying capacity, and produced the remarkable effect that with these modern boats and their fishing aids a group of less than 40 boats now has a catching capacity in advance of over 9,000 sail boats in the 19th century.

Both the pair trawl and the purse net were also from the 1960s increasingly deployed in the mackerel fisheries. This resulted in adjusted patterns of mobility, and as well as including working from West Coast bases such as Ullapool, it involved winter trips to South-West England to engage in the fishery from such ports as Plymouth (Coull 1991: 44, 45); and it might on occasion result in landings in Brittany as well.

There were far-reaching changes in the herring markets, and the overall result was a big contraction. The age-old food staple of salt herring for European markets fell much into decline with rising living standards, being replaced in diets by fresh fish and by expanded consumption of items like poultry and meat. The cured sector was effectively replaced as the leading market sector for herring of reduction to meal and oil, which was very much dominated by Danish and Norwegian processors. The reduction sector handled catches in the mass, and gave only a fraction of the onshore employment of the traditional curing; its development is very much related to the modern intensive livestock farming of pigs and poultry, for which it gives a cheap protein-rich feed. Prices of herring for reduction were low, and profitability for the fishermen had to depend on increased catches. Attempts by the Herring Industry Board to develop reduction of herring as a surplus market sector proved of limited effect, mainly because in the British market it paid fishermen better to turn to other things. Fish meal factories were built on Bressay in Shetland and at Fraserburgh. These were able to continue in operation from herring, other fish and fish offal: however the reduction sector in this country was always minor in scale compared to countries like Norway, Denmark and Iceland.

The other herring market sectors which persisted in Britain were the fresh, kippered and canned sectors, although these were of limited dimensions, and in any case tended to decline. Remaining herring processing capacity also took a severe blow when serious over-fishing resulted in a complete ban on fishing herring in the waters around Britain in the late 1970s and early 1980s.

In most of the modern period the importance of pelagic species other than herring has been very minor, although from the 1960s significant changes occurred. There was a period in the 1960s and 1970s when landings of sprats ran to tens of thousands of tons annually (fig. 12.3), as they were fished for the fish meal market in inshore waters by the mid-water trawl. However it is well known that considerable numbers of young herring were also landed with the sprats.

By the early 1970s the scarcity of herring was increasingly constraining the pelagic fleet, and by the middle of the decade there was the beginning of the rise in the purse-net fishery for mackerel as an alternative; and during the herring ban of the later 1970s and early 1980s this was to be the salvation of the pelagic fleet. This catered mainly for the low value fish meal market, although there have also been edible outlets in Britain and Europe.

The relative value of pelagic species has continued to decline, and the formerly dominant cured market in Western Europe had become but a shadow of its former scale. After the extension of fisheries limits by the EEC in 1977, there was a great increase in klondyking, which now became a main function for the fleets of distant-water fleets from the USSR and Eastern Europe, the fishing operations of which had become much constrained under the new regime of the International Law of the Sea. From that time the klondyking sector, though the prices paid are low, has become the main market outlet for the pelagic catches. This modern form of klondyking still involves mainly traditional salt curing, although it can involve freezing as well.

DIVERSIFICATION IN THE FISHERIES

The accelerated development of the period since World War I was associated with diversification at an unprecedented rate. This was achieved both by the bringing in of new fishing gears and methods and by fishing for species little previously sought. The increase in engine power that has been general in the fleet has allowed much of it to turn to trawling, the most important single fishing method of the modern world. This at first allowed the adoption of the light trawl, and of the mid-water trawl towed by two boats, and there have been various further developments. There were important developments in shell fishing (ch.14).

Nephrops fishing, in which there is also a considerable catch of white fish taken at the same time, became an important enterprise on both East and West Coasts, although as had been the case with the herring fisheries, much of the West Coast catch was taken by boats for which the home base was the East Coast. In addition, in the Clyde, and various other places on both coasts there was the development of dredging for scallops, mainly by smaller boats. In both of these fisheries success was related to new marketing initiatives for high-value species which had outlets in Britain and abroad in circumstances of rising disposable incomes.

While 'industrial' fisheries for reduction have never had the importance in this country that they have attained in Denmark and Norway, Scottish fishermen have also found it profitable to have some involvement in these modern fisheries, the main function of which is to supply feed for intensive stock farming, and now for fish farming. Material for reduction had long been supplied on a limited scale from fish offal, and in the pelagic fisheries for herring, mackerel and (more particularly sprats) have also given supplies to the reduction outlet, although on a much smaller scale than in Norway and Denmark. The species of Norway pout, sand-eels and blue whiting have all been fished specifically for the reduction sector: these species are all of low unit value, but various methods of mass handling limit the work for the fishermen. These fisheries can be profitable, but in Scotland they are of minor importance.

REFERENCES

Coull, J.R. (1991) 'Mobility in the Scottish Fisheries', *Scot. Geog. Mag.* 107, 40–46.

Coull, J.R. (1968) 'Modern Trends in Scottish Fisheries', *Scot. Geog. Mag.* 84, 15–28.

Coull, J.R., Goodlad, J.H., and Sheves, G.T. (1979) *The Fisheries of the Shetland Area. A Study in Conservation and Development*, Department of Geography, University of Aberdeen.

Reports of the Fisheries of Scotland (RFS.) 1938–48; 1955.

Steel, D. (1980) *The Fisheries of the Western Isles Area. A Study in Conservation and Development*, White Fish Authority, Edinburgh.

13

The Development of Mobility

While for the great part of known history, fisheries in Scotland have been prosecuted from permanent settlements, there have for long been elements of mobility associated with fishing. It is fairly certain that before the development of farming in Scotland, there was considerable mobility among a sparse population dependent on widespread living resources, many of which were irregularly available. This is clear with the original Mesolithic settlers, many of whose known occupation sites were in fact temporary encampments.

Even so, the advantages of permanent settlements must from a very early time been evident, and in the historic period any degree of mobility associated with fishing was rare. The main instances were the busses which were fitted out from the late 15th century in the towns to prosecute the offshore herring fisheries on both East and West coasts; and in both of these the busses ranged for distances of a hundred miles and more in a northward direction: this must, however have accounted for a minor part of the fishing effort and the production. Aside from this, the main other instance were the short-distance seasonal migrations which occurred in the Shetland Islands from the 18th century by which fishermen concentrated at the more outlying parts of the archipelago for the 'haaf' summer fishery for ling and cod.

INCREASING MOBILITY IN THE PERIOD OF THE INDUSTRIAL AGE

By the late 18th century, however, there was a new impulse of mobility in evidence, and there was to be a continued expansion of mobility over more than a century. This was associated with the expanding herring fishery and with the extension of shore-based curing to a range of bases around most of the British Isles; and to a lesser extent it was linked to the improvements in overland transport and to the delivery of more fresh herring to the home market. In the pattern that eventually emerged at the end of the 19th and the start of the 20th centuries, the great movement flows on the Scottish mainland to Caithness and the Aberdeenshire ports in the summer; and of primary importance also became the flows to Shetland in the summer and to East Anglia in the autumn. There were lesser movements to most of the coasts of the British Isles by boats of the Scottish herring fleet and by Scottish curers. Essentially these large scale movements of people were due to the development of a major fishery that was labour-intensive both for catching and processing; and as an oily fish like herring had to be processed with minimal delay, the bases for the fishery were regularly located at the nearest landfall.

The main movements to Shetland and East Anglia are in general well recorded; but for a range of the lesser centres it is difficult to gauge the scale of movement. The record is also conditioned by the fact that the jurisdiction of the Fishery Board from 1868 did not extend beyond Scotland at all, so that the herring fisheries engaged in at English and Irish ports were not covered by the arrangements for inspecting and branding barrels; and any record and statistics kept were in a different framework from that of the Board. This precludes the possibility of direct comparison.

By 1850 there was pressure to extend the herring season and this was to promote increasing mobility. With the growing fishery fishermen in many places, and especially on the East Coast, had achieved sufficient profits to reinvest in bigger boats which could carry more nets and make the fishery more productive. Initially the use of these boats, like others, was limited to the two-month summer season, and the usual pattern was for the men to go back to traditional inshore lining in smaller boats for the rest of the year; indeed for many fishermen seasonal inshore lining was an important part of their activity until the late 19th century. However, with the capital tied up in bigger boats, there was a substantial disadvantage in having them drawn up for the great part of the year, and this prompted leading fishermen to search for other outlets for their use.

In this Scotland, along with the rest of the British Isles, were in the long run to provide unmatched environmental conditions for expanding the herring fishery. Such is the range of herring stocks in the waters around the British Isles and the phasing of their life cycles that it was possible to fish herring from some part of the coasts at almost all times of year: the main difficulty was in the spring, although at that time of year it was possible to prosecute the deep sea longline fishery for white fish, or to maintain and prepare herring gear. The long-term development was in fact for many fishermen the lengthening of the herring year until it very much dominated their activity. As well as the mobility which developed among fishermen, equally important was that of the curers as essential processors, and indeed they might often take the initiative in the opening up of new centres. Also integral to the evolving pattern was the movement of shore labour, which was dominated in numbers by women gutters and packers; but it also included the coopers, who as well as being the makers of the barrels, often supervised the work of curing; and there were also numbers of male general labourers. While there was some hiring of local labour in different places, curers did tend to move with at least their key staff, and there was also considerable mobility on the part of other skilled labour. There was a varied degree of participation of local fishermen, curers and labour at different centres; but even so, there was a marked tendency for the industry to move as an ensemble: and although it would have been difficult to put a rigid line around the total personnel involved, the bonds of common interest held the whole together.

INCREASING MOBILITY ON THE EAST COAST

The success of the fishery at Wick had already by the 1790s attracted boats from other parts of the Moray Firth, and especially from the Buckie area. In addition to the establishment of Pulteneytown by the British Fisheries Society and the outlet offered by incoming curers, the peninsular location of Caithness meant it was closer to the main migration paths of the herring. Also a greater sea area could be reached from Wick than was possible from any of the centres on the inner part of the Moray Firth. Migration to Caithness increased as the 19th century progressed, and boats from Fife and other parts of the East Coast were drawn in. Already in the early 19th century Gaelic speakers from the West Coast might take a week or more to reach Wick in the search for seasonal employment (Sutherland 1983: 29). By mid–century Caithness was by a considerable margin the main herring producing area in the country, and the habit of mobility well established among herring fishermen, despite the inconveniences and additional overhead costs of the hiring of accommodation for the crews' time ashore. The pressure on accommodation was enormous, and incomers were housed in barns and sheds as well as other accommodation in Wick, and in the nearby villages like Staxigoe, Broadhaven and Sarclet (Sutherland 1983: 30). Many people were living in insanitary conditions, and outbreaks of cholera, typhus and diphtheria occurred. In the main centre of Wick the congestion in the season was especially great, and in 1840 it was stated that in the season there were 7,882 people in the town directly involved in the fishery: this included 3,828 fishermen, 2,175 women gutters and various others, the majority of whom must have been migrants (NSA. XV Caithness: 153). In the following year the Census gave the total population of the Parliamentary burgh of Wick with Pulteneytown as 5,522 and Wick itself had only 1,333. At the peak in the 1860s there were over 1,000 boats engaging in the Caithness fishery, the majority of them incomers from elsewhere on the Moray Firth and from Fife.

Other ports on the East Coast also developed the herring fishing, and Fraserburgh was established by 1820, with Peterhead about a decade later: and these Aberdeenshire ports gave another peninsular location from which the fishery could be prosecuted. They attracted in the first place crews from villages in their own immediate vicinity: in 1840 the villages of Cairnbulg and Inverallochy near Fraserburgh were 'almost deserted' in summer, when the men fished from Fraserburgh, and the women folk accompanied them to work in the curing yards (NSA. XII: 296). However the majority of visiting boats to the Aberdeenshire ports came from further afield, especially from Fife and the inner Moray Firth. By the 1840s upwards of 1,000 boats were gathering between Peterhead and Fraserburgh in the season, and these included hundreds of boats from Fife and elsewhere. After a reduction in the tempo of activity from the later 1840s with the decline in the Irish and Plantation markets, renewed growth saw the numbers of boats congregating at the Aberdeenshire ports at a peak of around 2,000 by the later 1870s and early 1880s: thereafter the number of boats fell with the increased concentration of catching power in fewer bigger boats, and the diversion of a big sector of the fleet to the booming Shetland fishery in the early 1880s.

There were other components in the developing pattern of mobility other than boat crews and curers from one fishing centre going seasonally to another. The curers from an early stage brought and engaged coopers as key personnel. Although there was a tendency to engage locally women for gutting and general labourers to help in the curing yards, in an expanding fishery pressure points (such as Wick) did develop where it was necessary to bring in extra labour for the season despite the extra expense this might entail. In especially the second half of the 19th century the main source of this labour was the crofting communities of the West Coast and Hebrides. This area in all its modern history has not had enough sources of employment within itself, and big numbers of both sexes have gone to various parts of the country to make extra earnings to balance household budgets. As well as supplying harvest labour on lowland farms, workers for railway and other construction, domestic service in cities and elsewhere, a main seasonal source of employment became the herring fishery on the East Coast. It was estimated by the provost of Stornoway in 1914 that from Lewis there were between 1,600 and 1,800 men went as hired hands to the herring fishing on the East Coast and Shetland, and that the numbers of women were between 2,000 and 3,000 ((RSDCNSFI. II 1914: 59). For the men this was in large part a reflection of the failure of the West Coast to keep pace in numbers and sizes of boats, with the result that a main opportunity became that of making up part crews on East Coast boats: this yielded a labour share, but did not give the returns to boat and net shares that many East Coast fishermen enjoyed. In addition, it became by the end of the century usual for thousands of women to go to East Coast ports to work in the curing yards. This necessitated the provision of accommodation, which was quite often in part built into the curing installations themselves. However in the later 19th century it was usual for the extra hands needed for the herring boats to live aboard, as the vessels were by now decked and had cabins.

INCLUSION OF THE WEST COAST

The extension of the operating season and the accompanying increase in mobility was to begin with the engaging in an early summer fishery in the months of May and June as a preliminary to the main summer season, and here the nearest opportunity lay in the Minch. By the 1840s men from the East Coast had started to go to places like Stornoway for an early summer fishery as a curtain raiser to the main season at Caithness: this was desired both by curers and fishermen, and already in 1844 curing had spread beyond Stornoway and stations were being set up at other places on the east coast of Lewis like Bayble Bay and Cromore, and there also were ideas of setting up others on the west side of the island. While giving an expanded opportunity that was wanted in the industry, it was to introduce a problem that was to persist for decades: that of catching herring too early in the season to be good for curing (SRO. AF 5/1: 114–115). Even so the numbers of participating boats in this early summer fishery swelled in the second half of the century. By 1865 the Stornoway district was capable of attracting 680 boats from all parts of Scotland to the early summer fishery (SRO. AF7/87: 216); and although participating numbers inevitably fluctuated in the fishery, with a fresh surge of development in the 1870s the number of boats recorded actually reached 1,381 in 1880 (SRO. AF7/90: 207), and this was due essentially to the expanded East Coast fleet coming in strength.

The herring fisheries of the Hebrides attracted visiting curers and fishermen in force and there were also locally based curers. Although the visiting curers brought coopers with them for making barrels and supervising curing, the great part of the labour force was recruited locally. It was a striking fact that in the late 1880s in the Barra district nearly all the gutting women, numbering around 1,500 were local (AF7/49: 1888: 122). There was a greater female labour force in the Stornoway district, and as well as being involved in the early summer and winter fisheries, many of them went to the East Coast and to Shetland to provide extra hands in the main summer herring season.

The development of the Shetland fishery siphoned off from the Hebrides a big part of the catching power for the early summer fishery from the 1880s, and there were also attempts, by agreement between fishermen and curers, to prevent the season starting too early, as otherwise the early caught herring would be poor for curing. Even so, the Minch was to continue as a significant early summer venue, with Stornoway and Castlebay as major ports: it was also to become the most important venue for winter fishing as the need to operate a longer working year intensified, particularly after 1900 with the steam drifters (see below).

One response to the efforts to delay the start of the season at Stornoway and other Minch ports was to be in the 1880s the flourishing of Stromness in Orkney as a base for the early herring season. This started in the mid 1880s and was engaging around 70 boats annually at the end of the decade; and by the mid 1890s a fleet of over 500 strong could be attracted for a fishery that lasted up to six weeks and in its peak year of 1894 realised 74,557 crans (ARFBS. 1896: 184). Stromness however was only temporarily prominent, as the fishing failed badly in a series of years at the start of the 20th century, and few boats came thereafter.

However, as the fishery built up to its peak in the early years of the 20th century, the availability of a fishery in the early months of the year became more important, and steam drifters especially made this part of their annual cycle as they reacted to the pressure to lengthen their working year. Several hundred boats could participate: in 1907 there were 300 drifters which landed an aggregate of 12,500 tons, and in the peak year of 1913 140 steam drifters caught a total of 20,000 tons (ARFBS. 1913: xxiv). Upwards of three quarters of the district annual total could in fact be landed at Stornoway between the months of January and March. Much of these herring were sent by steamer to fresh market destinations on the mainland, or 'klondyked' for the German market; a considerable fraction was kippered; and the remainder went for pickle curing.

MOVEMENT TO SHETLAND

There was a location where the opportunities for the early herring fishery were in fact better than in the Hebrides: this was the Shetland Islands, although this was relatively late in being fully appreciated. In fact there is an anomaly in the restricted part played through most of the 19th century by the Shetland Islands in the pattern of activity in the Scottish herring fishery. Although the archipelago had been the main base for the great Dutch herring fishery, attempts to develop the fishery with shore-based curing had limited success, and the 'haaf' fishery for ling and cod continued to be the main basis of the islands' economy. In these earlier phases catching was dependent completely on the islands' fleet although the bounties did attract a number of curers from the Scottish mainland in the 1820s and 1830s; and the main effort was in the late season after the 'haaf' season, when the general quality of herring was poorer, with a big percentage of fish spent after spawning (Coull 1983: 123–140). Even the Dutch fishery by its own strict rules did not begin until the end of June, and the fact that herring in quantity were available to the west and north of the islands from May was inadequately known.

With the momentum of expansion that there was in the Scottish fishery by the 1870s, there was a continued quest for fresh grounds,

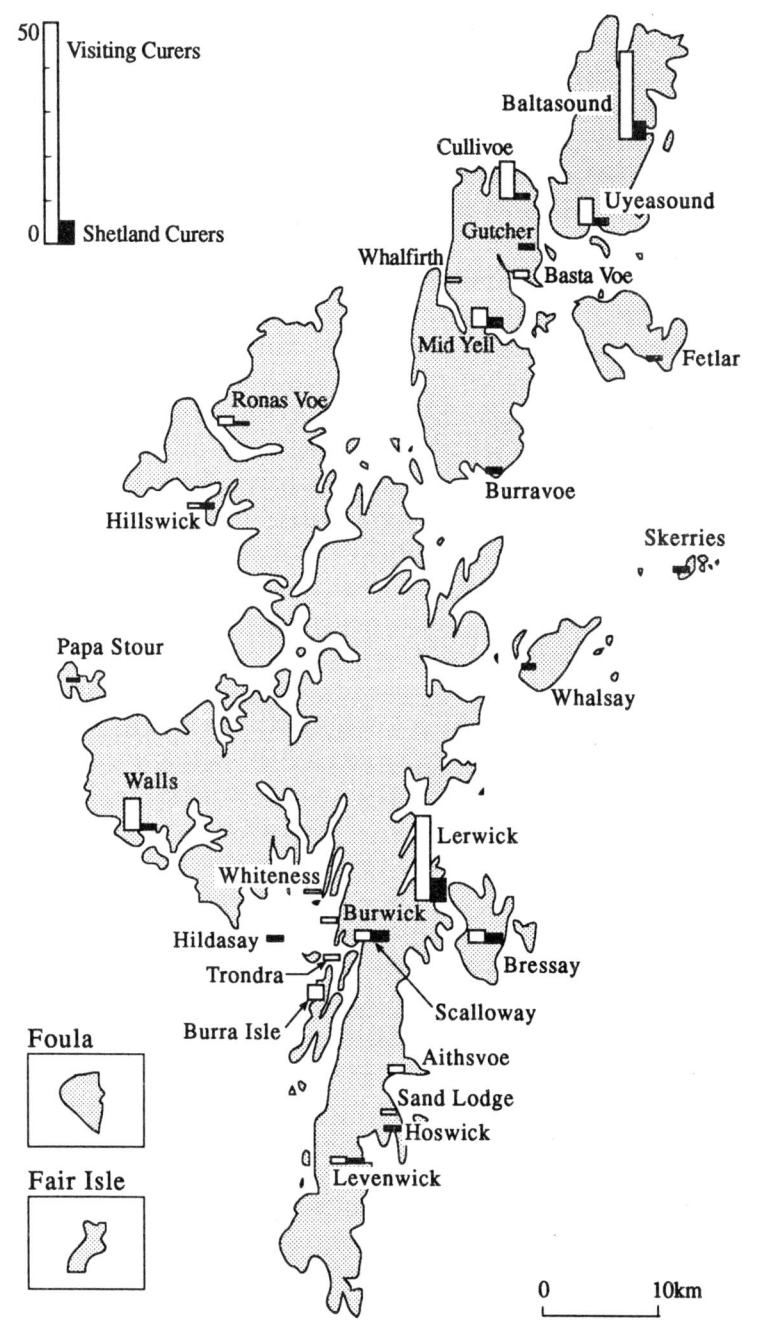

Fig. 13.1 Herring curing stations in Scotland, 1884.

and the increasing size of boats made longer voyages more feasible. Shetland came from the end of the 1870s to play a main role (Coull 1986: 25–38), especially as an alternative location to the Minch for the early summer fishery. The move to Shetland was triggered by the northward move of some mainland curers, accompanied by mainland boats, at the end of the 1870s for the summer season, and very good catch rates in 1880 kindled a widespread interest throughout Scottish fishing communities. The result was that a big proportion of the boats which had gone to the early summer Lewis fishery was diverted to Shetland. Curing stations were rapidly set up all around the islands, the big majority of them by mainland curers, although there were a significant number of island curers who also participated in what was in fact a great new opportunity in the setting up of a fishery which was to be the main economic basis of the archipelago for over half a century. By 1884 there were a total of 123 curing stations in 28 locations (fig. 13.1): this pattern reflected a situation in which herring were caught all around the islands by sail-boats, and the objective after catching was to make the nearest landfall with an adequate harbour. Particularly prominent was the concentration of stations at Balta Sound in the northern island of Unst: this was an extensive natural harbour which was the most convenient location for the early herring fishery in the north of the isles, and in 1884 it had 24 stations. With the big number of visiting boats in the early herring season, the total strength of the herring fleet in 1884 was 932 (fig. 13.2). Although there was a brisk increase in the islands' own herring fleet with this new development, in 1884 over 60% of the boats were strangers; and their proportion of the catch was considerably greater. The Shetland fleet at this stage was of lesser craft that included mainly smaller second-hand decked boats which had been discarded in mainland centres: and open sixerns were also used, especially for the late herring fishing after the "haaf" season had ended (fig 13.2).

Although in the recession of the later 1880s there was a big decrease in the numbers of boats coming to the Shetland fishery, interest again built up in the 1890s, and in the peak year of 1905 there were 173 curing stations operating in various locations, 148 of them by visiting curers (Manson's Shetland Almanac 1906: 96, 97). In the same year the number of boats included a total of 1,815 British vessels with a crew strength of 13,500 (ARFBS. 1905: 94), as well as a number of boats from other countries that is not precisely known but which would have run into hundreds. Visiting

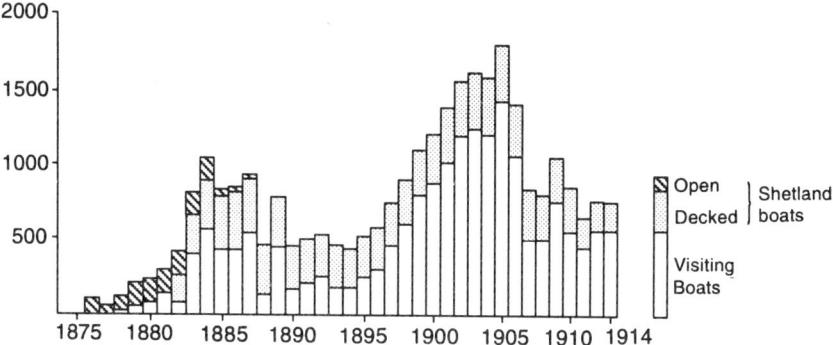

Fig. 13.2 Numbers of boats at the Shetland
herring fishery, 1875–1914.

boats at this time comprised over 75% of the active fleet in the early summer fishery and they included over 1,000 boats from the East Coast of Scotland; there were also hundreds of boats from East Anglia, and participants from the West Coast of Scotland and the Isle of Man.

In the early 20th century the distribution of activity within the islands changed markedly, with a great concentration of landings and curing capacity on Lerwick. This had in fact begun from 1893 with the establishing of auction selling of herring at Lerwick, with the reorganisation within the whole industry that followed the recession of the later 1880s. Very generally other centres had not the port facilities or scale of landings to make an auction market feasible, and curers at them continued with the engagement system, although there were attempts to modify it to attract more custom, especially at Balta Sound. When the steam drifter appeared on the scene, it rapidly came to dominate the Shetland fishery, and by 1907 steam landings had forged ahead of those of sail. Steam drifters could make for the Lerwick market from most of the grounds around the islands, and the 'country stations' were more and more by-passed: by 1913 out of the 82 curing stations operating, 50 were located at Lerwick and on its neighbour island of Bressay (Manson's Shetland Almanac 1914: 102) (fig. 13.3). As a very general rule these were also bigger and more elaborate than the earlier stations dispersed through the islands, and although the number of curing stations had decreased by more than half from the 1905 peak, curing capacity had been increased rather than reduced.

The movement to Shetland included thousands of both male

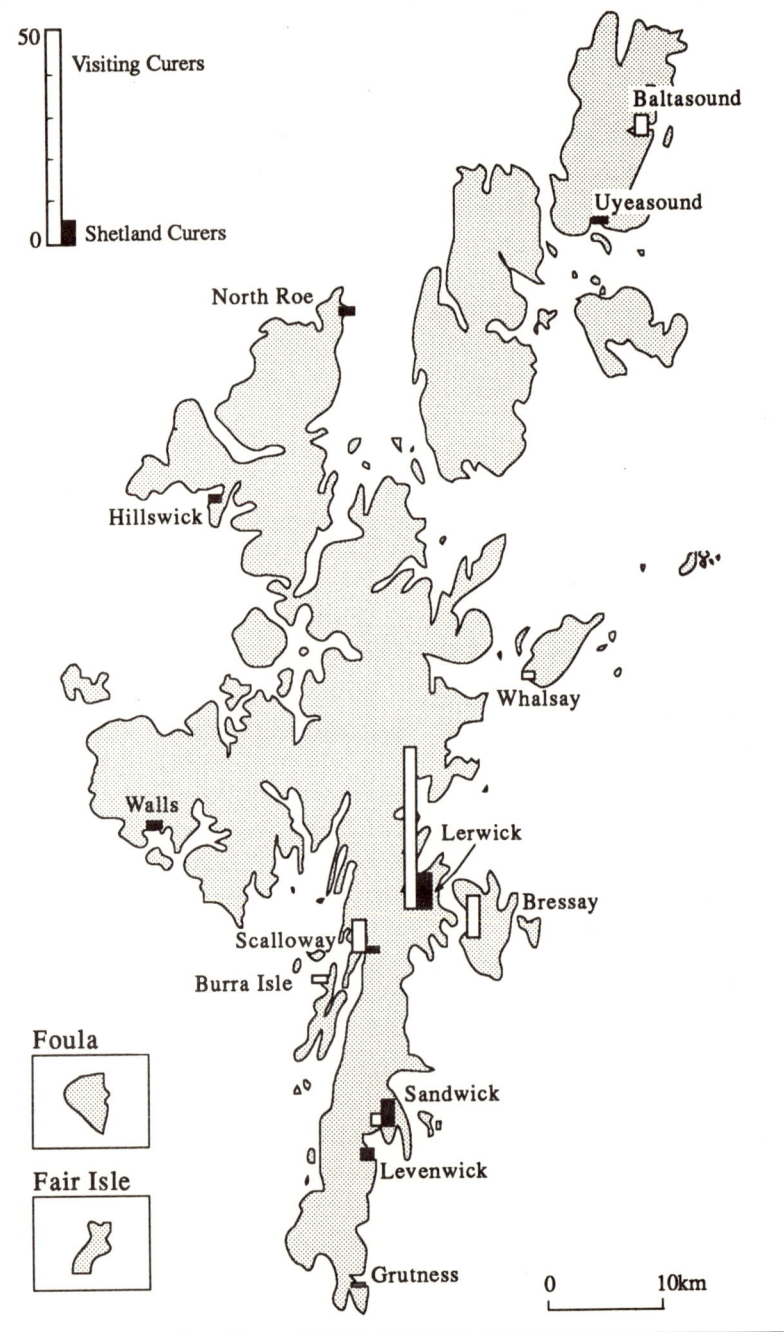

Fig. 13.3 Herring curing stations in Shetland, 1913.

and female workers for the curing yards, and the scale was such that at the peak of the herring season the numbers of visitors could approximate to that of the islands' own population. By this time visiting fishermen were generally on decked boats, and it was possible for them to live on their boats. For visiting workers there was a problem of accommodation, and in many cases, special accommodation was actually built by the curers at their curing yards, usually in the form of wooden huts. The great summer influx made a big impact in a variety of ways, and although most of the profits from the fishery left the islands along with the seasonal visitors, it did for a time lift the native community of Britain's most outlying location to a new level of prosperity, against a general background of decline in the marginal regions of an industrialising society with the accelerated movement to cities in Britain, and to new opportunities abroad. Within the islands the impact was of course greatest at the main seats of curing; and this meant above all the main centre of Lerwick, which always dominated the fishery from July onwards, and Balta Sound, which was the main centre in the early season until the more mobile steam drifter increased the concentration at Lerwick in the early 20th century. It was estimated in 1901 that out of about 20,000 fishermen and shore workers who came to Shetland for the early season, almost 10,000 came to Balta Sound alone, at a time when its off-season population was in the vicinity of 500 (Shetland News, 21st Sept. 1901).

WINTER HERRING FISHERIES: THE FORTH AND THE MINCH

While there had for centuries been some herring fishing in the winter, this had been a restricted operation carried out mainly on the Ballantrae Bank in the Clyde and at the entrance to the Firth of Forth; however bigger boats had at times ventured into the Minch at the same season. In the later 19th century there was to be an expansion in working in the winter part of the year to justify the increasing investment that was going into bigger boats. In 1886 it was noted that winter herring had recently become important along virtually the whole East Coast, and that this was associated with expanded market outlets for fresh and smoked herring (ARFBS. 1886: lv). While some herring might be caught in winter almost anywhere, the main opportunities were at the entrance to the Forth and in the Minch. On the debit side winter herring were leaner and of poorer quality, and the fisheries appear to have been even more variable than the summer ones, although this was at least partly because of more frequent bad weather interrupting fishing. On the home market prices tended to be good for winter herring, and there could also be a good continental demand for klondyking: however both of these market sectors could be compromised by herring coming from Norway, where the main herring fisheries were in the winter: in some years this brought prices down to levels of marginal profitability or worse.

At first in this phase the more important of these venues was that on the Forth: in 1886 between the Anstruther and Leith districts there were a total of some 320 boats at the fishery (ARFBS. 1886: xxxiii, xxxiv), which must have included a significant number from other districts. Even in the winter this fishery gave catches as high as 100 crans (20 tons) on a good day to individual boats, and the total yield of the fishery was 26,736 crans (*c.* 4,000 tons). In 1900 it was recorded that in the total fleet of 172 at Anstruther during the winter season there were crews from the Eyemouth, Arbroath and Banff districts as well as local crews, and the average catch was 226 crans per boat. Prices were very good at an average of 24/- per cran mainly because of the demand from klondykers plying to Germany (ARFBS. 1900: 218). With steam drifters coming on the scene this fishery increased at the start of the 20th century, and in 1904 the total catch approached 10,000 tons, although with the limited harbour accommodation available for bigger boats the numbers of visiting boats had declined (ARFBS. 1904: xxv; 196).

In subsequent years catches in this fishery went into a phase of long-term decline, and it became more and more a matter of local interest, and that mainly for boats of small and medium size. Herring might be caught in large measure by anchored nets close to the coast (ARFBS. 1913: 1890), and by the inter-war period the main fishery was by local motor boats, while the main catching power was still in steam drifters which fished elsewhere.

As the 20th century progressed it was the Minch fishery that became the main winter one. This was prosecuted mainly from Stornoway, and while local boats participated it was dominated by visiting steam drifters. Several hundred boats could participate: in 1907 there were 300 drifters which landed an aggregate of 12,500 tons, and in the peak year of 1913 140 steam drifters caught a total of 20,000 tons (ARFBS. 1913: xxiv). It was not unknown for these steam drifters to go to the Minch with as an insurance their great lines as well as their herring nets aboard, and to follow either fishery according to fish availability and prices. With this effort in the winter fishery, upwards of three quarters of the district annual total of herring could be landed at Stornoway between the months of January and March. Much of these herring were sent by steamer to fresh market destinations on the mainland, or 'klondyked' for the German market; a considerable fraction was kippered; and the remainder went for pickle curing.

This continued to be a main fishery for the steam drifters, even in the difficult inter-war period. In the peak inter-war year of 1924 there was a fleet of 245 steam drifters and 20 motor boats at the fishery, although this declined in the face of mounting market difficulties, which were again aggravated by plentiful supplies of winter herring from Norway (ARFBS. 1925: 22).

MOVEMENT TO EAST ANGLIA

In the longer term there were also opportunities in the other major herring fishery of the British Isles, which was the longest-standing of all as a great herring fishery – the autumn fishery at East Anglia (Coull 1992: 127–147): this was, however at a greater distance from home base, and was particularly challenging in the days of open sail-boats. The East Anglian fishery was useful in filling what had been largely a blank in the herring year, and the fact that the leading fishermen were now investing in full-decked boats with cabins also made it possible to reduce some expenses by living on the boat away from home. From the 1860s movement to East Anglia became an expanding activity, and the autumn fishery at East Anglia was to be a major part of the pattern of activity of the Scottish herring fleet for the great part of a whole century.

The first ventures in fishing for herring south of the Scottish border were made by Anstruther fishermen trying the summer herring from various points on the coast of North–East England in the 1850s (SRO. AF19/3/4, 177, 206, 242.277). Following attempts with indifferent success by the Fife men to extend their operations into the autumn on the Clyde lochs, two Anstruther boats ventured to Great Yarmouth in the autumn of 1864: they joined the local East Anglian fleet for the season, and their conspicuous success prompted more general interest and increasing numbers of boats made the voyage in subsequent seasons. Despite some variable results in early seasons, the number of boats engaging in the East Anglian autumn fishery could reach a hundred by the mid–1870s, and boats from Fife were being joined by ones from Leith on the opposite side of the Firth of Forth (SRO. AF28/89: 24, 57). At this stage too, Scottish curers began to take an interest in the East Anglian fishery by sending agents to buy herring to send to Scotland by rail for processing (SRO. AF28/89: 250). By 1880 boats from Buckie on the Moray Firth were also making the seasonal voyage to East Anglia, so that by the start of the 1880s the East Anglian season was a regular part of the yearly round of many leading Scottish fishermen, and there were upwards of 200 Scottish boats engaging in it. In these first phases of involvement in this fishery the catches of the Scottish crews went both to increase the supply of herring to existing markets in South-East England, and of cured herring that went mainly to the Mediterranean.

The effects of the post-1884 recession in the fishery were less serious in England than in Scotland, as at this stage the main markets for the East Anglian fishery were the Mediterranean and home British markets, rather than the market in Northern Europe, which was the main source of the problems. Total numbers of boats at the fishery, distinguishing locals and visitors, become available from 1884 (fig. 13.4). Although by the early 1890s there had been a fall in the numbers of boats at the fishery, the problems in Scotland did something to prompt wider interest in it, and most of the Scottish East Coast was represented by 1890, with boats from at least seven of the fishery districts taking part.

With the strong recovery from recession by the Scottish herring interests from 1893, the seasonal exodus to East Anglia swelled in remarkable degree, and it was reinforced by the linked interests of Scottish fishermen and curers. Scottish curers were already curing at East Anglia in the 1880s, and by the early 1890s exports of cured herring to Northern Europe were running at between 10,000 and 20,000 barrels annually. There was a rush of Scottish curers to East Anglia from the mid–1890s and a steep increase in the export of cured herring to the market in Northern Europe, which was now in its final phases of rapid expansion. By 1899 the export from East Anglia had reached 141,585 barrels (Wigg 1900: 52), and in 1913 the figure was actually to top one million barrels (TNNNS. 1914: 834). This market sector also became the main one for the whole East Anglian fishery. It prompted organisational adjustments and streamlining that included the measuring of catches by the

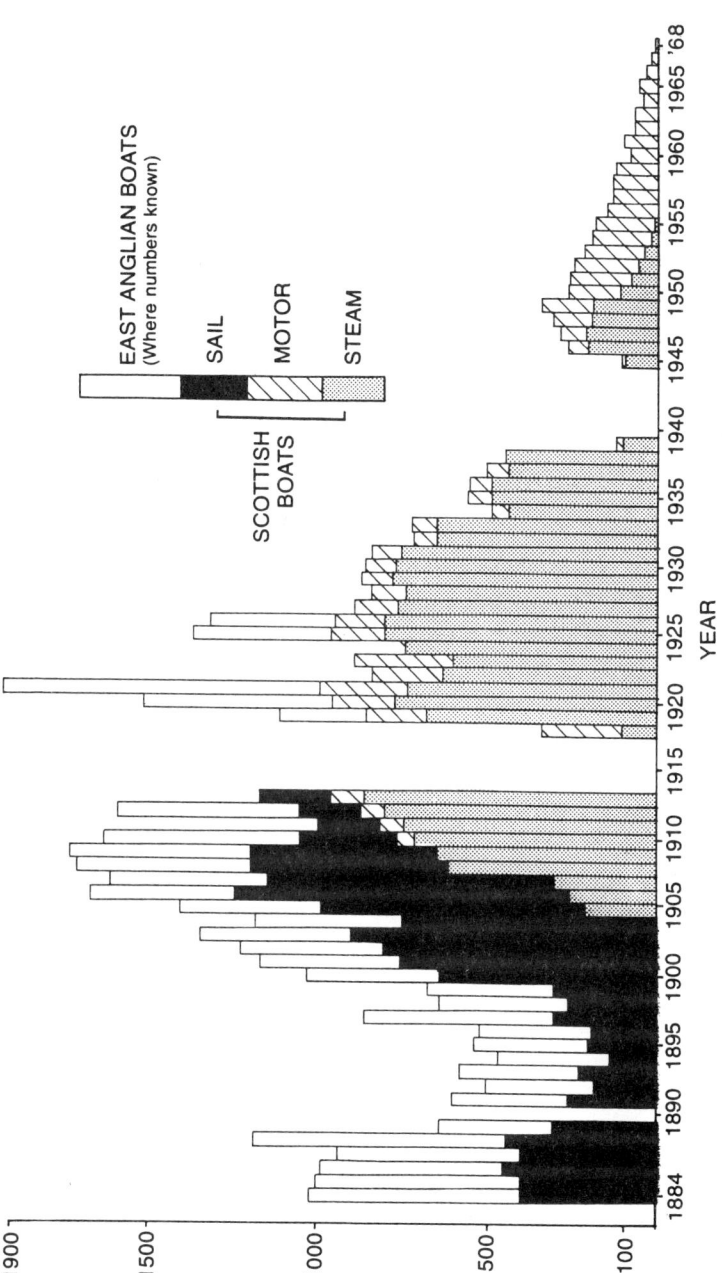

Fig. 13.4 Number of Scottish and English boats at East Anglian herring fishery, 1884–1968.

Scottish cran measure rather than the older method of actually counting fish.

The great boom at East Anglia was a magnet that attracted the bigger boats of the Scottish fleet in rapidly growing numbers in the 1890s, and by 1900 the number participating reached 634; and from then on it involved the vast majority of the Scottish herring fleet, and this regularly outnumbered the considerable home English fleet (fig. 13.4). From 1905 till the outbreak of World War I there were well over 1,000 Scottish boats making the autumn trip to East Anglia; and a strong driving force in the increase in numbers was the need to increase earnings for the big number of steam drifters: for them participation in this major fishery was a necessity rather than an option (Coull 1992: 138). The home bases of this visiting Scottish armada were almost all on the East Coast, and within it was much concentrated on the south shore of the Moray Firth: more than 80% of the boats came from districts between those of Findhorn and Peterhead. This emphasises how far the Highland area and West Coast had been left behind in the progress of the herring fishery. There was also a big-scale seasonal migration of personnel engaged in curing, who were very generally recruited by the dominant Scottish curers in Scotland: at the peak this migrant shore labour force was in the range from 6,000 to 7,000. Although there was not the same regular collection of data here as there was with boats, there are a number of estimates for particular years, although most of them are less than complete: these are shown on fig. 13.5. Some of these personnel came by normal service transport, but the chartering of special trains became the main means of travel, and it was known for over 20 special trains to come from the Scottish bases of the fishery (Coull 1992: 129–138). By the time this migration started on a major scale, Yarmouth was an important seaside resort and catering for summer visitors had become an important part of the town's economy. A consequence of this was that accommodation that was used to cater for summer visitors could also serve for the gutting women from Scotland, although they were provided for on a different basis: it was common for there to be as many as six to a room, with bunk beds being used.

As well as participating in the East Anglian fishery, Scottish boats might also operate from ports further north on the English East Coast, such as Scarborough or Grimsby: these could provide intervening opportunities en route to East Anglia, and when herring were available off the coast might be the venue of operation for

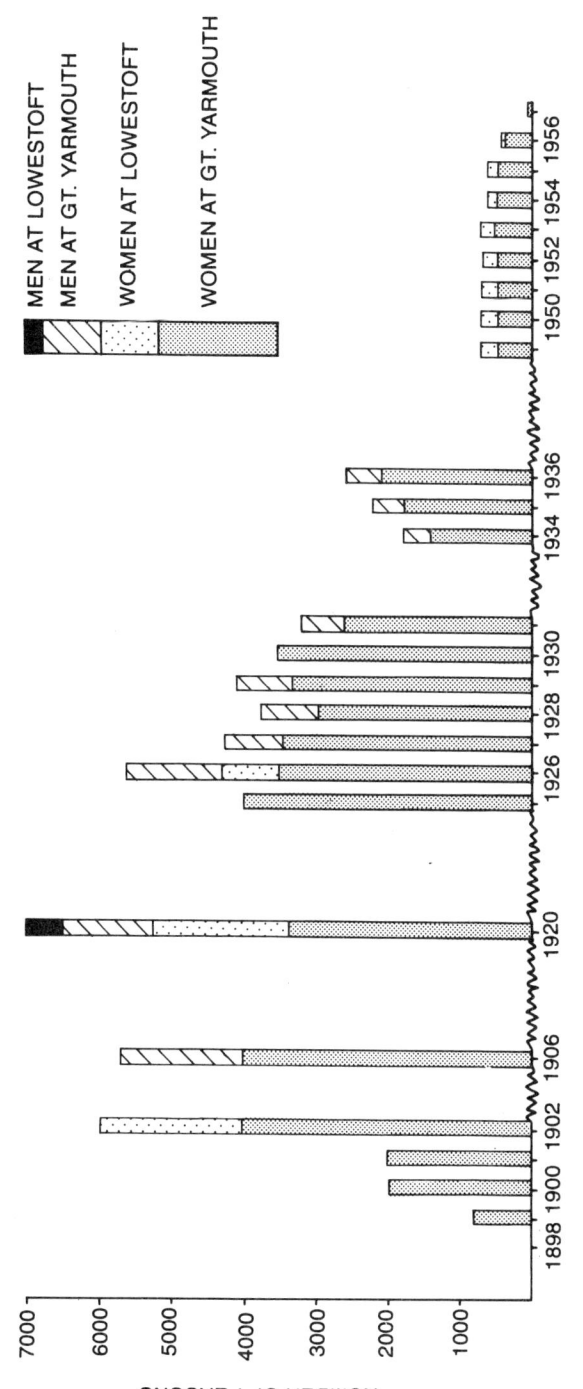

Fig. 13.5 Scottish migrant shore labour at East Anglian herring fishery, 1899–1957 (partial record).

weeks. It was also known on occasion for boats not to return home after the end of the East Anglian season, but to proceed to Plymouth to fish there in December and January. In addition there were fisheries in the Irish Sea which were prosecuted in late summer and early autumn. Peel on the Isle of Man was a main port here, although landings might also be made in Ireland at ports like Kilkeel and Howth, or on the opposite side of the sea at Whitehaven or Holyhead.

FISHERIES AT IRELAND

There had previous to the end of the 19th century desultory visits of Scottish herring boats to bases in Ireland, but from the 1880s fisheries at Ireland came to be a regular part of the activity of a sector of the Scottish fleet. This began especially with the men from Fife, who had been the advance guard in trying different new fisheries, and from 1893 the numbers of boats participating in the Irish fishery were regularly given by the Fishery Board for Scotland. By the early 1890s the Fife men were being joined in number by men from the Buckie district when for the winter fishery there could be 70 to 80 boats working at the south of Ireland from ports such as Kinsale; and in the first decade of the 20th century (before the big reduction in the size of the fleet as sail gave way to steam) the total participating in fisheries at Ireland could run into several hundreds, as boats from most Scottish districts made the voyage: in 1905 it actually reached 439 (ARFBS. 1906: viii). In the main the tendency was to make short rather speculative trips to Ireland in the attempt to find profitable fisheries in between main seasons elsewhere: these were mainly in late spring and early summer (from April to June) and in winter between December and February. It included working from ports like Buncrana and Downings Bay in Donegal as well as places like Dunmore and Waterford in the south. Even at the best the earnings of Scottish boats at these Irish fisheries did not reach a tenth of those made at East Anglia.

MOBILITY ASSOCIATED WITH STEAM TRAWLING

In the white fisheries, the development of mobility was less important than with the herring, although the scale of mobility was often greater in distances covered. Here the trawler fleet, especially from

Aberdeen by the end of the 19th century was ranging over a great area of the northern North Sea and had also extended operations to the West Coast. In the 20th century fishing was also extended to the Faroe Islands, to Iceland and the White Sea to the north of Norway; and in the case of the steam liners they ranged as far afield as Greenland. However white fish were only occasionally landed other than at the main trawl ports, and the shore-based personnel for the fishery continued to be concentrated in great degree at Aberdeen and to a lesser extent at Granton.

Mobility at these longer ranges inevitably involved considerable loss of several days in fishing time in making the passage between port and fishing grounds; and working conditions in high latitudes were often more arduous, especially in winter. However the better catch rates on these grounds was the attraction that counterbalanced the disadvantages.

THE CHANGING PATTERN OF MOBILITY FROM THE INTER-WAR PERIOD ONWARDS.

Mobility in the herring fisheries was at its peak in the years immediately prior to World War I. Although these fisheries continued to command the main interest of most Scottish fishermen in the inter-war period, they were struggling in a changed and more difficult situation of reduced and more uncertain markets. While the pattern of mobility already discussed continued, it was on a tapering scale as the fleet of steam drifters aged and declined in numbers. Although there was the installation of motors in most of the rest of the fleet by this time, and motor boats joined the steam drifters in the seasonal migration, it did not prove possible to develop an effective motor-driven winch to replace the steam capstan.

Shetland in the inter-war period continued to be the main venue for the early summer fishery, although participation in it inevitably declined. Numbers of boats at the peak season varied with the less prosperous and more unstable situation of the period, but could reach around 500. A feature was that the English steam drifter fleet of perhaps 200 boats generally made Lerwick their headquarters throughout the Scottish summer season, and indeed generally dominated the fishery after many of the Scottish steam drifters moved to home and mainland bases after the early summer fishery. With the difficult market situation, the early summer fisheries tended to become less important, as the herring from them were less good for curing, although some movement to the Scottish West Coast and to Ireland continued.

Throughout the 1920s the numbers of boats going to East Anglia held at around 900 (fig.13.4), but after the accentuated difficulties in the herring trade after the Wall Street crash of 1929 herring fishermen turned in increasing numbers to ground-seining for white fish, while increasing numbers of steam drifters were laid up or scrapped; by 1939 fewer than 500 boats were making the voyage to East Anglia. Participation continued to dwindle after World War II. This was related to the big decline in the traditional market for cured herring as living standards rose markedly in western Europe; and from the later 1950s the East Anglian herring stock itself dwindled as a result of the big fisheries for young herring developed by the Danes in the central North Sea for reduction to fish meal. By the end of the 1960s the seasonal migrations of Scottish boats and shore personnel to East Anglia was history.

In the inter-war period the most profitable sector of the British trawling industry was the distant-water one, involving fishing especially at Iceland and North Norway, and expansion continued at this time (Jüngst 1968: 43–58). This was however mainly an activity of the Humber ports of Hull and Grimsby, although German trawlers working at Iceland continued to contribute to the Aberdeen landings. The main development of the Scottish trawler fleet at this time was to increase fishing at the middle-water range at Faroe.

The decline in mobility related to the herring fisheries has to some extent been counterbalanced by new elements of mobility to prosecute other fisheries: these, however, have very generally involved only the boats and their crews and have entailed little movement of other personnel.

The fact that East Coast fishermen in general were already familiar with West Coast waters led them to look for new opportunities there with the fisheries which developed after World War II. The growth of the seine net fishery saw landings made at a variety of ports, especially on the East Coast; and although there was an obvious tendency for boats to operate from their home ports, their was also a tendency for strong-going crews to land at the main white fish port of Aberdeen, where there was more market competition. This was to result in a pattern of weekly commuting between Aberdeen and the home ports in locations like Fife and the south shore of the Moray Firth. With the rise of Peterhead to the position of leading white fish port from the 1960s, largely at the expense of Aberdeen, most of this commuting has been redirected to Peterhead.

From before 1950 the expansion of the seine net fishery was carried into West Coast waters from such ports as Lochinver, Kinlochbervie and Oban. On the West Coast it has overwhelmingly been prosecuted by fishermen from the East Coast, especially from the Moray Firth. While some boats went seasonally to the West Coast, and might spend the winter especially fishing from home, there were from an early date also boats which spent the whole year on the West Coast. At an early stage the West Coast seine net fishery gave rise to a pattern of weekly or fortnightly commuting. Although buses at times have been used to take crews to and from home, with the general spread of car ownership, this is very generally done by fishermen with their own vehicles.

The development of the nephrops fishery has had a comparable effect to that of the seine net in generating new patterns of commuting between the West and East Coasts. On the whole West Coast mainland north of the Clyde, landings have become dominated by visiting boats (Coull 1991: 42). In both white fish and nephrops the resources on the West Coast in total are less than on the East, but the availability on both coasts has been an asset that has spread the catching power more widely.

In the pelagic fisheries mobility is still an asset, but now involves only a fraction of the personnel of earlier in this century. With a very limited amount of processing, other than by the foreign klondyker vessels, and the catching much concentrated in between thirty and forty purse-net boats, based in the North-East and in Shetland; the numbers of men involved now is only about 500. Base ports now depend on formal designation for administrative purposes as klondyking ports, and this has been dominated by Lerwick in the north and Ullapool in the west, although in more recent times Peterhead has also had a significant share in the trade. Herring fisheries still mainly involve operation at Shetland and Peterhead, and the mackerel fisheries were dominated by Ullapool but now also include Lerwick.

REFERENCES

Annual Reports of the Fishery Board for Scotland (ARFBS.) 1886, 1896, 1900, 1904, 1905, 1906, 1913, 1925.

Coull, J. R. (1986) 'The Boom in the Herring Fishery in the Shetland Islands', *Northern Scotland* 8, 25–38.

Coull, J. R. (1983) 'The herring fishery in Shetland in the first half of the nineteenth century', *Northern Scotland* 5, 123–140.

Coull, J. R. (1991) 'Mobility in the Scottish Fisheries', *Scot. Geog. Mag.* 107, 40–46.

Coull, J. R. (1992) 'Seasonal Fisheries Migration: the Case of the Migration from Scotland to the East Anglian Autumn Herring Fishery', in Fischer, L. R., Hamre, H., Holm, P., and Bruijn, J. R. (eds.) *The North Sea. Twelve Essays on Social History of Maritime Labour*, Stavanger Maritime Museum, Stavanger, 127–147.

Jüngst, P. (1968) *Die Grundfischversorgung Grossbritanniens. Häfen, Verarbeitung und Vermarktung*, Selbstverlag des Geographischen Institutes der Universitat Marburg, Marburg, Germany.

Manson's Shetland Almanac 1906, 1914.

New Statistical Account (NSA.):- XII, 293–297 (Rathen Parish). XV Caithness 117–178 (Wick Parish).

Report of the Scottish Departmental Committee on the North Sea Fishing Industry (RSDCNSFI.) (1914) Vol. II

Scottish Record Office (SRO.) *Fishery Board Papers:- AF5 Secretary's Reports of Inspection; AF7 Letters and Reports from the Chief Inspector of the West Coast.* AF19 Anstruther District records. AF28 Leith District records.

Shetland News, 21st Sept., 1901.

Sutherland, I. (1983) *Wick Harbour and the Herring Fishing*, Camps Bookshop and the Wick Society, Wick.

Transactions of the Norfolk and Norwich Naturalists Society (TNNNS.)

Wigg, T.J. (1900) 'Report on 1900 Herring Fishery', TNNNS. VIII (1901), 263.

14

The Shell Fisheries

Shell fisheries have been important as long as any, as the archaeological record testifies, and in the modern period have come to yield
most of the highest value items in sea food. While in historical times
they have always been overshadowed in commercial importance by
pelagic and demersal species, they were from the earliest times a
significant dietary item for coastal populations, and in addition
were frequently of major importance to the poorer elements in the
population in times of dearth. There are obvious advantages in the
exploitation of species like oysters, cockles, mussels and razor shells
which can simply be gathered from the beds at low tide; and as well
as featuring in the diets of the earliest known settlers in Scotland
during the Mesolithic period, these have been collected right into
modern times. However for commercial development there were
additional difficulties in utilising shell fish, with their greater danger
of spoilage, and the food poisoning which could ensue.

At a wide range of places around the coast there is evidence in
archaeological sites of shell fish being collected for food, and in not
a few cases this occurs in heaps which consist largely of shells. In
one Mesolithic heap at Polmonthill, Falkirk it has been estimated
that there are actually between six and seven million oyster shells
(Stevenson 1948: 139), suggesting accumulation over a protracted
period. Around the Forth evidence for the use of oysters has also
been found in Neolithic and Roman times (Millar 1961: 5).

Investigations by modern archaeologists have shown other
important details of the use of shellfish by early coastal populations. Thus at Northton on Harris there was an alternation of the
dominant shellfish species in the diet between the Neolithic and
Iron Age periods which was in harmony with a changing coastal
morphology. In the Neolithic, cockles (a sandy shore species) were
most prominent, and this was succeeded by limpets (a rocky shore

species) in the Upper Beaker level of the Bronze Age. However in the following Iron Age the predominant species reverted to cockles. In Orkney, excavations at Knap of Howar (Papa Westray) have indicated that the shellfish in the Neolithic came from a more extensive sandy shore than now obtains; and evidence from the Buckquoy site suggests that during the Norse period there was an overexploitation of resources with a decreasing average size of limpets and an increasing proportion of cockles (Evans 1977: 21, 22).

Despite the early appearance in the archaeological record, for most of history very little is known about shell fisheries, although comments in various sources from at least the 16th century show that they featured in the traditional subsistence economy. They are referred to by both John Major and Bishop Leslie, and oysters, crabs, lobsters, crayfish, cockles and whelks are all specifically mentioned (Major 1521: 46; Leslie 1578: 140). It is also clear from sources such as the Statistical Accounts that by the early modern period fisheries such as those for oysters and lobsters had acquired considerable commercial importance. However comprehensive information about shell fisheries is not available before the end of the 19th century, when they eventually became part of the responsibility of the Fishery Board.

Commercial shell fisheries for long were for local markets, and there was no way of preserving their products by curing. In Scotland, the commercial fisheries for oysters, especially on the Forth, are of unknown antiquity: the fishery was of major importance in Newhaven, and the oysters were an important source of food for the urban poor, as they were to a greater extent in London. The redoubtable Sam Weller in 'The Pickwick Papers' remarks when going through east London, that the oyster shells to be seen lying about were a distinctive mark of poverty.

Despite the difficulties in preservation, improving transport in the early modern period considerably extended the market range of shell fish, and this was most prominent in the cases of oysters and lobsters. In the later 18th and in the 19th centuries commercial oyster fisheries expanded, and available beds were exploited all around the coasts. This resulted in a number of piece-meal attempts to formulate conservation restrictions, although with increasing market opportunities few of these were effective in the long term.

While shell fisheries have contributed in the modern period more and more to the luxury food market, their share of the total landed value of fish has always been small. By the late 19th and early 20th centuries the total value of shell fish was running at around £70,000 a year, but was only 2% or 3% of the total fish value. At the same time the social importance of the shell fisheries was greater than this would suggest, as much of the catch came from people at the lower end of the social scale like crofter-fishermen.

OYSTER FISHERIES

While the earliest exploitation of oysters would have included picking them at low tide, the main commercial method became that of dredging by boats rowed by oars; at the end of the 18th century this was done in water from four to fifteen fathoms deep, from September to late April, the summer being avoided as then the oysters were in poor condition (OSA. XVII: 69–71). While oysters were found in various' shallow bays and firths, the main beds of commercial importance were those in the Forth: and they were long the subject of contention between different interested parties. The free fishermen of Newhaven were at the centre of this contention: they claimed exclusive rights to the best grounds in the middle of the Firth under charters from James VI and Charles I, and they also had the sole right to operate the fisheries of the grounds of the city of Edinburgh under a charter of James IV. It was estimated that in the best years these oyster beds yielded around 30 million oysters annually, and fishermen from other places on the south shore of the Firth like Fisherrow and Cockenzie tried repeatedly to encroach across the unmarked boundaries (McGowran 1985: 125–127). In addition there was running friction with the city of Edinburgh on the arrangements for working the city grounds. The city claimed the right by a series of charters to control the oyster fisheries, and in both the late 17th century and the second half of the 18th made regulations that aimed to govern the minimum sizes of oysters which could be sold and also to define the close season (McGowran 1985: 128). The fishermen on their part saw this as interference from an authority that had no real knowledge of the practical problems of fishing. The period from about 1770 saw accelerated exploitation of the oyster beds in the Forth, and as well as oysters being supplied to the expanding market in Scotland, small oysters were taken to restock beds in England, France and Holland. In the 1790s a deteriorating situation was recorded, as a result of the expanded trade of the past twenty years, which was in considerable measure due to a Leith merchant contracting in 1771 with ten different firms to ship oysters to London. It was claimed that at the start of this period it was common for a boat in one day to dredge up 6,000 oysters. Forty boats were involved in the fishery and were paid around £2,500 yearly for oysters. Twelve vessels were engaged in carrying the oysters to the Thames in barrels, the average cargo being between 350,000 and 400,000 oysters; and although the intention was to exclude small oysters from this trade, this was poorly enforced (OSA. XVII, 69, 70). With declining yields from the early 19th century the city of Edinburgh intervened from 1814 to restrict fishing on the beds over which it had jurisdiction, and charged a rent from those allowed to fish. This for a time revived the fishery, and the rent was progressively raised. Transport by horse and cart by the late 18th century also brought the Glasgow market within reach of the fishermen on the Forth, and this was further enhanced by the coming of the railway in the 1840s. The market links with London were also reinforced as the century progressed with the coming of regular steamer transport; and there was also a considerable export across the North Sea to Holland (McGowran 1985: 125).

The increasing demand for oysters and the associated rise in value was to trigger the sort of dispute that illustrates the problem of administering a common property resource like the fisheries. After the Newhaven fishermen had secured permission to export oysters to London in 1834 and made the great profit of nearly £5,000 as a result, the city council of Edinburgh in 1839 refused to renew the lease to the Newhaven fishermen, but raised the annual rent to £600 and gave a ten-year lease to George Clark, an Englishman. The aggrieved Newhaven men continued to operate illicitly and sold oysters to Clark's English competitors. Clark, meanwhile brought up to 70 dredging boats north from Essex, which operated from morning till night taking oysters of all sizes, while Clark himself refused to pay rent to the council as they were unable to stop the local men from fishing. This in fact only occurred during a single season, after which Clark departed and the Newhaven fishermen reached an accommodation with the city fathers. Behind this scramble for the right to fish the oysters was a rising market driven by rising incomes: it was estimated that by 1870 the value of oysters consumed in England was over £4 million (RCOCUKF. 1870: 53). By the 1870s it is clear from the extant records of the Society of Free Fishermen of Newhaven that the fishery was dwindling towards extinction. In the 1874-1875 season the fishery was prosecuted from September to April, and at the spring peak as many as 34 boats were operating and 5,000 to 10,000 oysters were landed daily (SRO. GD 65/13/1); in 1878–79 daily landings were down around half on average although there were briefly as many as 44 boats involved in the spring (SRO. GD 265/13/2); by 1882–83 the maximum number of boats in autumn was 17 and in spring 10, and the peak daily landing had gone down to 1,600; and in the 1883–84 season there were no boats at all fishing oysters (SRO. GD 265/13/3).

Commercial oyster fishing also extended in the 19th century to various places on the Clyde lochs and also such locations as West Loch Tarbert in Argyll, the bays on the Solway, to Skye, Lewis, Orkney and Shetland, and the Moray Firth (Millar 1961: 10–13). With a strongly increasing demand there was an all too general tendency for overfishing to spread: it was recorded for example that at Gareloch Head in the Clyde lochs the beds were soon exhausted by big boats coming and dredging up 30 to 50 tons at a time (ARFBS. 1887: 51).

One result of this increase of pressure on resources was that there were some attempts to cultivate oysters. The most quoted example of this was in Loch Ryan in the south-west where William Wallace, Bart. held the oyster beds by charter and had a system of seeding the beds: in the 1860s he was charging boats £4 per month for the right to dredge oysters. Four-man boats were used, and the proceeds of catches were divided in six shares, with one to each crew member, and also one to the boat and one to the dredge; and each share was worth about £2 per week after the payment of the dues to Wallace. Even here, however, it proved difficult to administer an organised system to give a sustainable yield. The Scar in Loch Ryan had been recognised as having some of the best oyster beds, and Wallace had previously also leased these out, but after a dispute it was decreed that these were not covered by his charter, and the beds had become speedily depleted by local fishermen. Wallace discontinued permitting boats to dredge oysters in the rest of Loch Ryan in 1875, saying that the fishermen would not refrain form taking small oysters. He had succeeded in restocking the beds by 1885, but found that poachers soon depleted them again, and it proved very difficult to get adequate evidence to prosecute the offenders (ARFBS. 1887: 44–47). Despite the problems, Loch Ryan was to continue as the most important source of oysters into the 20th century. The Fishery Board also tried to encourage oyster farming under new powers under the act of 1885 which allowed it to recognise particular oyster fisheries as being operated by named individuals or companies, who could attend to restocking them. This appears however never to have had more than minor success.

In all, oyster fisheries in the modern period have been so valuable as to be all too prone to over-exploitation. By the late 19th century the leading oyster fisheries on the Forth were yielding only a fraction of previous times, and the fishing effort was dwindling. The fisheries in other locations generally lasted only a few years despite desultory attempts to manage and conserve them. It was something like an ironic 'coup de grace' when with the establishment of regular steamer links across the Atlantic, there was a disastrous price fall with the coming of cheap oysters in quantity from the USA.: modern transport which started by giving a great boost to the market ended by killing it.

THE RISE OF LOBSTER FISHERIES

In lobster fisheries, by the late 18th century, market links by fast sailing ship were sufficiently reliable for English merchants to come on the scene, and Newcastle became an important market. More important was the expansion of the orbit of the London market: by 1785 this was being supplied with hundreds of thousands of lobsters every year: lobsters were being collected in quantity from Norway as well as Scotland, and it was difficult to satisfy the market demand (RCHC. vol. X 1785: 23–24). To get the lobsters to market, well smacks were frequently used as, like shellfish generally, lobsters deteriorated relatively quickly if they could not be delivered alive.

A limited amount is recorded on early methods of lobster fishing, and nets and hooks both seem to have been employed, although the main method in the earlier 19th century was to sink on a rope to the bottom an iron ring with a pouch net attached; bait was put in the net and the ring quickly pulled up when a lobster took the bait. This appears to have been most widely used in the north, in Orkney, Caithness and Sutherland. The method is best described in the New Statistical Account of Durness in north-west Sutherland. Here there was a summer fishery that lasted from May till August operated by two-man boats of 14–foot keel. Each boat operated 20 or more nets, and the men spent the whole day lowering and hauling them (NSA. XV: 99). By the middle of the century the lobster creel or trap had been devised, by which the lobster could be lured by bait into a trap left on the sea bed: to do this it had to pass through a narrow tapering net funnel which made it very difficult for it to return once in the trap. It became the practice for individual lobster fishermen to work dozens of creels at the same time; commercial fisheries developed all around the Scottish coasts and on the most prolific lobster grounds of the West Coast and Western Isles it became a fishery of major importance for many crofter-fishermen: its requirements in capital were small, and as the fisheries were close inshore it was still feasible to work with small boats.

There was a buoyant demand in city markets, especially in London. Despite increasing levels of production, the price was driven up in the later 19th century and between the 1860s and the 1880s the price rose about three times. By the 1880s hundreds of thousands of lobsters were being caught annually: in 1887 the total value of the catch was £26,647, of which more than two-thirds came from the West Coast (Fulton 1887: 190). Obviously the availability of rapid transport by railway was an asset, although there was always a problem of some lobsters not surviving the journey, especially in the warmer weather of summer.

There were already fears that this upsurge in catch had caused overfishing: although comprehensive catch statistics only began in 1884, it was known from the experience of fishermen and merchants that lobsters were becoming more scarce and decreasing in size. These fears appear to have been in part premature, as later lobster investigations have seen fluctuations of stocks that are only partly explicable by fishing pressure, and the catches from 1892 to 1954 are shown in fig. 14.1. One result of the developing scarcity was that lobsters were being imported from Norway in quantity, and by the 1880s these imports were numbering between 300,000 and 500,000 annually. However despite the considerably greater stocks of lobsters in Norway, there too catches were declining and signs of overfishing were in evidence.

While a minimum landed size of six inches carapace length had been imposed in Britain, it was clear that this was inadequate for conservation purposes. Enlarging the minimum landed size and the banning of landing berried lobsters were to help conservation, and in the years before World War I the Northern Isles too had started to become more important in contributing to the supply: at this period there were around 700,000 lobsters being landed annually, and the value was between £30,000 and £40,000. They accounted for around one half the total value of all shellfish landings. With the increasing interest in lobster fishing in the Northern Isles, especially in Orkney, where it proved a good complement to small holdings for men with limited capital, these became established as the location of the most intensive production, although the West Coast and the Long Island continued to be the leading area in total production.

From the inter-war period there was an increase in the effort in lobster fishing on the East Coast (fig. 14.1), which was related to the difficulties being experienced in the main herring fishery, although the West Coast continued to dominate the fishery. In the inter-war period West Coast catches were on a par with the catches around 1900 and were running between 500,000 and 600,000 annually; on the East Coast they had risen from around 100,000 to the 300,000 level, while in the North Isles catches continued around 150,000 (Thomas 1958: 14).

SHELL FISHERIES SINCE WORLD WAR II

The period since World War II can be seen as that in which what might be termed the cycle of development in shell fisheries turned towards completion. Earlier shell fishing had been concentrated on oysters, which were mainly a food of the poor, and the low-value bait mussels, which were essentially ancillary to traditional white fishing. Increasing purchasing power over the period of the Industrial Age has been associated with a general rise in value of shell fish, and increased incentives to catch increased numbers and a greater variety of them. The rise in demand for oysters and lobsters has already been discussed over this period, and the more rapid increase in living standards after World War II has led to a burgeoning market for shell fish not only in Britain but in the developed world generally. Important here has been an increasing capacity to preserve by freezing: shell fish are particularly prone to rapid spoilage and a great part of the catch is now frozen. Also important has been improved transport which allows for the quick transport of fresh shell fish, or the less urgent transport in the frozen form; and this transport can include air freight as well as transport by surface modes. To the continued importance of lobsters has been added a major fishery for nephrops and a valuable fishery for scallops; and the modern facilities for freezing have also given a new impetus to the fisheries for crab, even if these are of lower value.

In the early post-war period the lobster and crab fisheries

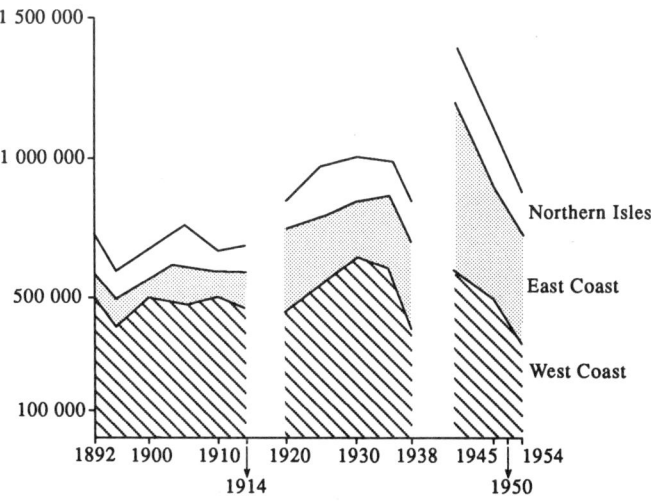

Fig. 14.1 Lobster landings, 1892–1954
(*After H. J. Thomas*).

continued to dominate the activities of shell fishermen. After the reduced activity of war-time lobster landings for a time in the late 1940s approached a total of 1.5 million (fig. 14.1), while the more numerous crabs could reach well over 3 million. These fisheries show a markedly seasonal rhythm of activity. There was considerably less fishing effort in mid-winter when operating conditions for small boats were at their most difficult, and the danger of loss or damage to gear is greatest. Landings were much higher from April to October, with a definite peak in the late summer and early autumn. The seasonal variation was particularly marked in the lobster fishery which is closer inshore, with the late summer peak being more than 15 times the mid-winter low. With crabs, in contrast the difference was of the order of a factor of 4. In 1954 the number of boats on the East Coast at the peak season approached 200, and on the West Coast was well over 100 (Thomas 1958: 8). By the 1960s lobster landings could reach 20,000 cwt. The Northern Isles came to account for around 30% of national production. Although landings have generally been lower since the 1960s, demand has continued to drive the price up, and after the 1970s the value was regularly over £1 million.

From 1958 landings of all shell fish have been recorded by weight as well as value, and the trends in landings from 1958 to 1974 are shown in fig. 14.2. The increasing importance and effective widening

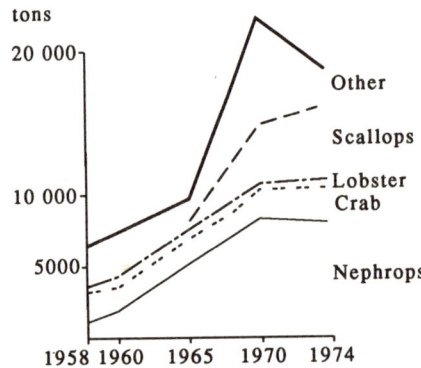

Fig. 14.2 Landings of shell fish, 1958–1974.

of the resource base with shell fish has been in many ways opportune. It has effectively compensated for the continued contraction of the herring fishery, and more recently has done something to ease mounting pressure on decreasing stocks of white fish.

NEPHROPS FISHERIES

What was to become the most important of Scottish shell fisheries had to wait till after World War II. The stocks of Norway lobster (nephrops, or 'prawns' to the fishermen), are extensive off both Scottish coasts, and are especially abundant in the Moray Firth and the Minch: unlike the lobster and crab this species lives offshore on areas of muddy bottom, and although its presence was known and it did appear to an extent in trawl and seine net catches, it had been thrown overboard, as there was no market for it. In the 1950s freezing plants were established at a small number of points on the coast, and fishermen started to land nephrops that had come up in seine nets and to fish systematically for them in between main seasons. However the fishery has always been a mixed one in the sense that white fish are caught at the same time. To begin with the main landings were on the Forth, with landings at Anstruther and Leith: but the fishery soon spread to the Moray Firth with landings at Lossiemouth and Buckie, and to the West Coast with landings at Mallaig and Oban. It was soon taken up also on the Clyde. By the early 1960s the light trawl was recognised as the most effective gear for catching nephrops, and there were crews who were turning to it as a full-time occupation. There were relaxations of the by-laws earlier laid down by the Fishery Board which had restricted trawling in the firths, so that nephrops trawling could be operated in them for at least part of the year, provided that not more than 25% of white fish were caught with the nephrops. In the case of the Moray Firth it was allowed for seven out of the twelve months, but with no restrictions on the gear on the West Coast, the boats concerned were able to work the whole year if they wished. Nephrops trawling became the major recourse of a big group of fishermen, many of whom had previously depended mainly on the herring fishery, which was now in decline. Early processing was mainly at Berwick and Annan, but by 1965 processing plants had also been established at Stornoway, Oban and Inverbervie. From the end of the 1960s the balance of production had switched to the West Coast.

From being of negligible dimensions in 1950, by the mid 1960s nephrops were dominating the shellfish catch. In 1965 the landings had risen to over 5,000 tons, and by 1973 to over 10,000 tons. By the later 1960s the aggregate value regularly exceeded £1 million, and the value since has been five to ten times that of lobsters. From the later 1960s nephrops have constituted up to a half or more of the total shellfish catch by weight.

OTHER SHELL FISHERIES

Directed fishing for scallops was largely an initiative of men in the Campbeltown district in the south-west from the mid–1960s when they began dredging for them on both sides of the peninsula of Kintyre, From landings of 7,456 cwt. in 1965 landings rose to 110,700 cwt. in 1969. While the value of landings has regularly topped the £1 million mark from the early 1970s, this has partly been achieved by extending the fishery to queen scallops. Since the early 1970s scallops landings have markedly fluctuated. This is due to the all too easy overfishing of a sessile species, and although new beds have been found over a period off various parts of the coast, they have tended to be intensely exploited. The main protection is the establishment of minimum landed sizes, which does allow beds some respite for a period once mature scallops are removed.

While crab fisheries have continued all the while, and their landings are usually several times of those of lobsters by weight, their unit value is much lower, and they contribute only a fraction of the value of lobsters. Crab fisheries have, however, benefited from the establishment of freezing plants. Various other shell fish contribute to the catch in fluctuating amounts. On of the more important in the modern market is shrimp, while periwinkles are still collected for the edible market. Mussels now enter this category also, although part of the production now derives from fish farming in the Hebrides.

REFERENCES

Evans, J. G. (1977) 'The Palaeo-Environment of Coastal Blown-Sand Deposits In Western and Northern Britain', *Scottish Archaeological Forum* 9, 16–26.

Fulton, T.W. (1887) 'The Scottish Lobster Fishery', *Annual Report of the Fishery Board for Scotland* (ARFBS.) Scientific Investigations, 189–202.

Leslie, Bishop (1578) ' History of Scotland', in Brown, P.H. (ed.) (1893) *Scotland before 1700 from Contemporary Documents*, David Douglas, Edinburgh, 113–183.

McGowran, T. (1985) *Newhaven-on-Forth. Port of Grace*, John Donald, Edinburgh.

Major, J (1521) 'On the Boundaries of Scotland, its Cities, Towns and Villages, of its Customs in War and in the Church; of its Abundance of Fish, its Harbours, Woods, Islands, etc., in Brown, P. H. (1893) *Scotland before 1700 from Contemporary Documents*, David Douglas, Edinburgh, 41–61.

Millar, R. H. (1961) *Scottish Oyster Investigations 1946–1958*, DAFS. Marine Research no.1, HMSO., Edinburgh.

New Statistical Account (NSA.) XV (Sutherland), 82–104 (Durness Parish)

Old Statistical Account (OSA.) XVII, 61–88 (Prestonpans Parish).

Report of the Commission on Oyster Culture in the United Kingdom and France (RCOCUKF.) (1870) HMSO., Dublin.

Reports from Committees of the House of Commons (RCHC), (1803), Vol. X, Miscellaneous Subjects, 1785–1801

Scottish Record Office (SRO) *Papers of the Society of Free Fishermen of Newhaven* GD 265/13/1, 265/13/2, 265/13/3.

Stevenson, R. B. K. (1946–47) 'A shell-heap at Polmonthill, Falkirk'. *Proc. Soc. Antiquarians Scot.* 80, 135–139.

Thomas, H. J. (1958) *Lobster and Crab Fisheries in Scotland*, DAFS. Marine Research no.8, H.M.S.O., Edinburgh.

15

Fishing Boats

While there are several well known traditional types of boats used for fishing in Scotland, for most of them the early history is largely a matter of conjecture, and it is well into the 19th century before there are any systematic descriptions of any of them. Fishing in Scotland in the main involved operation from open exposed coasts, and even for inshore working, robust and seaworthy boats were needed. Like vernacular housing, traditional construction skills for boats were passed down through the generations by craftsmen. At the same time the fact that from the earliest settlement of the country it was possible to move from mainland to island shows that boats of some kind must have been available. However for most of history the subject is largely a closed book. This contrasts with the more recent situation in which, especially during the 20th century, there has been continuous development in the design of fishing boats and of their gear and equipment.

EARLY AND TRADITIONAL BOATS

The earliest boats may well have been vessels of hide stretched on a framework of wooden spars of the type that has survived into the 20th century in the curragh of Western Ireland. These would be easier to build that wooden hulled boats, although they were less robust; and coracles of this general style were used into modern times on Scottish rivers.

It has been claimed that the skill in the construction of clinker-built open boats that was developed in Scandinavia in the centuries prior to the Viking era was spread by them around the coasts of the North Sea and to other shores of Western Europe. However, proving the lineage of traditional boat styles in Scotland is near impossible other than in the North Isles, where the traditional double-ended boats have a clear family relation to the boats used in Norway, the

Faroe Islands and Iceland. The original Norse faering style is best preserved in the modern Shetland Ness yole, and indeed this type of craft was in very general use for fishing in Shetland until the development of the bigger sixern (plate 8) in the 18th century (Henderson 1977: 52–55). However the sixern as well as the Ness yole were lightly built, although both were very seaworthy craft: such boats incorporated an important element of elasticity in their construction and had strength-through-resilience rather than strength-by-solidity (Morrison 1978: 61).

There is a fairly clear regional variation in traditional Scottish boat styles. The traditional style on the East Coast as far north as about Kinnaird Head was the 'fifie' with its vertical stem and stern posts. The name strongly suggests that it was originally from the Fife area, which was certainly prominent in fishing and other sea-faring from Medieval times onward. In the Moray Firth area the characteristic type was the 'scaff' or 'scaffie', with its curved stem and raked stern. On the Clyde was the 'nabbie' or 'Loch Fyne skiff' which had a curved stem and rounded and raked stern.

In the case of the double-ended Shetland model with its sheared bow and stern, boats were actually imported into the 19th century to the treeless islands in kit form from Norway, and there was an effectively unbroken tradition in style of building. There is evidence that the scaffie was derived from boat types of the North Isles, and there are certainly traditions of settlers from Orkney coming into the Moray Firth area.

In addition to the small boats which were always much greater in number, there was some use of bigger boats for the herring fisheries from at least the 15th century. The 'crears' from Fife and the 'busses' from the Clyde were decked or half-decked boats, although there are indications that they were something of general purpose vessels which were also used for trade. Little is known of the details of early vessels of this type, although the evidence from involvement in the Norwegian timber trade suggests that even these were mainly of modest dimensions, and that some were as small as 4 to 6 lasts (i.e. 8 to 12 tons) burthen (Lillehammer 1986: 102–103).

When the fisheries expanded in the 18th and 19th centuries, boats became bigger and improved, but largely through development of traditional designs. When the Shetland 'haaf' fishery developed in the 18th century, and there was an emphasis in working further offshore, this was achieved essentially by employing the bigger six-oared 'sixerns' rather than the smaller four-oared 'fourerns'. At the same time on the East Coast mainland there was evidently a greater use of six-man fifies and scaffies for the great line fisheries. Traditional boats always had oars for propulsion, but the bigger ones also used square or dipping lug sails on a single mast. In Shetland, boats retained the traditional Viking square sail into the 20th century. While this simple arrangement was good for running before the wind, it was less suitable for tacking; and Shetland 'haaf' fishermen often preferred to row sixerns into the wind rather than make distance by tacking into it (March 1970: 55–57). However in most other places the use of lug-sails was general by the 19th century. An important advantage of the lug-sail was that it involved no bowsprit or spars projecting out beyond the hull on the beam, and this was a definite asset in the congested harbours that were the norm in the herring fishery. It was better for tacking into the wind than the square sail, but was actually less satisfactory for this purpose than the ketch and other rigs: even so with the harbours that continued to be congested it was the dominant rig as long as sailing fishing boats lasted in Scotland (March 1952: 239–252).

BOAT DEVELOPMENT IN THE INDUSTRIAL ERA

It was the expanding herring fishery in the 19th century that accelerated the development of boats. From an early date it had sufficient success for fishermen to invest in bigger boats which could fish in more demanding weather and also carry more herring; even so the cost of the boat itself was in the vicinity of £50. These boats were generally built in small boatyards in or near the places where they were based, and in detail there were often local variations in style. Increase in boat size was slow in the first half of the 19th century, but accelerated in the second half of the century as the fishery came to dominate European herring fisheries. By mid–century the Fishery Board was trying to promote a move to decked rather than open boats as a safety measure. However this also was very slow to come, partly because of the greater capital investment it entailed, but also because the fishermen saw the early decked boats, which generally had a low rail, as less safe. However there was a tendency from the 1850s to build new boats with a small forecastle in the bow which gave covered bunks; and at Eyemouth in 1856 the first full-decked boat was built at the relatively high cost of £130. This first full-decked boat was still only 40 feet in length, and it was not until the later 1860s that with increasing size of boats to over 50 feet that decked boats became generally accepted: by 1872 the leading Buckie district alone had 400 decked boats, and advance was effectively promoted not only by the capability of such boats to venture further out to sea for herring, but also by their suitability for winter line fishing (ARFB. 1872: 4–5). In the 1870s too there was the first installation of man-powered winches to aid the heavy task of hauling the main rope, and also the beginning of carvel rather than clinker construction; and by the end of the 1870s new boats were costing £200 to £220 (SRO. AF34/7: 20). A limited amount is known about the types of timber that had been used in earlier times in boat construction, but by the 1870s the general practice was to employ mainly oak and larch. The oak was used for the frames, and larch for the hull and deck planking, although there could be use of other timbers, such as beech for keels and Scots pine for planks.

The bigger boats had been developed on both the fifie and scaffie designs, but in 1879 came an innovation that was to lead to a new word in the vocabulary of Scottish fishing. William Campbell of Lossiemouth pioneered a new design in his 'Nonesuch' (March 1952: 252) that in effect incorporated some of the best features of the fifie and the scaffie (Wilson 1965: 26–27): it had the vertical stem of the fifie but the raked stern of the scaffie. The traditional scaffie with its relatively short keel had less a grip of the water when running before the wind, but the new design improved this while also giving a relatively long deck for the size of the boat. This development started at the time of the Zulu War in South Africa, and the new boat became known as the zulu. One remarkable result of this is that the word 'zulu' is the last word in the Scottish National Dictionary. Continued increase in vessel size saw new boats reach lengths of 60 and even 70 feet, and by the 1890s they might cost £500 or more ready for sea. The 'zulu' became a very popular design, and by the end of the century there were 400 of them in the Buckie fishery district alone (March 1952: 274). Another development of the late 19th century was the building of the 'baldie', an improved small boat on the fifie design for inshore fishing, which acquired its name because of the prominence at the time of Garibaldi on the political stage in Europe. The baldie was decked, and carvel rather than clinker built. It was of around 25 feet in length, had the mast well forward and also used the lug sail (Wilson 1965: 26).

Parallel with the increase in length in boats there were necessary adjustments in other dimensions and in masts and sails. The hulls became broader in beam and deeper in draught, and an important consequence was a great increase in the demand for piers and harbours for craft that could not be pulled up out of the water on a daily basis. Masts became bigger and main masts reached over 50 feet above the deck, while main yards could be 30 feet long. Such masts, yards and sails in proportion were necessary to propel the bigger boats, and it became a really major effort to raise mast, yard and sail by block and tackle and muscle power. This was especially so as it was necessary to lower the mast when the boat at sea was riding at her nets to cut down rolling; for this reason the raising of the mast was a daily and very strenuous task.

There were also developments in the traditional craft of the Clyde, largely associated with the rise of ring-netting. Here the increase in size of boat was more modest, and this is much related to the need of vessels to be manoeuvrable in this fishery and to be able to operate in narrow and sometimes shallow water. The usual size of skiffs was increased from the 20–25 foot range to around 35 feet, and small forecastles built in the bows which allowed the fishermen to live and sleep on their boats (Martin 1981: 76–78). They continued to use lug sails, but a bowsprit and small foresail were added (plate 19) (Martin 1981: 76–85). With the smaller size of boat that continued on the Clyde and in the south-west there were some variety of local designs and sizes that developed in the 19th century. In the middle of the 19th century there were on Loch Fyne four sizes of boats costing from £25 to £60 each, and their net gear for the herring fishing cost from £25 to £40 (Marsh 1970, I: 297, 298). Also on the Solway coast there came into use a type of boat similar to those used in similar conditions of shallow water on Morecambe Bay. They were about 30 feet overall, were cutter-rigged, and they were used for beam trawling for flounders, shrimps and prawns (Marsh 1970, II: 294–295).

STEAM POWER AND STEEL BOATS

Although there was the development of vessels driven by steam power from early in the 19th century, it was decades before there was any serious thought of installing steam engines on fishing boats. This was essentially due to the cost of such craft being of another order of dimensions from the cost of fishing boats. With the increasing efficiency of the marine steam engine the spectrum of use for it during the century extended from the specialist packet and high value freight trades to general trade, and ultimately it was being installed aboard custom-built trawlers by 1880. Even for trawlers it meant a great jump in capital costs of several times; and the increase in running costs was also substantial with the price of fuel and other requirements. However, such was the demand for fresh fish in the city markets of the Industrial Revolution in Britain that steam trawling quickly proved profitable through greatly enhanced catch rates. The extra and dependable power for pulling the trawl over the sea bed, and the steam winch to haul it at the end of a tow, proved decisive advantages. For rapid distribution to market, steam power was also the key, with markets reached by the railway network.

SCOTTISH STEAM TRAWLERS

There had been some development of trawling in Scotland with sail boats before 1880, but it was the advent of steam trawling from that time that was to work a revolution in the white fisheries. This was essentially an offshoot of developments in England, and indeed much of the early trawling off the Scottish coast was by English trawlers seeking fresh grounds. However there was also rapid investment in trawlers in Scotland, and by 1893 trawlers landing in Scotland were dominantly Scottish. It was thus that a major new departure arose in Scottish fishing fleets: steam trawlers from the start were steel built decked craft, generally of at least 60 feet; and they were generally company-owned. The earliest trawlers had an aft wheelhouse, with the wheel in a position copied from sailing craft, and the funnel and engine were forward of this. However this soon gave place to the more efficient lay-out by which the wheelhouse was brought forward to the middle of the ship and connected to the rudder by steering chains; and this allowed the skipper a clear view of the working foredeck from the bridge (plate 30). The trawl winch was mounted immediately forward of the wheelhouse, and under the foredeck was the hold or fish room, while the cabin was in the stern. In the fashion of the time the steam trawler had a vertical stem. The first steam trawlers were 'beamers', i.e. used a trawl in which the mouth of the bag was held open by a heavy wooden beam along the headrope, but this was swiftly replaced at the end of the 19th century by the more efficient otter trawl, with which the net is kept open under water by vertical trawl 'doors' or otter boards. The trawl had originally been worked with ropes, but these were replaced by stronger steel warps. Fore and aft 'gallows' were installed at the rail; and over pulleys attached to the gallows the trawl warps were let over the side, and the trawler was to acquire the style which persisted for over half a century. Construction in steel was concentrated at much fewer yards than that in wood, and the big part of the Scottish trawler fleet was built in yards in the main trawl port of Aberdeen.

STEAM DRIFTERS

Although steam trawling was proved profitable in the early 1880s, it was not apparent that it could be extended to other fisheries. While the new degree of independence of steam power from wind and tide was an advantage for any fishery, sundry early attempts to use steam boats for the great drift-net fishery for herring were less than successful. For drift netting there was limited advantage in having steam power for the fishing operation itself: although a powered winch could be used to heave in the main rope, the main part of the work still consisted of the manual hauling of nets and of the shaking the fish out of them. Also by the mid–1890s sailboats were availing themselves of the advantage of a steam capstan with a much smaller and more economic engine than was required to drive a boat. The early experience in herring drift-netting on steam boats was that any increase in catch was more than swallowed up in increased costs.

However it was also the case that most of these early experiments at herring fishing with steam boats were made at a time when the whole trade was depressed and profit margins slim in the period between 1884 and 1893. As the herring fishery at the end of the century recovered vigorously form this recession, there were fresh ventures with steam boats immediately prior to 1900, and these generated sufficient profit to arouse a new interest. By 1902 steam drifters were being custom-built, and their average catch rates were more than double those of sail boats; they were also able to make a quicker and more dependable return passage to port, and to get the best prices in the auction markets in which herring were now sold.

The earliest steam drifters, like the early trawlers had the wheel-house aft and the funnel forward of it, but there was a swift transition in lay-out parallel to that of the trawler, with the wheelhouse in the middle of the boat. The cabin was now built in the stern, and the rope locker in the bow, as it became practice to haul the main (or 'bush') rope on to the capstan placed in the bows (plate 20). The stem was vertical and the big majority had elliptical counter sterns. The hold was aft of the rope locker and the engine room under the wheelhouse. Custom-built steam drifters at between £3,000 and £4,000 were three or four times as expensive as the biggest sailing drifters; and although there was a great fever of construction of them in the years between 1900 and 1914 for the leading seats of the herring fishery on the East Coast, there was always a question over the economics of them in the herring fishery. The building of steel steam drifters was concentrated in relatively few yards: many were built in Aberdeen, but places like Montrose also became involved. An important development as the drifter fleet expanded was the building of a big number of wooden vessels at this time, which were around £1,000 cheaper than the steel drifters; and many of them were built in local yards, especially in the North-East.

The viability of steam drifters had been questioned before World War I, and the problems of operating them economically became much worse in the difficult inter-war period. A number of new government-financed 'standard' boats were built in the period immediately after World War I to replace boats that had been requisitioned for war purposes and had been lost. After this, however, there was very little new construction for an industry in prolonged difficulty.

MOTOR BOATS

Although the steam engine was the great motive power of the industrial revolution on land as well as at sea, it had its disadvantages in being of relatively low mechanical efficiency, and in requiring a considerable amount of the restricted space in the hull of a boat being given over to a boiler as well as to an engine. For fishing boats too it was not always feasible to get fresh water for the boiler, and the use of salt water inevitably led to more rapid deterioration of the boiler.

In the longer term the superior source of motive power on sea as well as on land was motor and diesel engines. As well as being more efficient mechanically these required no boiler. Marine motor engines began making their appearance in the early years of the 20th century at the same time as the internal combustion engine was making its first impact on land transport. The early motors were mainly restricted to small and middle sized boats, although they were able to effect a speedy revolution in a country like Norway in which the fisheries were still very much dominated by small boats. As well as the problem of getting a motor engine sufficiently powerful to drive the bigger boats, there was also the problem of devising efficient motor winches, which held up development in Scotland for a generation. Even in the very difficult inter-war situation in the main herring fishery, the superior efficiency of the steam winch greatly retarded the move to boats with the more efficient and economic motor engine.

There was none the less an accelerating movement in the installation of motor engines in boats in Scotland from the early 20th century. As well as being suitable for smaller craft in fisheries like the ring-net and the lobsters, the installation of motor engines in former sailboats was a way forward in the herring fishery for men who could not command the capital for a steam drifter, even if the same boat might still need a small steam engine to power the capstan.

In the years immediately prior to World War I there was a notable increase in the numbers of motor boats, very largely through the installation of motors in existing craft; but they were conspicuously concentrated in areas and centres which were not in the van of the fisheries. From slow beginnings the rate of motorisation accelerated markedly after 1909 and by 1914 the total number of motor boats was 694, 397 of them in the 18–30 feet ('second class') range. In this period there was a great deal of experimentation, and this is reflected in the variety of makes of engine that were used: in the period up to World War I this totalled 22 (Hawkins 1984: 8). The main impact was on the Clyde with 230 motor boats, as ring-net men swiftly took up the new opportunity, while the rest were spread thinly over the rest of the coast of the country, the one other sizeable concentration being at Eyemouth; the men in this southeast corner of the country were pioneers in using the Gardner diesel engine (Wilson 1965: 37–38) which was, as it developed, to be a mainstay of the Scottish motor fleet for two generations. The first really successful bigger boat with a motor engine was the 65 foot 'Maggie Jane' of Eyemouth (Hawkins 1984: 26).

During World War I it was usually easier to get fuel oil than coal, and many coal-burning fishing boats had been requisitioned by the Admiralty. In these circumstances there was considerable progress in motorisation of boats: by 1919 there were 1,830 motor boats and in 1922 the total had risen to 2,020. Thereafter the number fell back somewhat with the scrapping of old vessels in the difficult inter-war period. The numbers of sail boats dwindled even faster, and by the later 1930s there were only a handful left on the register. For the installation of motors the traditional fifie style was better than the zulu with its deeply raked stern, and in the inter-war period there was the construction of a considerable number of boats on a modified fifie design, custom-built with motors. These were less deep in the water aft, as they did not need so deep a keel to counterbalance the high mast and sail; and they had lighter masts as well as a wheelhouse. During the interwar period there was also the improvement in the design of ring-net boats: Robert Robertson of Campbeltown, who had previously taken an initiative in the area in having an engine installed in 1907, designed a bigger boat in his revolutionary 'Falcon' that was to trigger a series of changes. It was fully decked and had a canoe stern which allowed the propeller to be mounted forward of the stern post and mini-mised the danger of the net getting into it. Other boats began to be built fully decked rather than half decked for the first time, and to have steering wheels rather than the traditional tiller. (Martin 1981:21–221). With the sheltered waters in which these boats operated, it was also feasible to have the engine aft under the wheelhouse, the cabin forward in the bows, and the fish hold in between.

From the 1930s on the Moray Firth new construction for seine net purposes began to incorporate the cruiser stern, which also brought the propeller forward into a position less likely to foul the net. Also the original position of the winch immediately forward of the wheelhouse was changed to a position just aft of the fore mast: this was due to the adaptation of the original Danish method of winching in the net while at anchor to the technique of 'fly drag-ging' in which the boat maintained forward motion. The Danish method was better calculated for a fishery which caught a high proportion of flat fish and involved a slower movement of the net. With more haddocks and other round fish being caught in Scotland, fly dragging meant that the net moved forward with the boat and was more rapidly winched in. A main development of the 1930s was the extension of motorisation to many smaller boats of less than 18 feet with the result that by the end of the inter-war period the total number of motor craft approached 2,500. In all, by World War II, the effective fishing fleet consisted of the bigger sections of the steam-powered trawler and drifter fleets, and the rest which were motor powered.

DEVELOPMENT OF BOATS AFTER WORLD WAR II

The renewal of the fleet with government financial help after World War II resulted in regular work for more than two decades in traditional boat-yards located in the main fishing areas, especially on the East Coast. The renewal of the fleet was begun in 1946 along lines that had been worked out at the end of the inter-war period. Of the motor-powered MFVs (vessels built by the Admiralty during the war for naval purposes for later conversion to fishing use), 107 were allocated to Scotland. Grants and loans for new construction were provided from 1946 through the Secretary of State for Scotland and the Herring Industry Board. Most of these were originally for small boats, but from the start vessels of over 45 feet were included; by 1948 238 of these bigger boats had been built, the big majority of them for the established leading fishing areas of the North-East and of Fife (RFS. 1948: 22, 23). From 1953 fleet rebuilding was accelerated by the grant and loan scheme of the White Fish Authority. Through these government schemes the main catching power became concentrated in boats of lengths between 60 and 75 feet, most usually powered by 150 h.p. Gardner engines, and capable if desired of alternating between seine-netting and drift-net herring fishing. By this time the cost of building a new boat was of the order of £15,000, although with the government financial help available there was a spate of new construction. The remaining steam drifters were rapidly phased out, and the last of them sailed in 1955 (RFS. 1955: 9).

The great part of the new generation of boats were wooden built with cruiser sterns and diesel powered (plate 32), although when the trawler fleet was renewed under the government 'scrap and build' scheme in the years around 1960 construction here continued in steel. The general move to diesel power, as well as being more efficient both mechanically and economically, had the important result in saving hull space. The elimination of the boiler saved ten or more feet in hull length, while diesel oil needed less bunker space than coal. The general tendency also continued of the concentration of catching power in fewer bigger and more modern boats. In the rebuilding of the trawler fleet its numbers were reduced by more than half. Although the new diesel-powered trawler fleet was more economic and efficient, the boats were still the traditional style of "side winders" which hauled the net to the side of the vessel, and the crew worked on exposed open decks. There was no construction at this stage of the safer but more expensive stern trawlers, which hauled the net by a stern ramp, and in which most of the work was done under cover. Such stern trawlers first made their appearance in Britain in the distant-water fleets of Hull and Grimsby, and it was only later that a handful were built for Aberdeen owners for middle-water grounds like those at Faroe.

The later move to steel hulls for the bigger seine net and purse net boats meant that construction was often displaced, usually to bigger shipyards and often outside the areas where their owners lived: construction could be in such locations as the Clyde, the Humber and the Thames. In the longer term too there has been an increasing component of international competition in the construction of fishing boats, and a considerable number have been built abroad in countries like Denmark and Holland, and for the biggest purse-net boats construction in Norway has been the norm. All these developments led to a progressive escalation in the costs of boat construction, and the rate of increase was regularly considerably more than that of fish prices. By the 1970s new boats were costing over £100,000, and more recently the costs of boats in the main section of the fleet have been over £1 million; and in the case of the bigger purse-net boats the cost of the individual vessel rose to several millions. In later years there has also been some use of other construction materials for smaller boats, especially of fibre glass.

The changing emphasis on different fishing methods can be illustrated by the statistics which have been published since 1963 which aggregate the numbers of boats according to main fishing method employed. Fig. 15.1 shows that the most numerous sector of the fleet has continued to be the small craft which fish for shell fish, mainly by creels, but also using in some cases such gear as scallop dredges or shrimp trawls. Between 1963 and 1974 this sector of the fleet declined only marginally despite the small size of the production units, and this is essentially due to the high value of shell fish in the modern market. The traditional method of lining continued to decline, and the traditional pelagic methods of drift and ring net both declined to insignificance. The use of the seine

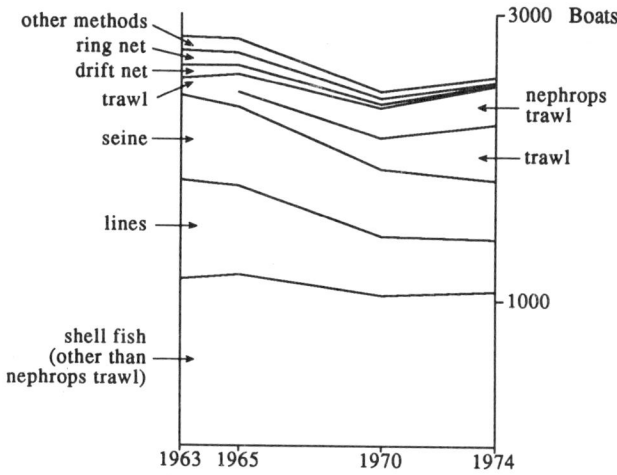

other methods
ring net
drift net
trawl
seine
lines
shell fish
(other than
nephrops trawl)

3000 Boats
nephrops
trawl
trawl
1000

1963 1965 1970 1974

Fig. 15.1 Fleet composition by main fishing method,
1963–1974.

net, though still important, fell off to some extent; in contrast, although trawling had earlier declined with the contraction of the fleet of steam trawlers, it has proved the method best suited to the age of power operation of fishing gear, and numbers of trawlers expanded during the period by almost seven times. While the nephrops trawl was the main single trawling method in the later 1960s, this has given way to a combination of other methods including both single boat and pair trawling, especially for white fish.

The great part of the catching power has continued to concentrate in fewer bigger boats, and this can be shown by the trends within the group of vessels over 40 feet. The number of these has risen significantly since the 1950s despite the general contraction in numbers of the fleet (fig. 15.2). After 1965 a more detailed break-down of the fleet structure is available, and of the districts in which it is based. The very prominent concentration of catching power in the North–East, in the districts from Aberdeen to Lossiemouth, is evident. While the rest of the country in total has a bigger share of the boats in the 40–60 feet range, in the 60–80 feet range the North–East marginally increased its share of the fleet to nearly 80% between 1965 and 1974; while in the over 80 feet range it remained in the vicinity of 85% over the same period. Here, although the Aberdeen trawler fleet was declining, the fleet of its only rival Granton was actually declining more rapidly. At the same

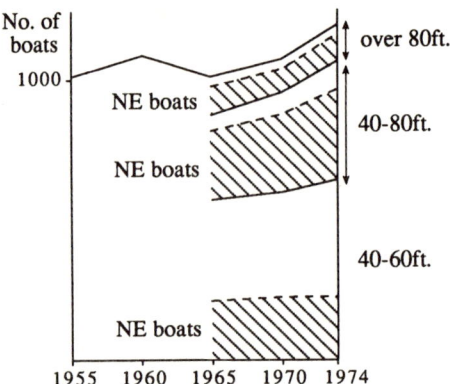

Fig. 15.2 Development of the section of the fleet over 40 feet, 1955–1974, with numbers registered in the north-east, 1965–1974.

time there was the beginnings of investment outside the trawler companies in bigger boats, especially for the purse-net pelagic fisheries.

THE REVOLUTION IN EQUIPMENT

There have also been other developments which have added substantially to catching power. Engine power has increased substantially, and this is related much to the increase in use of the trawl method for which capability to tow at some speed is a marked asset. There has also been the general adoption of hydraulic equipment, especially for winches. On seine net boats the addition of automatic coilers to winches effectively saved the work of one man; and with the adoption of the power block on the same section of the fleet, the last phases of hauling the net have been eased. As a general rule now little of the heavy work is done manually. The durability and strength of gear has also been enhanced with the use of synthetic fibre materials for ropes, nets and lines. While there is still inevitable wear and tear on fishing gear, the age-old problem of deterioration from repeated submergence in sea water has been effectively eliminated. Generators driven off the engines and backed up with batteries give the substantial advantage of electrical power for lighting and other purposes; and the bigger boats now have auxiliary engines for this which are as powerful as the engines which drove the boat in the early post-war period.

During the post-war period fishing has been revolutionised by improved equipment and fishing aids, not a few of which were originally developed for naval use during the war. Previously while steam and motor power could haul fishing gear, navigation and position finding out of sight of land had depended on the compass and dead reckoning; and there was no ship-to-shore communication. The most important other advance had been that trawlers in the inter-war period had begun to install radio receivers, which had the substantial advantage of keeping up with weather forecasts.

During the past half-century fishing boat construction has become a considerably more sophisticated operation, with the range of hydraulic, electrical and electronic equipment that has become normal. An early innovation after World War II was the installation of the echosounder, which had developed for naval purposes during the war, and as well as showing the depth of the sea bed and eliminating the use of the lead-line, this had the great advantage for pelagic fisheries of showing shoals swimming underneath the boat. The echosounder was later refined for the white fisheries to show fish near the sea bed; and it was also to be supplemented by the sonar, which could effectively sweep the sea for a distance around the boat.

Improvements in radio technology soon meant that nearly all boats had radio receivers and many also transmitters, which were obviously of great consequence in the event of emergencies. The radio direction finder was a useful aid installed during the early post-war period, but this was soon to be displaced by more sophisticated navigation aids (particularly in the seas around Scotland the Decca system), which could pin-point position at sea by the intersection of radio beams. This has in turn been supplemented by satellite navigation systems.

While there are necessary limits to the improvement of conditions of life at sea, there have been substantial additions to comfort and amenity in the later phases. There is often now more space in cabins, galleys are routinely equipped with such equipment as refrigerators, and such amenities as shower baths fitted carpets and washing machines are not unknown.

REFERENCES

Annual Report of the Fishery Board (ARFBS.) 1872.

Hawkins, L. W. (1984) *Early Motor Fishing Boats*. Published privately by L. W. Hawkins, Sprowston, Norwich.

Henderson, T. (1978) 'Shetland Boats and their Origins' in Baldwin, J. (ed.) *Scandinavian Shetland. An Ongoing Tradition?*, Econoprint, Edinburgh, 49–55.

Lillehammer, A. (1986) 'The Scottish-Norwegian Timber Trade in the Stavanger Area Sixteenth and Seventeenth Centuries', in Smout, T. C. (ed.) *Scotland and Europe 1200–1850*, John Donald, Edinburgh, 97–109.

March, E. J. (1970) *Inshore Craft of Great Britain*, vols.I and II, David and Charles, Newton Abbott.

March, E. J. (1952) *Sailing Drifters*, Percival Marshall and Co., London.

Martin, A. (1981) *The Ring-Net Fishermen*, John Donald, Edinburgh.

Morrison, I. (1978) 'Aspects of Viking Small Craft in the Light of Shetland Practice', in Baldwin, J.R. (ed.) *Scandinavian Shetland. An Ongoing tradition?*, Econoprint, Edinburgh, 57–75..

Scottish Record Office (SRO.) AF34 (Records of Peterhead Fishery District).

Wilson, G. (1965) *Scottish Fishing Craft*, Fishing News Books, London.

16

Fishing Piers and Harbours

Piers and harbours built to serve fishing boats are very much a thing of the industrial era. Previously boats were generally small enough to be pulled up on a beach after use, and in any case with the cost of pier and harbour works there could be little question of providing them from revenue that could be raised from use by fishing boats. However there were various estates and burghs for which the needs of fishing did contribute significant stimulus for harbour construction from at least the 17th century (Graham 1987: 361). At ports generally the expanding economy of the 18th century prompted activity in the construction of quays: despite this, however, even on the East Coast with its long history of trade in the North Sea theatre, there had been surprisingly little improvement of ports as late as the end of the 18th century (Lenman 1975: 54). While there were some bigger boats used for fishing in pre-industrial times, any harbourage available was primarily for trading vessels, and fishing boats might be fortunate enough to share in its use.

There had been some harbour construction in Scotland from the Medieval period, and harbour works from the pre-industrial era are still to be seen, mainly at places on the Firth of Forth such as Crail and Dunbar. These appear from the beginning to have catered for fishing as well as for trade, and there are a fair number of harbours built in succeeding centuries elsewhere to which this also applies: these include, for example Arbroath, where the harbour was originally built by the abbey before the Reformation (Graham 1976–77: 335); Stonehaven where the construction was due to the Earl Marischal at the end of the 16th century (Graham 1976–77: 339); and Portmahomack in Easter Ross, where the harbour was built by the Earl of Cromarty before 1690 (Graham and Gordon 1987: 284). Coastal works like harbours require substantial structures, and even these may suffer serious storm damage; marine

251

engineering had to become fairly advanced before much construction was technically possible. In addition such works are very generally costly, and were difficult to finance, especially in earlier times.

In Scotland trade has always been the leading activity at the major harbours, apart from a very small number of naval bases. However fishing has operated from more places on the coast than has trade, and with the great improvements which have taken place over the modern period in overland transport, shipboard trade has become concentrated at a very few main ports. Although parallel forces of concentration have to an extent operated in fishing, fish landings have always shown considerable dispersal, and fishing has often continued as the main activity of the smaller ports where harbours were originally built mainly to serve trade. Exceptionally fishing has continued as a major activity at ports of the status of Aberdeen and Lerwick. The development of harbours in general accelerated through the 19th century, especially after 1860, and the Fishery Board was active in building and improving fishing piers and harbours from 1828; and indeed the Board was to play the leading role in developing these for fishing purposes. Many burghs, local councils and other authorities also took initiatives in getting their own Parliamentary piers and harbours bills: well over a hundred of these were passed between 1860 and World War I, and over half of them were for places where the main interest was fishing. This in itself is significant: the improvement of overland transport, particularly the spread of the railway network, from about 1840 onwards ate into coasting seaborne trade. One result was that for the harbours of a number of places like Eyemouth, Peterhead, Fraserburgh, and Lossiemouth, for which the main early function was that of trade, the expanding fishing industry became their main business. The general tendency was for these harbours to centralise operation at the local level, as they became used by fishermen from neighbouring villages, as well as from the towns themselves. For most such harbours the Fishery Board was also at some stage to make contributions to the work of harbour improvement, especially after Word War I when it functioned as something of a co-ordinating body although resources were provided mainly by other public agencies.

While the building of piers and harbours was not defined as a main objective of the Fishery Board, it recognised that they had a vital part to play in improving the fisheries, and it played a leading role in their construction at over a hundred places (fig. 16.1). The harbour programme contributed significantly to the Board's achievements of promoting the Scottish fisheries to the point that they became world leaders.

While commercial demands and the problems of finance were main issues in the building of fishery and other harbours, there is ample evidence that there were other matters of great moment – those of minimising dangers of shipwreck and loss of life. Dangers have in general always been greater for seamen than for landsmen, even when landsmen have been involved in high-risk occupations like mining. Among seafarers risks have always been particularly high among fishermen, both because of the nature of their work, and because much of their activity has always been conducted in relatively small boats. There are many records which illustrate the hazards of life for fishermen, and these can be encapsulated in instances from a late 19th century review of the dangers to Scottish fishermen: in one big storm in 1848, 100 fishermen were lost on the East Coast; in a seven-year period in the 1870s the one community of 300 at Cellardyke lost 30 men; and in the worst ever fishing disaster in the country, 191 men were lost in one day in October, 1881 in South-East Scotland, 129 of them from the one village of Eyemouth (Young 1883: 69–70).

THE CONTEXT OF THE FISHERY BOARD'S WORK IN HARBOUR DEVELOPMENT

The character and requirements of Scottish fishing changed greatly during the period of the Fishery Board's existence. At the start of it, fishing in the great majority of places was from open boats which operated from coastal villages without harbours and which were small enough to be pulled up on the beach when not in use. However a long-term trend was for the boats employed to get bigger and more elaborate, and for the needs for harbourage to expand.

The need for harbourage was accentuated by the fact that the main fishing effort in Scotland was based on the East Coast, the alignment of which is characterised by long unbroken open stretches, with virtually no good natural harbour other than the Cromarty Firth. This of course contrasts with the West Coast, which is much more indented with its sea lochs.

During the period the fisheries became progressively more dominated by the herring fishery, with its emphatic peak in the summer

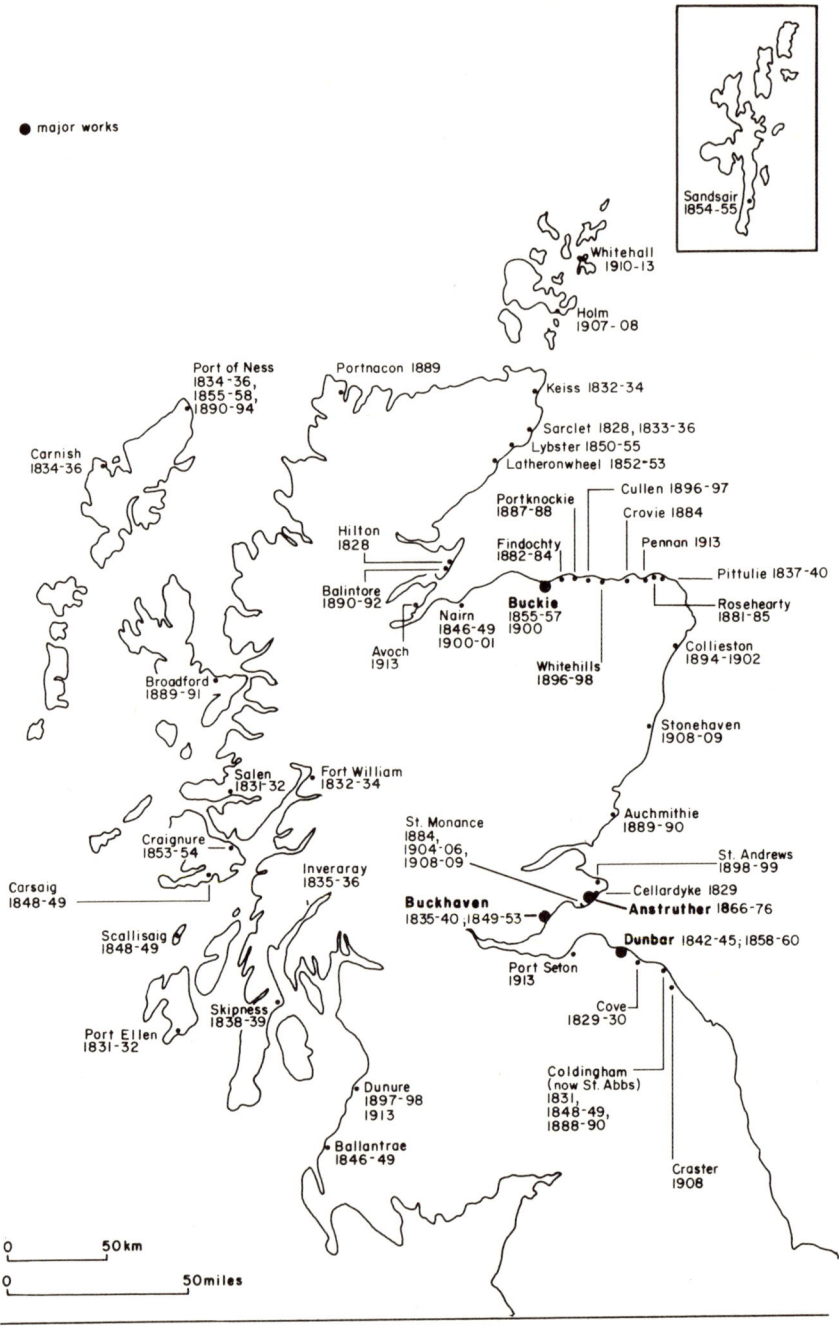

major works

Sandsgir 1854-55

Whitehall 1910-13

Holm 1907-08

Port of Ness 1834-36, 1855-58, 1890-94

Portnacon 1889

Keiss 1832-34

Sarclet 1828, 1833-36
Lybster 1850-55
Latheronwheel 1852-53

Carnish 1834-36

Cullen 1896-97
Portknockie 1887-88
Crovie 1884
Pennan 1913

Hilton 1828

Findochty 1882-84

Pittulie 1837-40

Balintore 1890-92

Buckie 1855-57 1900

Rosehearty 1881-85

Nairn 1846-49 1900-01

Collieston 1894-1902

Avoch 1913

Whitehills 1896-98

Broadford 1889-91

Stonehaven 1908-09

Salen 1831-32

Fort William 1832-34

Auchmithie 1889-90

Craignure 1853-54

Inveraray 1835-36

St. Monance 1884, 1904-06, 1908-09

St. Andrews 1898-99

Carsaig 1848-49

Cellardyke 1829
Anstruther 1866-76

Buckhaven 1835-40 ;1849-53

Scallisaig 1848-49

Dunbar 1842-45; 1858-60

Port Seton 1913

Skipness 1838-39

Cove 1829-30

Port Ellen 1831-32

Coldingham (now St. Abbs) 1831, 1848-49, 1888-90

Dunure 1897-98 1913

Ballantrae 1846-49

Craster 1908

0   50 km

0   50 miles

Fig. 16.1 Pier and harbour works undertaken by
the Fishery Board, 1828–1913,

season. At an early stage there was a definite tendency for this to concentrate much of its activity at relatively few ports, like Wick, Fraserburgh and Peterhead. Hence while there was the need to improve the harbours of such ports, the forces of centralisation were not uncomplicated, as congestion led to effective overspill from an expanding fishery to a variety of other ports. In addition there was also a demand for better facilities at the villages where many of the fishermen continued to live, and where their boats needed accommodation outside the summer herring season.

While there were considerable achievements in developing piers and harbours for fishing boats, there were periodic reverses, sometimes of a serious or even devastating kind. These were mainly caused by storm damage to structures, and this could retard plans for extending improvements by the Board having to commit funds from its limited budget to repairs. The early effort to build a harbour at Sarclet in Caithness, the 'grand object' of the major harbour of Dunbar in the 1840s (ARFB. 1840: 3), and Port of Ness in the Isle of Lewis at the end of the 19th century, were all cases in point, where it was judged essential to lay out considerable funds for repairs after the original construction in order to maintain the harbours in working condition.

The involvement of the Board in piers and harbours began fifteen years after its founding, when in 1824 it took the decision to divert the sum of £3,000 which it had previously committed annually to encouraging the 'open sea' herring fishery to other uses. The open sea fishery involved operation with relatively big boats which cured their catches aboard, in the manner that had long been the basis of the successful Dutch herring fishery; and this reappraisal of priorities was linked to the realisation that in the Scottish situation, where the herring were caught within a few miles of the coast, greater success could be achieved by the simpler and easier expedient of on-shore curing. With this mode of operation the herring could be caught by smaller and less expensive craft; and for curing the inevitable constraints of limited working and storage space which applied on vessels were effectively absent.

Even for the boats of 25 to 30 feet used in the inshore herring fishery there were big advantages from being able to berth at a pier, rather than have to land on an open beach. In addition from 1820 the Board had been given the function of developing the cod fishery as well as its main responsibility of promoting that for herring; and piers could be used for this fishery outside the herring season.

In addition to simplifying landing, piers and harbours also provided an important measure of shelter, which meant that the arduous task of pulling boats up out of danger had to be performed less often.

The original intention of the Board in 1824 for the use of the annual sum of £3,000 was to commit it to both the building of piers and the help of poor fishermen for the repairing of damaged boats. After some delay, caused by the requirement to get approval from the Treasury and the Board of Trade for the regulations proposed by the Board for both purposes, the first advances to poor fishermen were made in 1827 (ARFB. 1827: 23), and the first help for piers in 1828 (ARFB. 1828: 2). From the start the intention was to commit most of the available money to piers, and from 1832 a pattern crystalised by which £2,500 was committed to these, while £500 was devoted to repairing the boats of poor fishermen (ARFB. 1832: 19, 38–41). Considerable problems were in fact encountered in assessing the claims from the big numbers of fishermen requesting aid for boat repairs, and in administering the proper use of the many small sums allocated, few of which were as high as £5. From 1839 onwards the Board was advocating instead the encouraging of friendly societies among local groups of fishermen as an insurance against boat damage, while it wished to use the whole of the £3,000 to help meet the proliferating requests being made to help the building of piers. From 1850 onwards it was eventually permitted to do this (ARFB. 1849: 10).

THE GREAT PROBLEM OF FINANCE

For the great part of the time the Board had very restricted resources at its disposal: in the main it was limited to the earmarked annual grant of £3,000 for the construction, improvement and maintenance of fishing harbours. None the less it became recognised that the aid it provided in building piers and harbours had a great positive effect in the development of the fisheries at many places. The general policy of government until the 20th century was that harbour development should in principle be self-sustaining, and that the only schemes seen as warranting special help from public funds were harbours of refuge and naval harbours. In effect, the great developments in harbours for shipboard commerce were regularly funded directly or indirectly from the profits from trade.

Until the 20th century, the capital for fishing harbours came mainly from coastal burghs or their harbour authorities; some came from coastal estates and other bodies; in rather exceptional cases groups of fishermen combined to raise necessary finance; and of the sources of government finance the main one was the Fishery Board, although it could be supplemented by other sources for special projects. In the special case of Wick, which was the leading fishing port throughout the first half of the 19th century, the British Fisheries Society were responsible for the earlier phases of harbour development; and this also applied for the smaller developments at Tobermory and Ullapool. There was also the national decision to construct from the 1880s at Peterhead a harbour of refuge which was to enclose 200 acres in the bay. This was actually part of a wider strategy for the building of safe harbours, and it in fact contributed minimally to the development of fishing at the port.

Although the Board did occasionally succeed in securing extra public funding for special projects, only in the period after 1909 did a change in government policy give any substantial enhancement of funding. The practice of the Board was to insist that local interests should contribute to costs of construction; it generally paid the major share of the cost, up to a maximum of three-quarters of the total. The remainder usually came from local landlords or burghs, or sometimes from groups of fishermen. In the first 40 years of the programme, a total of £75,000 came from private funds (ARFB. 1869: 2), which was equivalent to over 60% of the Board's own outlay.

While it was exceptional for fishermen to have the requisite resources to contribute much towards harbour projects, there were a number of cases in which go-ahead fishermen were the instigators of developments. Cellardyke in the early 19th century was one of biggest and most prosperous fishing communities, and in 1829 the fishermen agreed to subscribe £500 towards the cost of a harbour improvement costed at £1,720 (ARFB. 1829: 21). Coldingham fishermen took the initiative in applying for a pier and offering one quarter of the cost in 1831 (ARFB. 1831: 2); and this had no sooner been built and paid than they were collecting money for harbour deepening and other improvements (ARFB. 1833: 2). However with the Board's limited finances and long list of priorities, these were not undertaken until 1848 when a more substantial improvement was undertaken by deepening the harbour, building an additional wharf, and providing boom gates at the entrance to keep out severe storms (ARFB. 1848: 15). At Fort William in 1832 the fishermen combined with curers to raise one third of the cost of a new pier (ARFBS. 1832:19); and when a major improvement at Buckhaven harbour was started in 1850 the Board recognised that the £3,000 collected by the fishermen towards the cost was 'infinitely larger' than any previous sum contributed by fishermen to any harbour works (ARFB. 1850:7). Undoubtedly, however, the main achievement in fishermen providing their own harbour was at St. Monance in Fife, where they spent £15,000 in developing it with no help from any source, and did not apply to the Board for assistance until 1883 when bigger works had become necessary (ARFBS. 1883:xxxiv).

Far more frequently the local contribution came not from fishermen but from the landlord. The perceptions of different landlords differed widely in this. There were cases, such as at Gourdon in Kincardineshire and Dunure on the Clyde where landlords unaided laid out big expenditures on harbours for fishermen: in the latter case the cost ran to £12,000 (SRO. AF 5/1:282,55). However for other landlords the building of piers helped by the Fishery Board was a subsidised way of improving their properties, despite the ban imposed by the Board on charging dues on fishing boats. A number of coastal burghs also were able to secure the aid of the Board in improving their harbours.

Thus in the early years Sir James Miles Riddell of Ardnamurchan and Sunart paid one third of the cost of a pier at Kentra in North Argyll, and Walter Campbell of Islay a similar proportion of one at Port Ellen (ARFB. 1831:19). At Keiss in Caithness in the following year, although the estate was at the time in trusteeship, the trustees were sufficiently convinced of the potential benefits of a pier that they agreed to contribute the usual one third of the costs (ARFB. 1832:19). At Scallisaig on Colonsay and at Carsaig on Mull, the Board undertook to rebuild and extend existing piers built by local proprietors which had become seriously deficient (ARFB. 1848: vii; 1849:30; 1850:6). In both of these cases the proprietor paid half of the cost of the new works.

Burgh harbours tended to be bigger and more costly, and with its strictly limited resources the Board helped in the development of few of them till finance became available on a more generous scale in the 20th century. For what was to be the first phase of the major project at Dunbar, built in the early 1840s, the Board recognised at the start that it was likely to require almost its whole annual grant for a three year period (ARFB. 1840:3); and it was quite typical that by the time it was complete, the cost had risen and the Board's contribution of £11,262 was very close to four times its whole annual grant. The town paid the remaining £4,500 (35%) of the total cost of £15,762 (ARFB.1844:7).

There were also town harbours built in the later 1840s at Nairn and Ballantrae. At the royal burgh of Nairn, the town paid half of the total cost of £3,925 (ARFB. 1848:15), but Ballantrae was a burgh of barony and the proprietor of the burgh paid the town's contribution, which amounted to almost half of the total of £5,648 (ARFB. 1849:31). At a later date the 'enlightened liberality of Lord Saltoun' was recognised as crucial in the advance of Fraserburgh to the status of a major fishing port (Young 1883:81–82).

The great problem of raising necessary finance was shown nowhere better than at Buckie, which by the middle of the 19th century had the greatest number of fishing boats of any locality in Scotland. In 1855 it had a total of 849 boats and 3,111 fishermen (ARFB.1855:6). Here the Board undertook a major project in the later 1850s, but it had to recognise from the start that it was unable on this open exposed coast to construct an all-tidal harbour. A total of £14,000 was committed to the construction of a basin extending to 4 acres; and of this the Board paid £10,000, and the proprietor, Mr. Gordon of Lettfourie contributed £4,000 (ARFB.1848:6,48). However as boats continued to increase in size, the need for an all-tide harbour became more pressing, and the Board got its engineers to survey possible sites in the Buckie vicinity in 1864, and they concluded that even the cheapest site would take between £22,639 and £34,639 to develop (ARFB. 1864:6). This was beyond its resources, and the position was also complicated by the fact that by this time at Buckie the boats did most of their fishing elsewhere, and the main need for the harbour was for accommodation when the boats were at home between seasons. The Buckie problem was eventually resolved by the greatest ever harbour project undertaken by a landlord, when at the end of the 1870s Mr. Gordon of Cluny built the harbour that still bears his name: at the cost of £50,000 on a new site he built a harbour with two basins of total extent 8 acres which could accommodate 400 boats (Buckland *et. al.* 1878:49).

By dint of vigorous lobbying the Board did get extra resources for a small number of further special projects by getting the main burden of their cost transferred to other government agencies. For the repairs and improvements needed at Dunbar in the late 1850s, after more than a year of negotiations, the Public Loan Commissioners agreed to advance £20,000, while the Board committed itself to finding £10,000 (ARFB. 1858:32); and the big half of the costs of the Anstruther Union Harbour built between 1866 and 1876 were met by an advance from the Public Loan Commissioners and a special Parliamentary grant (see below).

The general increase in the size of fishing boats which took place with the vigorous growth of the herring fishery after 1860 inevitably compounded the problem, as they needed not only harbours, but harbours which extended into deeper water. In 1874 the major authority Thomas Stevenson, in making the case for greater government involvement, claimed that smaller piece-meal efforts at harbour construction in effect slowed economic development as well as leading to greater long-term costs (Stevenson 1874: 270–272). By the late 19th century, both needs for better harbours and pressure to secure that end were mounting, and eventually it became recognised in the 20th century that fishing harbours warranted public investment on a new scale.

In addition to the basic problem of raising finance for capital works, there was also the problem of maintenance. The material gathered from the fishery officers of the day in the Washington Report of 1849 shows that on the East Coast where fishing boats paid harbour dues at all, the general level was low (Washington 1849:56, appendix 19). The majority of places recorded charged no dues, although this itself was a comment on the facilities available at these places which were non-existent or rudimentary. However, even in the middle of the 19th century the position could be complicated by landlords providing some facilities, but no separate harbour dues were itemised as these might be effectively included in wider arrangements. In the Portgordon (later Buckie) district each fisherman simply paid to the proprietor an annual sum that covered both ground rent for his house together with boat accommodation. At East Clyth, Clyth, and Occumster in Caithness fishermen paid no dues, but in leasing their holdings they bound themselves to fish for the curers leasing the curing stations at a rate for herring of 1/- per cran below the average price at Wick, which must have brought down their gross earnings around 10%.

The bigger harbours generally did charge dues, but at considerably variable rates, and in most cases the revenue generated could only have been modest: it was certainly very rare for revenue to be generated at a rate that could be the basis for significant capital works. There could also be preferential rates for local boats, and in the case of Wick for feuars of the burgh. The highest rates charged were nearly all on the Forth: from the point of view of fishing, however these dues applied in the main at the less important harbours. This must reflect the more advanced level of development of the region, including its harbours. The top rate quoted was 15/- per season at Burntisland and North Berwick, but this seems to have reflected in part the lack of safe harbourage on the inner part of the Forth rather than heavy demand for the facilities for fishing purposes. At Dunbar was the next highest rate at 11/6 per season: like the other two rates already mentioned, this appears to have been calculated to maximise profits when boats congregated for the herring season to help recoup the town's major outlay on harbour development, undertaken with the help of the Fishery Board. Evidently too Dunbar got an exemption from the usual requirement of the Board that no dues be charged for fishing boats in the schemes it assisted, and this could have been justified on the basis of the dues levied for the old harbour before the new one was begun. At Fisherrow, where the main function was to cater for the boats of the village, the rate was 7/6 annually. Contrasted was the situation at Eyemouth, where the level of 10/- per year was a historical carry-over of a teind payment to the minister. However at Anstruther, Pittenweem and St. Monance in Fife the only boats recorded at paying dues were strangers, and here the modest rates of 1/2 per year suggests that such visitors were wanted to increase the participation in the winter herring fishery.

The top levels of harbour dues levied on the Forth were only occasionally met elsewhere. Portsoy, Lossiemouth and Findhorn were three of the better harbours on the south shore of the Moray Firth: these had been built mainly to cater for trading ships, but were on the part of the coast which vied with Fife for leadership in fishing by this time. At Portsoy fishing boat dues were 10/- per year: at Lossiemouth 10/- per year for white fish boats and 7/6 for herring boats; and at Findhorn the rate of 10/- per year included the right to draw up a boat during the winter.

A rate of 5/- per season was recorded at several places: this was the rate levied at Peterhead, Rosehearty, Banff, Whitehills, and Stonehaven; and it was also quoted as the rate in the Cromarty district. The Peterhead rate may well have been calculated to draw custom away from Wick, where any who were not feuars of the burgh paid 8/- per season; and in any case with then numbers of boats coming to Peterhead at this time its yearly income from fishing boat dues would have been at the relatively high level of over £100. Fraserburgh, the rival of Peterhead, charged dues on a different basis, it being stipulated that a boat should pay the price of one cran of herring caught at the start of the season. This appears to have been calculated as a method of creaming off a share of the early season profits before the date at which boats entered on engagements, and when prices were generally high but catches small; it was stated that this amounted to about £1 per boat. While the port stood to make over £400 from this, it can only have been a factor diverting boats to fish elsewhere. The only other port recorded as operating on a comparable basis was Gourdon in Kincardineshire, where on the basis of dues of the price of half a cran of the first herring caught in the season, and this was stated to have a value of *c*.15/- per boat. This appears to have been an arrangement by the local landowner to recoup his investment on the harbour, which is unlikely to have been used by any boats beyond the locally recorded fleet of 26.

In the leading fishing area of Caithness harbourage was at a premium: only Wick itself (Pulteneytown), Lybster and Sarclet had effective harbours. At the main port of Wick, feuars paid 3/- and non-feuars 8/- for the season, and both paid 2/6 for laying up for the winter. Discrimination against boats which did not contract for the season was greater, as they had to pay 1/-every time they entered the harbour, and 5/- if they chose to lay up for the winter. The other data supplied in the report about numbers of boats suggest that the total harbour dues from fishing boats at Wick would have been between £150 and £200 annually. At Lybster the rate was 8/- annually, which would have raised about £40: but Sarclet had been built from the start by the Fishery Board, and charged no dues. Over the border in the county of Sutherland, Helmsdale, which at this point had a significant share in the herring fishery, charged 3/- per year.

Where there was a separate rate for visiting boats this was often charged on a daily or short term basis, and could be much heavier. This practice was almost confined to the Forth, and appears to have been geared to making revenue of boats making occasional calls, and this would have included shipping consignments of cured fish from other ports on the East Coast. Thus at North and South Queensferry the daily rate was 4d; at St. Andrews stranger boats paid 1.75d every time they entered the harbour, and at Crail the rate was 3d. Occasional visits to St. Monance and Burntisland were charged at 2d per time, which meant that if a visiting boat came for the winter season that leaving the harbour six times incurred the same expense as a year's fishing. The only other district in which stranger boats were charged higher dues is that of Cromarty where the rate was from 3d to 1/- per visit, according to the size of the boat, and this was a matter of much complaint.

THE BEGINNINGS OF THE FISHERY BOARD'S INVOLVEMENT IN PIERS AND HARBOURS

The policy which the Board followed until after it was reconstituted in 1882 was to assist in the construction of piers and harbours on condition that no pier dues were to be levied on fishing boats using them, although it did allow the charging of dues on other vessels. While this was of course appreciated by the fishermen, by the later part of the century it was clear that this carried the substantial disadvantage of creating situations where there was no specific provision for maintenance. The Board was approached with increasing frequency with requests for help with repairs, which threatened to be a serious drain on its limited funds, and in its later projects it made a practice of making aid conditional on the setting up of appropriate trust bodies to maintain the works with the help of pier dues.

When the Board embarked on its programme of aiding in the construction of fishing piers in the 1820s, its policy was at first eclectic, and aimed at providing facilities at localities where needs were particularly pressing. In general the bulk of its activity was on the East Coast, where fishing settlements and boats were more numerous, and where natural harbours were very few. However from the start it did help construction at a number of places on the West Coast and in the Hebrides, and at a later stage also at a number of places in the North Isles.

The regular method was to put works out to tender and to have the construction supervised by the Board's own consulting engineer. In 1828 Joseph Mitchell, Engineer to the Parliamentary Commission for Highland Roads and Bridges, took on the extra work of pier engineer to the Board at a salary of £100 per year (ARFB. 1828: 1). Mitchell, in addition to his main work as a road engineer did have some previous experience of harbour works in places like Findhorn and Lossiemouth (Mitchell 1883, I: 142–144); he realised that the scale of finance available could improve conditions for the fishermen considerably, but could not provide really safe harbours (Mitchell 1883, I: 292). Mitchell continued as the board's engineer until 1851, when with increasing work and problems the leading marine engineering firm of David and Thomas Stevenson of Edinburgh were appointed as consulting engineers (ARFB. 1850: 6).

The main initial function of the Board was to help provide piers at places where none existed, like Cove and Burnmouth in the south–east, Hilton in Easter Ross, Salen in North Argyll and Carnish in the Isle of Lewis. However one of its first projects was also an improvement to the existing harbour of Cellardyke, and the building of a harbour at Sarclet in Caithness. The latter was to go through several rather exasperating phases as they sought to improve it and to repair storm damage. On the West Coast at several places piers built or improved with the help of the Board, as well as helping local fishermen, also proved useful for the developing freight and steamer trade: this occurred at Fort William, Inveraray, Scallisaig on Colonsay, and Skipness on the Clyde.

It was fairly frequent for the works to take longer than anticipated, as well as for costs to rise. In some of the projects in remoter locations, the difficulty in getting enough skilled labour could be an additional cause of delay. In the case of the early projects of Callicot (Port of Ness) and Carnish in Lewis, the contracts for which were awarded in 1834 to a contractor from Wick, it was stated in the following year that they had got behind schedule through shortage of skilled men for the work of construction (ARFB. 1835: xx). Similar problems were reported again 16 years later at Carsaig (Mull) and Scallisaig (Colonsay) (ARFB. 1851: 8). The work at Ballantrae Harbour was started in 1844 (ARFB. 1844: 7), but after the contractor ran into difficulties in 1847 it was being continued under the Board's own superintendent (ARFB. 1847: 50); and the following year the contract was relet to another firm (ARFB. 1848: 15).

After the severe damage to the first works at Sarclet, the Board agreed to their rebuilding and improvement when strongly pressed by the curers and merchants in the leading port of Wick, who wanted more harbour facilities on the Caithness coast to take pressure off their own overcrowded port. However in this case it was stipulated that the contractor would be bound after completion to uphold the works for three years at his own expense, and half the payment was to be held back until then (ARFB. 1833: 3). None the less even these improved works were to succumb to the force of storm waves, and the Board later concentrated its efforts in Caithness to improving the harbour of Lybster.

One of the Board's greatest embarrassments occurred with the improvement of the major harbour of Buckhaven at mid–century. Here the contractor ran into severe problems in 1851, through the inadequate work of the engineer who had prepared the initial survey, and who had reported rock at a point where the contractor found quicksand in the foundations. The contractor threatened to go to law, and the Board settled the matter by agreeing to pay him almost £2,000 above the contract price to complete the work. After this the Board insisted that preparatory surveys should all be prepared by its own engineer (ARFB. 1851: 7–8).

THE NEED FOR BIGGER HARBOURS

The great majority of the piers until the late 19th century were in fact inaccessible at low tide, and some indeed barely reached to the low tide mark. While this meant that they were relatively cheap and simple to build, it inevitably meant that their use was restricted. The herring fishery gained in dominance and in general commercial success during the 19th century, and as well as the expansion in numbers of the fleet, there was a tendency for boats to get bigger: and this accelerated notably from around 1870. Such bigger craft were only drawn up out of water at the end of the season, and the need for bigger harbours extending into deeper water became more pressing. The ideal was obviously to have secure all-tide harbours which boats could enter and leave at all states of the tide, and have safe berthage afloat while they were in the harbour. Improvements in harbourage were needed especially at major fishing ports like Wick, Peterhead and Fraserburgh with the development of serious congestion in the herring season, when the fleets from most of the coast concentrated in them. There was also an increasing demand for harbours, or better harbours, at the many villages where the fishermen who owned herring boats lived, even if the boats were inactive for most of the year while the men prosecuted the line fisheries for white fish with smaller craft.

The biggest early project was that at Buckhaven, which was then one of the most important home bases for men engaged in the herring fishery, and was unusual in having (in addition to the summer fishery) a winter fishery in the Firth of Forth. This was begun in 1835 with the decision that finance for it be allocated from the annual budgets of two years (ARFS. 1835:2); it became obvious, however that it could be substantially improved by the building of an additional protective breakwater, for which there were not available funds at the time that construction began. This was approved in 1838 (ARFB. 1838:2), and the harbour was finally completed in 1840. A total of over £6,000 was spent in building this harbour, of which the Board contributed three quarters.

With the proved success of the Buckhaven harbour, the general policy of the Board became that of concentrating its resources on relatively few bigger harbour projects, and this was an effective recognition of the success and needs of the expansion and needs of the dominant herring fishery. It was thereafter generally at intervals between major projects that the Board reacted to the continuing demand for aid for smaller works.

Building of small piers and harbours was suspended in 1840 for 'the one grand object' of the construction of a considerably bigger harbour at Dunbar on the south shore of the Forth, which was also important for both summer and winter fishing. This was to be a harbour of safety and refuge for all the boats on the Forth taking part in the fishing (ARFB. 1840:3). It involved the building of a wharf and sea wall nearly 1,000 feet long, and the excavation of over 43,000 cubic yards of material, to enclose a basin 4 acres in extent. It cost £15,762, of which £4,500 was paid by the burgh of Dunbar, and the remainder by the Board (ARFB. 1844:7). A harbour of this size was at this time one of high capacity in fishing boat terms, it being reckoned that 85 to 115 boats could be accommodated per acre (Stevenson 1874:153).

The building of a proper harbour for the Buckie area of Banffshire, as already stated was by the middle of the 19th century a major project, and when this was completed in 1858 emphasis reverted to Dunbar, which had suffered serious storm damage, and which also now required extension and improvement. This came to dominate the Board's harbour programme until 1865, after which attention was focussed on the other leading fishing area of the Fife coast with the decision to undertake the major development of the Anstruther Union Harbour. This was to provide jointly for the boats of Cellardyke and Anstruther, and also to provide for boats from other ports in Fife, and indeed from elsewhere in Scotland at the time of the important winter herring fishery. The Union harbour was begun in 1866, and commanded the Board's almost undivided attention for a whole decade, partly because of serious set-backs through storm damage in the course of construction. Towards the costs of this project an advance of £16,500 was obtained from the Public Works Loan Commissioners in 1866, and this was supplemented by a special Parliamentary grant of £9,000 in 1873, in addition to the Board committing its own annual grant of £3,000 over the period. The eventual total cost was in excess of £55,500.

By the 1870s the gap between the resources of the Board and the needs of the fleet had widened to the point of being a major issue and a running sore. The Buckland, Walpole and Young Report on the Herring Fisheries of Scotland of 1878 reviewed the situation at an important juncture, and it recognised that the shortage of adequate harbours had become a main problem in the further development of the fishery. This was also echoed by one of the three authors Sir Archibald Young, in another major survey of 1883 (Young 1883:65–92). The Buckland, Walpole and Young Report recognised that the one harbour with a major involvement in the fisheries which was adequate was that of Aberdeen (Buckland *et.al.* 1878:xxxix), and this itself was hardly due to fishing but rather to trade. The report amassed an impressive collection of evidence form all around the coast of the inadequacy of the harbours and the problems that ensued: not only had the fleets expanded in numbers but many boats were now between 40 and 50 feet keel, and drew from 5 to 7 feet of water (Buckland *et.al.* 1878:xxxvii). Included in the evidence submitted to the Buckland committee was the suggestion made by William Boyd (the agent for the trustees of the Peterhead harbour) that the surplus income which the Fishery Board received from branding fees should be added to the funds allocated for harbour improvement (Buckland *et.al.* 1878:27). This became accepted in part, and from 1882 a variable sum was added to the £3,000 statutory grant allocated to the newly reconstituted Board for piers and harbours: however this sum over the next decade averaged little more than half the annual statutory £3,000, and its effects were marginal. The situation was also complicated by that of a rival priority: by this time several of the remoter but important locations from which the fishery was prosecuted were conspicuous in the national context for being underprovided with telegraphic connection, and the Board also used surplus income from branding fees to subsidise telegraphic extension.

Developments in fishing accelerated towards the end of the century, and the Board was reconstituted in 1882. The first annual report of the new Board eloquently echoed the findings of the Buckland Committee on the urgent needs for harbour improvement (ARFBS. 1882: xv–xviii). However its expressed intention of making a detailed survey of these requirements was stillborn when a Select Committee of the House of Commons was appointed to make a survey of the national needs for harbour accommodation (1883: xxxiii). It was recognised that harbourage at Aberdeen, Peterhead, Fraserburgh and Buckie were adequate for contemporary needs, although before recent improvements at Peterhead it had been estimated that the limitations of the harbour had effectively resulted in a loss of £60,000 in a single season (ARFB. 1882: xv). In addition the major effort by the British Fisheries Society to enlarge the Wick harbour with a substantial breakwater had earlier come to grief in a series of severe storms between 1870 and 1872, and this actually culminated with the bodily movement of a great part of the breakwater of the remarkable mass of 1350 tons (Stevenson 1874: 46–47); and this is still one of the most outstanding events ever recorded in the history of marine erosion. At Wick over a period of 13 years £140,000 had been spent in the expectation of substantial increases in revenue. In 1878 not only was the structure in ruins, but debris was scattered over the anchoring ground, and there was a 'dead weight' of debt over the port of Wick (Buckland *et. al.* 1878: 65) which continued to dog its development, and was partly responsible for the Aberdeenshire ports of Peterhead and Fraserburgh emerging as the leading fishing centres in place of Wick.

In making the case for greater help from public funds for harbour improvement, the board noted that the main herring fishery had reached a pitch of importance such that the average value of landings in the now leading fishing area of Aberdeenshire approximated to the total annual rental value of the county (ARFBS. 1882: xv). In its second report it stated that pressing improvements were needed at 23 harbours, but that its limited budget had compelled it to select only four of these for assistance (ARFBS. 1883: xxviii). By the early 1890s the new Board were pressing with renewed vigour for improved funding, linked to long-term planning to meet the needs of the fishing fleets. In 1891 it made the case in its annual report that the Government should 'abandon the present hand-to-mouth system, and fix definitely on a plan of harbour extension all around the coast, to be carried out systematically over a series of years' (ARFBS. 1891: xviii); and in the following year it commented that several Royal Commissions and Select Committees had recommended more funds for harbours, but to no avail (ARFBS. 1892: lii). A new act passed in 1894 gave limited additional powers to the Board by allowing it to underwrite loans contracted for by local harbour authorities from its grant allocation. From the 1890s too it became policy to insist that aid from the Board be conditional on arrangements being made for the imposition of dues on boats using harbours, and on fish landed and exported; and there can be little doubt that this was justifiable, as the turnover of the industry had much increased, and there was an obvious need to provide for maintenance of harbours. It was also of some help that pier and harbour works in the Highlands and Islands could get aid under the Western Highlands and Islands Work Act of 1891, and from 1897 under the Congested Districts Act of that year. From 1894 till after World War II the Fishery Board withheld its aid from the Highlands and Islands, apart from help at a very few major harbours like Stornoway and Lerwick.

The need for better fishing harbours reached a new pitch in the years after 1900, as the steam drifter rapidly replaced the sailboat as the main catching vessel in the herring fishery. The Scottish herring fishery had now attained the unquestioned status of the world's leading fishery, and in addition to requiring all-tide harbours of adequate depth, the steam boats needed a variety of new facilities, including bunkering and engine-room stores. Provision was at length made for increased funding under the Development and Road Improvement Act of 1909. Under this legislation what became known as the Development Fund was instituted; this did take over three years to become effective, but in 1913 and 1914 a series of 18 improvement schemes were approved at a total estimated cost of £354,669; and of the total assistance of £309,030 to be provided from public funds by the Board and the Development Commissioners, two-thirds were to be in loans and the remainder in grants (ARFBS. 1914: 247). The timing of this could scarcely have been more unfortunate, as with the outbreak of World War I in the following year, harbour development was largely put in abeyance, although in fact work did not entirely stop.

THE INTER-WAR PERIOD

The hopes that World War I would prove but a brief interlude in orderly development were soon proved false in fisheries and harbours,

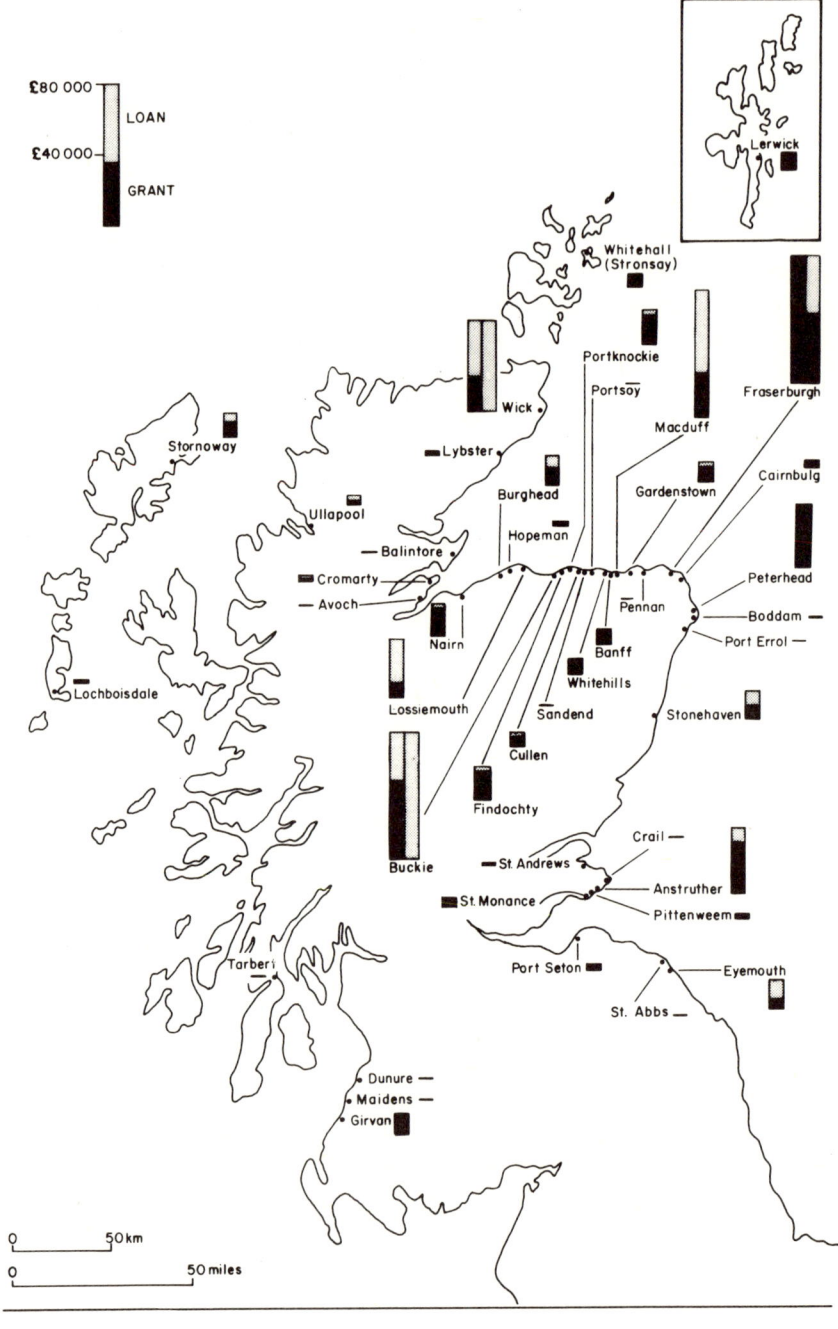

Fig. 16.2 Grant and loan expenditure by
the Fishery Board, 1910–1938.

as well as in many other fields. In the much changed climate from 1919 onwards, despite the new and improved financing arrangements, harbour development proceeded in a somewhat stuttering manner. There was also a change in priorities, and there was a new emphasis on concentrating development at main harbours which in earlier times had been left largely to their own resources. Funding for smaller projects still tended to come from the Board's Piers and Quays Fund (which was still limited to £3,000 annually), with bigger schemes being financed principally from the Development Fund; by 1938 a total of £734,063 had been advanced from the Development Fund, of which 52% was in grants (ARFBS. 1938: 77). On top of this additional funding was also obtained in a number of cases from the Public Works Loan Fund; this tended to be given when benefits beyond fishing were envisaged, as (for example) at Lerwick and Stornoway, where harbours provided essential transport links for island communities. In addition, the policy became more flexible in gauging the scale of aid appropriate to different projects: some (especially of the smaller ones) were completely covered by grants, while others got widely varying proportions of grant and loan aid. While in general the major part of costs was still covered by grants and loans, in a number of cases the assistance given covered a minority part of the cost. Regionally aid was very much concentrated at the main bases of the fleet, which meant that it was concentrated on harbours on the south shore of the Moray Firth: of the total expenditure under the new policy between 1910 and 1938, 61% of the total aid of £822,000 was spent on the 80 miles of coast between Nairn and Fraserburgh (fig. 16.2).

However, the revised arrangements for helping fishing harbours were now set in a new environment which was in many ways less favourable. The fisheries, from being in a generally prosperous position of world leaders, were now in a position of prolonged depression for the two interwar decades, and the 'knock-on' effect of this was to lead to a serious drop in harbour revenues and profitability. At the same time there had been 'an enormous increase in the costs of labour and materials' (ARFBS. 1920: 61). Also superimposed on the problems of fisheries and harbours were various measures of national stringency: and cuts in public expenditure resulted in cut-backs in harbour funding, especially in the financial year 1922–23 and in the early 1930s. In all, the objectives of harbour improvement became seriously handicapped by the great burden of debt on the harbour authorities. In fact the debt burden escalated in the 1920s, when a series of major expenditures were approved in the expectation of a fairly rapid post-war recovery. This millstone of debt was to persist through the interwar years, even although in the trough of the Great Depression considerable sums were written off as a special concession. By 1929 total outstanding debts to public funds had reached a figure of about £900,000, of which almost two-thirds were owed to the Public Works Loan Fund, and the bulk of the remainder to the Development Fund (ARFB. 1929: 49); and between 1929 and 1930 debts to the total value of £241,000 were remitted (ARFB. 1930: 51).

Despite this the new framework of finance did allow some harbour improvement during the period, even if much of the available funds were committed to maintenance and repair. The situation might indeed have deteriorated further had it not been for the very hard times that the fishing communities were undergoing, and the political pressure that this generated for ameliorative measures. Harbour projects were public works which generated some employment; and this included unskilled work which fishermen who had perforce been rendered idle could undertake. A high proportion of the harbour projects were on the south shore of the Moray Firth; and this stretch of coast had the highest concentration of fishing boats and men in Scotland. One specific example of the recognition of a project as a relief measure occurred in 1927, when the work of deepening one of the basins of the Buckie harbour was undertaken to help relieve local unemployment (ARFBS. 1927: 49), after work there had been suspended the previous year for shortage of funds (ARFBS. 1926: 71). In fact out-of-work fishermen got employment in various other public works, and one that is specifically recorded is that of removal of silt from the Portsoy harbour in 1934 (ARFB. 1934: 81), a stage when prospects in the herring fishery were at their bleakest.

The main fishing harbours, which had previously been left to their own resources but which now received substantial help, included ports like Peterhead, Eyemouth and Lossiemouth; and the greatest outlay was at the major centres of Buckie, Fraserburgh and Wick which between them received a total of almost £400,000 (ARFBS. 1938: 77), towards works for which the total cost exceeded £600,000. The Board at the same time recognised its continuing responsibility to help harbour works at the villages where many of the fleet were still based (ARFBS. 1929: 49). These included many places on the East Coast, and again places on the south coast of the Moray Firth, such as Portknockie, Cullen, Findochty, and Gardenstown were prominent. In the cases of Cairnbulg (Aberdeenshire) (ARFBS. 1926: 71) and Broonies Taing (Shetland)(ARFBS. 1931) the building of single new piers were undertaken: and such projects at this late juncture would certainly seem to embody a relief element.

A matter in harbour maintenance which received enhanced attention during the inter-war period was that of dredging, which was essential in nearly all harbours to keep them in operating condition. The great part of this work was done by the two dredgers the Board had acquired, and in principle its intention was to charge harbour authorities the cost of the work. However with the parlous state of the finances of many harbours, a considerable amount of this service was in fact provided free of charge.

HARBOUR DEVELOPMENT SINCE WORLD WAR II

After World War II the Fishery Board for Scotland was wound up and the administration of the fisheries passed to other government agencies. In a new political climate, and with the help of government grant and loan schemes the fisheries entered into a more prosperous era. The general trend was for the average size of boat to increase and become more sophisticated; and there was accordingly continued need for port development. It was now accepted that government should make major inputs into capital projects, although local contributions were still required: public funds continued to be the main source of finance for fishing harbours, and the scale of the work expanded to the point that a full-time post of harbour engineer was created in place of the former arrangement of depending on consulting engineers. In the fisheries the centralisation of operation at major harbours continued, and the coming of general car ownership made it much easier for fishermen to have the option of continuing residence in traditional villages, while keeping their boats in harbours some distance away. However the long-term trend for fishermen to leave the villages for towns with main harbours has continued, and the towns have superior amenities and educational facilities for fishermen and their families. Any modern expenditure at village piers and harbours is now mainly to improve their appearance for tourists.

REFERENCES

Annual Reports of the Fishery Board (ARFB.), 1827, 1828, 1831, 1832, 1833, 1835, 1838, 1840, 1844, 1847, 1848, 1849, 1850, 1851, 1855, 1858, 1864, 1869.

Annual Reports of the Fishery Board for Scotland (ARFBS), 1882, 1883, 1891, 1892, 1914, 1920, 1926, 1927, 1929, 1930, 1931, 1934, 1938.

Buckland, F., Walpole, S., and Young, A. (1878) *Report on the Herring Fisheries of Scotland*, HMSO.

Coull, J.R. (1992) 'The Development of the Fishing Districts of Scotland', *Northern Scotland* 12, 117–131.

Graham, A. (1976-77) 'Old harbours and landing places on the east coast of Scotland', *Proc. Soc. Antiq. Scot.* 108, 332–366.

Graham, A. and Gordon, J. (1987) 'Old harbours in northern and western Scotland', *Proc. Soc. Antiq. Scot.* 117, 265–352.

Lenman, B. (1975) *From Esk to Tweed. Harbours, Ships and Men of the East Coast of Scotland*, Blackie, Glasgow and London.

Mitchell, J. (1883) *Reminiscences of my Life in the Highlands*, I, Gresham Press, Chilworth and London.

Stevenson, T. (1874) *The Design and Construction of Harbours* (2nd edn.), Adam and Charles Black, Edinburgh.

Washington, Capt.J. (1849) *Report on the Loss of Life and Damage to Fishing Boats on the East Coast of Scotland*, British Session Papers LI 177–297.

Young, A. (1883) 'Harbour Accommodation for Fishing-Boats on the East and North Coasts of Scotland', in Herbert, D. (ed.) *Fish and Fisheries. A Selection of the Prize Essays of the International Fisheries Exhibition, Edinburgh*, 1882. William Blackwood and Sons, Edinburgh and London, 65–92.

17

The Development of the Administrative Framework: Fisheries Districts

The development of a system of administration for the fisheries is very much a thing of the industrial era. While in earlier times there were various measures taken by the national Parliament, most of what practical administration there was in fisheries was in the hands of the Convention of Royal Burghs. Essentially there was freedom for everyone to fish, but trade was legally possible only for merchants. The herring fishery was always the sector of the sea fisheries that dominated and was the subject of most regulation. With the privileged position of the Royal Burghs in trade, such matters as the use of standard barrels for the export of herring or salmon, or the charging of export dues, were in their hands. While such functions were discharged, the regularity and dependability of this is questionable, in the light of the limited facilities available for inspection at the national level.

With the general quickening of the economy from the late 17th century, there were increasing attempts to promote the fisheries; this included the setting up of a number of companies and government bodies, and the latter were charged with the disbursement of funds for fisheries development. One of the results of this was that administration actually became less co-ordinated during much of the 18th century. The Convention of Royal Burghs in 1727 resolved to pass over control of the herring fisheries to the new Board of Manufactures: a main reason for this was that the Board had available £2,000 per year under the terms of the 1707 Union; it was also envisaged that the Board should administer curing and packing of herring and appoint inspectors to that end. However, Government legislation in 1750 giving bounties for fitting out herring busses (ch. 5) also entailed that customs officers were given responsibilities to check their nets, salt and barrels. When the Society for the Free British Fishery was set up in 1750 it was given the powers to regulate and manage trade and commerce in herring for 21 years; and these powers included the right to regulate the curing, sorting and packing of herring (RCHC. 1803, X: 100). This appears to have had very limited effects, and the charter of the society was not renewed when it expired in 1771, but the tonnage bounty arrangements were still in force until 1785, so that the customs officers continued to have a function in the administration of the fishery. There were in 1785 the most exhaustive government reviews up to the time of the state of the industry. While the outcome was the extension of tonnage bounties for herring busses, there was also the beginning of the payment of barrel bounties for which the small boat fishery could qualify: and within a few years this had produced a noticeable response in the inshore fisheries in Caithness and on the Forth and Clyde.

It was the decision 1809 to give more attention to the inshore herring fishery which required officers to be placed around the coasts to inspect the cure, which was a necessary concomitant to the paying of the barrel bounties (ch. 7). This gave rise to a system of fisheries districts, each overseen by a fisheries officer, whose work was supervised and co-ordinated by the Fishery Board from its office in Edinburgh with the help of inspectors who made regular tours of inspection on both East and West Coasts. This was the foundation of the system that is actually still in operation.

In the second decade of the 19th century, then, the coast was divided up into sections, each supervised by a fishery officer. The evolving pattern of districts which ensued reflected the needs of a growing industry, although the trend was also punctuated at intervals by important government decisions and policy changes. Relatively soon, in 1820, the responsibilities of the Fishery Board were enlarged to cover other fisheries (ARFB. 1820: 1), and from 1824 it also gave attention to the improvement of fishery piers and harbours (ARFB. 1824: 1) (ch. 16). In the early stages the policy involved financial incentives as pump-priming for expansion, and this was allied to quality control; with the later removal of financial incentives, quality control continued, linked to comprehensive record keeping in the hey-day of laissez-faire in the nineteenth century; and encouragement to export, direct or indirect, was always part of the policy.

As the system expanded in the early stages it was relatively frequent for new offices to be established and for districts to be sub-divided. However, once the system was established, it was also the practice to rationalise and streamline it at times by incorporating less active into more active districts, although new offices could still be instituted to cater for new developments. There are also a number of cases of offices being closed and then re-opened at a later stage.

THE BEGINNINGS OF FISHERY DISTRICTS 1809–1820

It was mainly to administer the barrel bounties for herring that fishery officers were originally appointed, and their most important function was to ensure that herring were put in salt in barrels within 24 hours of catching; the officers were empowered to put the crown brand on the barrels as a guarantee that this had been done. This was to continue to be a main function of Scottish officers well into the twentieth century.

The fishery officers at the start might have to deal with three types of operation in the herring fishery. In the early stages there were still a limited number of busses fishing on the tonnage bounty, and these had to be 'cleared' at an approved port to record the quantities of salt and barrels aboard, and to ensure that their barrels and nets met legal specifications. They also had to be cleared at the end of a voyage to ensure the herring had been properly cured and to record the numbers of barrels filled. Secondly there was the 'coast fishery' which was given special encouragement in the early stages. This was a fishery operated by smaller vessels, defined as being between 15 and 30 tons (ARFB. 1809, Appendix 2, 35). These like the busses also cured herring aboard: they could be given fitting out premiums, and their barrel bounties could be enhanced, although in both cases this was linked to their performance in catching. Despite its name, the boats in this fishery were required to fish out from the shore, and this reflects the accepted principle of the time that a successful fishery demanded off-shore operation by mobile vessels that could seek out herring in different places, on a pattern that had been successfully operated by the Dutch for centuries.

In the third place there was the inshore 'boat fishery' operated by small open boats within about 15 miles of the coast, in which the herring were cured on shore; and it was this section of the fishery which, to the surprise of officialdom, came to the forefront within half a dozen years. The critical and inadequately appreciated point was that in the herring seasons, the shoals were generally available in quantity within 15 miles of the Scottish coasts; and curing on shore did not have the severe constraints on space and storage that always prevailed aboard. Moreover, as boats and equipment improved in this growing fishery, the distance range from the coast expanded so that in the second half of the century shore-based curing could be linked to fishing at a range of 50 or even 100 miles offshore. While curing aboard was to continue, the key to the success of the Scottish herring fishery lay in curing on shore.

The barrel bounties which played the vital pump-priming role were given at the rate for 2/- per barrel from 1809 to 1815; this represented a supplement of *c.* 10 per cent by value, and moreover could be enhanced at a rate of 2/8 per barrel for herring exported. From 1815 to 1826, this system was amended and simplified by removing the export premium while giving 4/- on all approved barrels; thereafter between 1826 and 1830 the bounties were phased out (ARFB.S. 1882: 15) despite the protests of the trade, and production continued to expand.

While the system was based on having officers at particular locations, there were important elements of flexibility in it, and this always included some movement of personnel. Officers generally had to cover a number of places in their own particular districts, and there were always transfers between districts to cover seasonal peaks of activity at different places. There are various records of officers hiring gigs or horses to get around their districts, and from the early days of the Board they might have to move from Wick or even Lerwick, where activity was concentrated in the summer season, to help cover the winter fishing on the Forth or Clyde.

When the Fishery Board was set up in 1809, it inherited a situation where the main location of enterprise was on the long-established Firth of Clyde where, in addition to on shore curing, bigger boats were fitted out for longer-range operation, much of which was in the Minch. The other long-established area of the Firth of Forth was mainly involved in on shore curing, as was the more recently developed area of Caithness; and Yarmouth in East Anglia was a main base for the buss fishery. The siting of the original offices faithfully reflected this pattern of activity: in the first year five were established on the Clyde and two on the Minch; two were placed on the Forth, one in Caithness and one at Yarmouth, in addition to the head office in Edinburgh (ARFB. 1809: 3).

In the subsequent expansion there was extension of coverage to all parts of the Scottish coast by 1820, although with a variable spacing which was mainly related to the scale of local production. Offices were also established at a handful of locations in England adjudged to have potential for the herring fishery; and in addition offices were placed at Liverpool and Bristol (to oversee export to Ireland) and at London, to allow close links with other government bodies and agencies.

As the system expanded there were various requests submitted to the Board from local officials and from curers at a range of places wishing to have an officer established so that they could avail themselves of the bounties. The Board's usual reaction was to decide whether there was enough activity to justify a new office, and the working of the system can be exemplified by the way that the towns of Dumfries (SRO. AF 1/1: 292; AF 1/2: 117) and Stranraer (SRO. AF 1/1: 114; ARFBS. 1821: 1) were both turned down on first application, but were later provided with offices on the grounds of a growing trade. It was also known for officers to feel themselves overburdened, as occurred with the men at Leith in 1816 when they successfully argued that the inclusion of Eyemouth (50 miles away) in their district added unduly to their work, and Eyemouth was provided with an office of its own (SRO. AF 1/3: 301–303: ARFB. 1816: 1).

From the start the fishery expanded more rapidly on the East Coast, and it was on this side of the country that there was the main proliferation of offices, although they were also reinforced on the west. By 1820 the addition of three offices on the Clyde and four further north brought the West Coast total up to 15. The main growth area on the east was the Moray Firth where six new offices were placed, and the overall East Coast total rose to 13, stretching the length of the country from Eyemouth in the south to Lerwick in the north.

EXTENDING THE SCOPE 1820–1830

The early 1820s saw the peak of the pump-priming role of the Fishery Board. It was no doubt a recognition of the considerable success in expanding the herring fishery by means of barrel bounties that a parallel system was brought in 1820 for the other most important commercial fishery, that for cod and ling. It was decided to give a bounty of 4/- per cwt on fish that had been properly dry cured with salt, and given a distinctive punch mark by an officer on approval of the quality of the cure; and 2/6 per barrel was to be given for cod cured in barrels to an approved standard (ARFBS. 1882: 19). In this fishery the bigger part of the catch came from onshore curing linked to an open boat fishery, although effort was more spread around the coasts than had been the case in the herring fishery. There were also bigger cod smacks involved in this fishery, and the Board showed its faith in their potential to make a greater contribution in production by making the bounties available to them, irrespective of whether they caught the fish themselves or bought them from smaller boats. As it turned out, the cod and ling bounties were to run for ten years, and were abruptly terminated in 1830 and not phased out like the herring barrel bounties.

It was decided that the cod and ling bounties could be administered with a limited increase in the establishment of fishery officers. This was because there was a considerable overlap with the herring in the bases from which the fishery was operated, and also because it was less subject to the pattern of seasonality which dominated the herring. However, the increased work meant not only an increase in staff at existing offices, but the establishment of new offices and districts. While the cod and ling fishery was operated all round the coasts, the best locations were towards the edge of the continental shelf, which gave greater opportunities to the Hebrides and Northern Isles, and indeed the Shetland Isles were by far the leading centre of the fishery. At the start of the 1820s Shetland was divided into three districts, with new offices being established at Walls on the west side, and on the northmost island of Unst; Orkney was divided into two districts with the placing of an office on Stronsay for the north isles of the archipelago (ARFB. 1820: 2) to supplement the one at St. Margaret's Hope on South Ronaldsay; and further offices were sited on Barra (ARFB. 1820: 2) and at Dunvegan (ARFB. 1821: 1) on Skye. Although the relative importance on the East Coast was less, the cod and ling fishery was sufficiently important in the area of the Scotland-England border that the Eyemouth district was sub-divided with the creation of a new office at North Sunderland (ARFB. 1821: 1). The cod fishery was also part of the justification for the placing of a new office at Stonehaven (ARFB. 1820: 2), but the creation of an office at Burghead for the new Findhorn district (ARFB. 1820: 2) was related more to the continued momentum of growth in the herring fishery.

In the cod smack fishery, the main interest was on the part of merchants in the Northern Isles, but it was also important at the

English ports of Yarmouth, Whitby and Gravesend, and offices were established at the latter two ports (ARFB. 1820: 2; 1826: 1) in addition to the one already existing in Yarmouth. By the later nineteenth century the cod smack fishery, originally mainly exploited around the Northern Isles, was extending to grounds at Faroe and Iceland. However, in the 1820s it was found that there was insufficient justification for retaining the offices on the south coast of England – those at Plymouth, Portsmouth and Dover were closed, as was Oban in Scotland (ARFB. 1826: 2; 1822: 2; 1824: 1; 1827: 1).

CONSOLIDATION CULMINATING IN RECONSTRUCTION 1830–1850

The period from 1830 to 1850 was not a static period, but it was one of limited change until the mid-century year. Though the bounties were discontinued, a system of inspection was maintained for quality control of curing, but lower proportions of both herring and of cod and ling were presented for inspection. In the 1830s the withdrawal of bounties led to a sag in production of cod and ling, while towards mid–century there was a hiatus in the expansion of herring production, largely due to shrinkage in two important markets – those of the Caribbean following slave emancipation, and of Ireland following the Potato Famine.

In 1830, the offices at Yarmouth, Whitby and Gravesend were closed (ARFB. 1830: 2) and inspection provision for the cod and ling fishery concentrated on locations where open boats were the main source of production. There was also the closure of a number of offices in Scotland before mid–century, although this was done largely by taking the opportunity to amalgamate districts on the retirement of men in post. The removal officers from Barra and Islay (ARFB. 1837: 2) took away a coverage that was directed mainly to the cod and ling fishery. The closure of offices at Ayr, Loch Gilphead, Tobermory and Fort William (ARFB. 1833: 2; 1835: 2; 1839: 1; 1836: 3) reflected the continuing displacement of the main field of activity in the herring fishery towards the east and north; and the opening of offices at Peterhead and Tongue (ARFB. 1834: 2; 1832: 1) was the counterpart of these closures. However, the closure of the Burntisland office (ARFB. 1845: 6) on the inner part of the Forth also reflected the increasing emphasis on bases fronting the open sea.

The most abrupt change ever experienced in the pattern of the districts came in 1850, when there were severe cuts in public expenditure as the laissez-faire philosophy and the associated move to Free Trade became the accepted political orthodoxy. The inspection of cured cod and ling was abandoned in 1850, but the importance of a guarantee of quality in the curing of the more easily spoiled herring species was still judged essential, although after 1858 branding fees were introduced (ARFB. 1858:2) and this was followed by a further fall in the proportion of barrels presented for inspection. By this time the Scottish herring had an established reputation on the main continental markets, and although the availability of branding continued to provide important underpinning to the export trade, there were curers who preferred to depend on their own market contacts and reputation.

In the stream-lining of the system in 1850, all five offices on the English mainland were closed (SRO. AF 3/1/36: 372), and only that on the Isle of Man was retained outside Scotland. The pruning of the establishment in Scotland saw the closure of nine offices (SRO. AF 3/1/36: 373; AF 3/3/1: 51–54). As well as further contraction in the south-west with the closure of Dumfries and Campbeltown, there was the removal of the offices at Thurso, Tongue and Portgordon, and of that at Montrose which had become the centre for the Stonehaven district. The main function of these offices had been to deal with the herring fishery, but the closure of three in the North Isles (Walls, Unst and Stronsay) was essential due to the discontinuation of inspection in the cod and ling fishery.

During this enforced contraction, the Board took the opportunity to codify other important things in their organisation. It was realised that there was likely to be an interval of years before any new officers were appointed and that there would be an inevitable slow-down in promotions. From the inception of the Board, with the importance of herring curing, officers had generally been recruited from coopers, and it was now formally laid down that a candidate for appointment must have served an apprenticeship as fish curer and cooper, and must be under thirty years of age (SRO. AF 1/17: 197–199).

There was also the official and comprehensive laying down of precise boundaries between districts. The early strategy had been to locate officers at points where herring fishing was active, and to have them cover also neighbouring places within reach. Boundaries between districts were a subsidiary issue and evolved in an ad hoc manner. However, as early as 1811 it was necessary to define the

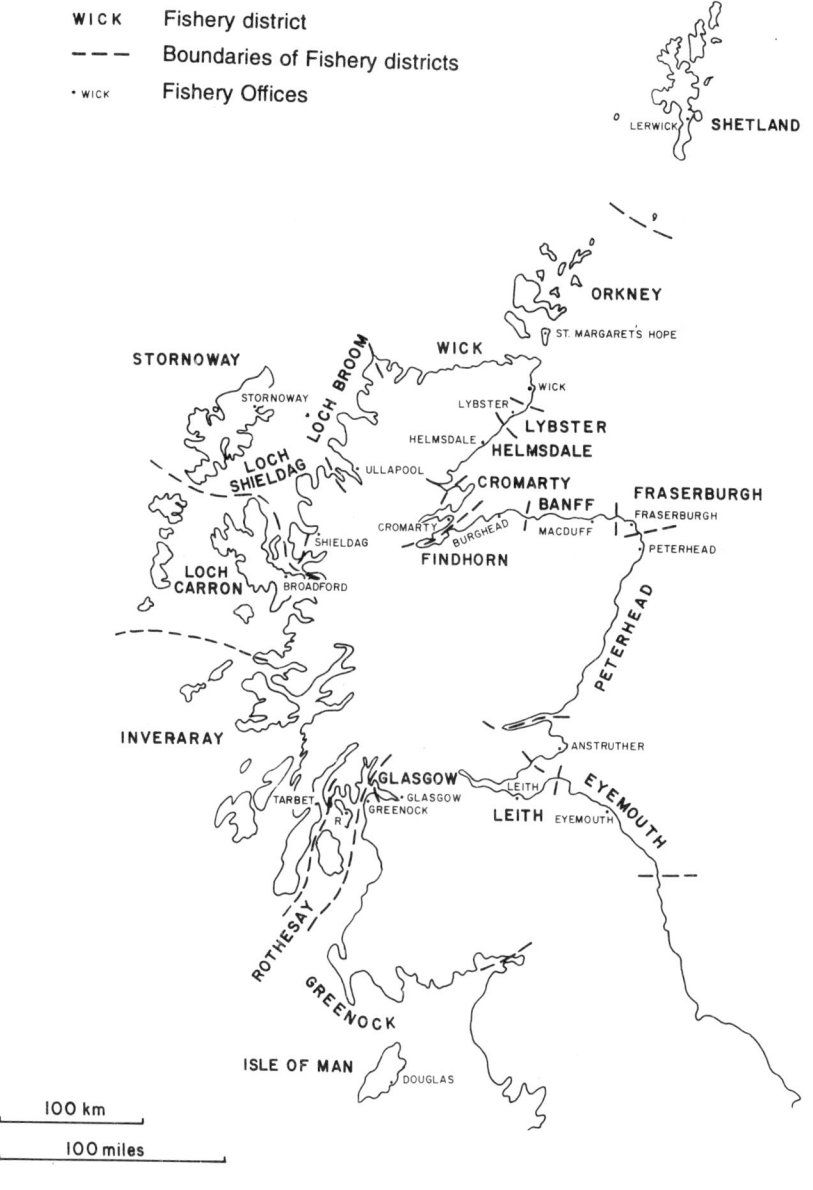

Fig. 17.1 Fishing districts and fishery offices in 1850.

boundaries of the three far-flung districts of the north-west mainland of Loch Broom, Loch Shildag and Loch Carron, and when the Loch Gilphead office was created in the same year, it was deemed necessary to define the boundaries between it and the neighbouring districts of Inverary and Campbeltown (SRO. AF 1/1: 229). The development of boundaries can be more fully illustrated by events on the south shore of the Moray Firth. The first office opened in this area was at Portgordon in 1816, and the district was defined as stretching from Banff westwards 'as far up as possible' (SRO. AF 1/3: 279), which meant that although there was a definite boundary with the Fraserburgh district on the east, that with the Cromarty district on the west was deliberately left indistinct. Offices were established on the intervening stretches of coast at Banff in 1818 (ARFB. 1818: 11) and at Findhorn in 1820 (ARFB. 1820: 2), and although the boundaries on the south shore of the Moray Firth were revised in 1824, it was still necessary for the Banff officer in 1825 to get a ruling from the head office that the village of Crovie was in his district and not in that of his neighbour officer at Fraserburgh (SRO. AF 1/7: 45).

The revision of boundaries which was effected in 1850 was to lead to the laying down of a pattern (fig. 17.1) in a document that became a reference base-line for future years (SRO. AF 3/3/1: 51–54). There were now a total of eight districts on the west coast and thirteen on the east; and while in the busiest areas a district might still be as little as 10 miles in coastal extent, as was the case at Lybster and Fraserburgh, some districts were now of such a size as to require extensive travelling on the part of incumbent officers, and the efficiency of the system can only have suffered as a consequence. The greatest problem was in the Inveraray district: as well as including both sides of Loch Fyne and all of Kintyre, the mainland as far north as Ardnamurchan was included along with the offshore islands (SRO. AF 3/3/1: 52). The Wick office too must have been under heavy pressure, as in addition to covering the busiest single port, its jurisdiction stretched right along the north coast to Cape Wrath.

1850 - 1953 RENEWED EXPANANSION, FOLLOWED BY STABILITY

During the second half of the nineteenth century, a number of new developments were to extend the work of the Fishery Board. From the end of the 1850s the herring fishery entered its main expansion phase which was to carry it to the undisputed position of the world's leading fishery; and by the 1880s the demersal fisheries were being radically changed by the rise of steam trawling as the dominant catching method. In addition, other duties and more detailed record keeping were added to the responsibilities of the fishery officers. From 1855 three classes of fishing boats were recognised based on size lengths (ARFB. 1855:3), and from 1863 there was fuller recording of the values of boats and gear. A new act of 1868 established a system for the registration of all fishing boats, and made the officers responsible for all sea fisheries apart from salmon (ARFB. 1868:4); and this act removed the Isle of Man from the Board's responsibility. Finally, as expansion and change gathered momentum towards the end of the century, the Board was reconstituted under a further act in 1882, its title changed to the Fishery Board for Scotland and its powers extended to include superintendence of salmon fisheries (ARFBS. 1882:xiii).

It had always been part of the Board's function to superintend good order at the fishing: off various parts of the coast large numbers of boats could be fishing in close proximity and disputes were not infrequent. These could be exacerbated by the participation at particular bases of stranger boats from other areas, and in the earlier decades this mainly involved the boats on the 'coast fishery' on the West Coast. However, the greatest problems the Board encountered in maintaining good order came from two innovations in the second half of the nineteenth century. The first of these was the beginning of the use of the ring-net on the Clyde, the use of which expanded in the 1850s (Martin 1981:6–25) although the method at the time was called the trawl: it allowed the surrounding of shoals of herring, and could be operated relatively easily on the sheltered waters of the Clyde, but it met strong hostility from traditional drift-net fishermen. It was legally banned for 16 years before becoming accepted (Martin 1981:10–25).

The other innovation was the use of the trawl, in its accepted modern sense of a drag net pulled over the sea-bed to catch demersal fish (ch. 9). Its greater efficiency was proved in the southern North Sea in the early nineteenth century, and its use spread northwards till it was being employed off the Scottish coasts by the 1860s; and its adoption sharply increased with the installation of steam power aboard trawlers in the 1880s. This method provoked hostility from the traditional line fishermen who saw their livelihood being undermined, and in the 1880s trawling was banned inside the three-mile limit, and in the Moray Firth and Firth of Clyde (ARFBS. 1889:xlii–xliv).

The expansion of the second half of the nineteenth century required adjustment in the pattern of offices and districts, and most especially reflected the increasing concentration of operation and catching power in the North-East. It is significant that in 1860 the Loch Shildag office was closed, and the officer transferred to a new office at Buckie (SRO AF3/3/9:366,455), the home area of the biggest single concentration of the herring fleet. In 1863 an office was re-established at Montrose; and when increasing congestion in the harbours of Peterhead and Fraserburgh helped direct a sizeable part of the herring fleet to Aberdeen at the start of the 1870s, a new office was opened there in 1873 (SRO. AF7/53:264) while Stonehaven was re-established in 1881 for similar reasons.

Matters were not, however, static on the west coast, and in 1863 there was the re-establishment of the Campbeltown (SRO. AF3/3/11:7) office, along with the setting up of one at Ballantrae (SRO. AF3/3/11:13) on the opposite side of the Firth of Clyde. These could be justified by increasing production, but were also related to the need for the presence of officers during the stormy controversy related to the rise of the ring-net fishery. Further north there was enough activity to justify the re-establishment of the Fort William office in 1868, although the closing of the Glasgow office in 1875 shows the decline in its suitability as a base for the coast fishery with the strong competition of other port uses. In the Outer Hebrides the importance of the herring fishery at the south end of the chain eventually led to the re-establishment of an office in Barra in 1892. However, despite the great scale of the herring boom in the Shetland Islands from 1880, the whole archipelago was maintained under the Lerwick office, although for a period of 30 years Balta Sound in Unst was a major herring port by any reckoning.

The opening of an office in Aberdeen proved to have fortunate, if unforeseen, consequences. When in the early 1880s steam trawling for demersal species was proved to be amply profitable, Aberdeen rapidly developed as one of the half-dozen main trawl ports delivering fish to the British market via the railway network, and by the early twentieth century fishing rivalled granite as the main economic basis of the city, while the Aberdeen district became the most important in Scotland in value of its fishing fleet and of landings. Trawling had less impact elsewhere in Scotland: only

Granton, covered by the Leith district, was to become another significant trawl port.

For 60 years, from the creation of the Barra district in 1893 until 1953 the pattern of offices and districts (fig. 17.2) was unchanged; this represented the peak of Scotland's relative importance as a fishing nation in the late nineteenth and early twentieth centuries. The pattern also survived the essentially difficult inter-war years, a time when the pattern of production changed relatively little, although the herring fishermen especially had a great struggle at this time. In this stable pattern, there were 15 districts on the East Coast and 10 on the West, with two in the Northern Isles.

DEVELOPMENTS SINCE WORLD WAR II

Administrative adjustments had to be made to cater for the accelerated changes there were in the fisheries after World War II. There was the continued run-down in the importance of herring, the rise to dominance of the white fish sector, and a substantial increase in the importance of shell fish. In addition there was the development of new fishing methods and new market sectors. However there was not initially any change in the structure of administration or in the pattern of districts; but there was considerable reorganisation between 1954 and 1957, and in subsequent years a number of further adjustments were made.

In 1954 significant streamlining was achieved by the elimination of five districts in which activity had dwindled to minor proportions by merging them into neighbouring districts: the Stonehaven, Cromarty, Lybster, Rothesay and Greenock districts thus disappeared (RFS. 1954: 6), and the following year the Long Island was made into a single district by the merging of the Barra district with that of Stornoway (RFS. 1955: 6). In 1954 also several new names appeared on the list of districts. The substitution of Macduff, Lossiemouth, Arbroath and Oban for Banff, Findhorn, Montrose and Fort William respectively reflected the ports that had come to the fore as the main fishing harbours in their own sections of coast. The replacement of 'Loch Carron and Skye' by Kyle was another instance of the same kind. The new Ayr district resulted from the merging of the former Greenock and Ballantrae districts, while the new Tarbert district was an amalgamation of the former Inveraray and Rothesay districts.

In 1956 there was a structural reorganisation, with the old system

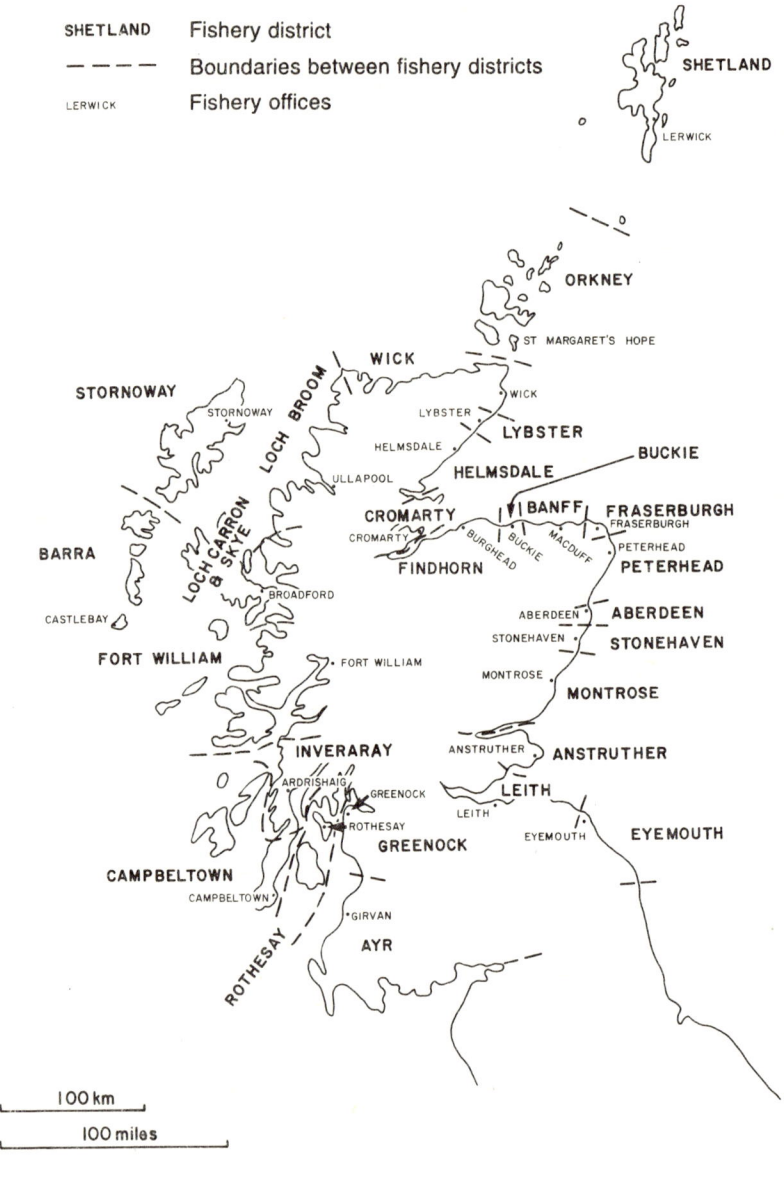

SHETLAND   Fishery district

– – – –   Boundaries between fishery districts

LERWICK   Fishery offices

SHETLAND

LERWICK

ORKNEY

ST MARGARET'S HOPE

WICK

STORNOWAY

LOCH BROOM

WICK

STORNOWAY

LYBSTER

LYBSTER

HELMSDALE

HELMSDALE

BUCKIE

ULLAPOOL

CROMARTY

BANFF

FRASERBURGH

BARRA

CROMARTY

BURGHEAD

BUCKIE

MACDUFF

FRASERBURGH

PETERHEAD

LOCH CARRON & SKYE

FINDHORN

PETERHEAD

CASTLEBAY

BROADFORD

ABERDEEN

ABERDEEN

STONEHAVEN

STONEHAVEN

FORT WILLIAM

FORT WILLIAM

MONTROSE

MONTROSE

INVERARAY

ANSTRUTHER

ANSTRUTHER

ARDRISHAIG

GREENOCK

LEITH

ROTHESAY

LEITH

EYEMOUTH

CAMPBELTOWN

GREENOCK

EYEMOUTH

CAMPBELTOWN

ROTHESAY

GIRVAN

AYR

100 km

100 miles

Fig. 17.2 Fishing districts and fishery offices, 1893–1953

of inspectors for East and West Coasts giving way to a pattern whereby the coast was divided into seven sections, each consisting of a group of districts under an inspector with a chief inspector in Edinburgh (Coull 1992:128,129). At the same time there was further consolidation with the merging of the Helmsdale district into that of Wick; in the Clyde area the Tarbert district was merged into that of Campbeltown; and in 1957 the Orkney district was merged with that of Shetland. This series of changes had entailed a reduction from 27 to 19 districts: and of the 19 districts the fact that 11 of them were on the East Coast showed that it still had the main concentration of bases and activity.

CONCLUSION

During the period of over 180 years during which the Scottish system of fishery districts and offices has been in existence, it has had to adapt to far-reaching developments in the fisheries. While they have greatly changed in scale and emphasis, for almost a century and a half the predominant position of the herring fisheries did give a substantial element of continuity and stability.

Integral to the changes has been a series of developments in regional emphasis which has been reflected in the evolving pattern of districts. Early in the nineteenth century, the East Coast outpaced the West Coast in being in the lead; on the East Coast itself, the leading area of Caithness was displaced in the third quarter of the nineteenth century by the north-east shoulder of Aberdeenshire; and in the peak days of the herring fishery around the turn of the century the Shetland Islands emerged as the predominant single district. At the same time the rise of steam trawling for white fish made the Aberdeen district preeminent in value of landings. While the leading position of East Coast districts has been maintained during the diversification of activity which has occurred since World War II, there has been a rise in the relative importance of the districts of the north-west, especially since the 1970s.

Set amid this pattern of regional change, there has been a transition from a largely dispersed system of operations from a multitude of villages and landing points towards a concentration on a relatively small number of specialised ports; and these ports are very generally the location of the fishery offices. The herring fishery, prosecuted from the early nineteenth century even in the 'boat fishery' with relatively big vessels which needed harbours rather than landing beaches, was a powerful initiator of this change; and the continuing general increase in the size of boats, together with the development of auction markets and the range of specialised facilities and services which the fleet have come to require, all combined to increase the forces of centralisation towards the main ports.

The system of administration originally established in 1809 has been continuously adapted to changing circumstances up to the present. An organisation originally formed for expansion and quality control in the herring fishery serves to enforce the policies and directives now emanating not only from Westminster but also from Brussels under the Common Fisheries Policy.

REFERENCES

Annual Reports of the Fishery Board (ARFB) 1809, 1816, 1818, 1820, 1821, 1822, 1824, 1826, 1827, 1832, 1833, 1836, 1837, 1839, 1845, 1855, 1858, 1868.

Annual Report for the Fishery Board for Scotland (ARFB.S.) 1882, 1889, 1890.

Coull, J. R. (1992) 'The Development of the Fishery Districts of Scotland', *Northern Scotland* 12, 117-131.

Reports from Committees of the House of Commons (RCHC), 1803.

Scottish Record Office (SRO.): AF 1 Fishery Board Minutes; AF 3 *Fishery Board Letter Books.*

Martin, A. (1981) *The Ring-Net Fishermen*, John Donald, Edinburgh.

Reports on the Fisheries of Scotland (RFS), 1954, 1955, 1957

18

Epilogue: Conservation and Extended Regulation

In modern times change has accelerated in the fisheries as in so many spheres of life. While fishing boats have in general become much more efficient and their gear and equipment much more sophisticated, life for fishermen has become greatly complicated by increasing pressure on the resource and by the plethora of regulations that have developed in the interests of long-term conservation. Conventional fisheries have now become something of a test case in the global use of resources, as there are ample indications that the yield of fisheries has approached or reached the biological ceiling. In many areas, including the waters around Scotland, the point of maximum sustainable yield has been passed.

It is now clear that the rapid increase in catching power of the decades after World War II was poorly matched by the development of conservation measures to maintain healthy fish stocks. Catching power built up in nearly all countries, and Scotland developed a fleet of boats mainly of the 60 to 80 feet size that was capable of exploiting the waters of the continental shelf around the country and beyond, and that had few peers at the world level for boats of this size class. These boats were still generally owned by working fishermen: horizontal integration of enterprise has still made limited progress in fishing, in contrast to what has happened in many economic sectors. Fisheries modernisation and enhanced catching efficiency was generally seen by officialdom as the way to rising living standards and economic survival in Britain and indeed in other European countries. The rebuilding of the fishermen-owned sector of the fleet meant that, among other consequences, it became more than a rival to the company-owned trawler sector in the catching of white fish, and exceeded it in production.

The increased intensity of expolitation inevitably made greater demands on fish stocks, and by the later 1960s there was increasing concern about symptoms of overfishing. It was also ironically aggravated by the measures the government had brought in to help the fishermen under the grant and loan schemes: when fishermen had a good year there was added incentive to reinvest their profits in bigger boats rather than pay a higher level of tax. Such concern had to an extent been present for some demersal species from early in the 20th century, and it was becoming clear that there were now bigger proportions of immature fish appearing in catches – a key indication of a deteriorating fish stock structure. Although there was agreement to set and increase minimum mesh sizes under the new North-East Atlantic Fisheries Convention, these essentially were confined to demersal stocks, and in any case proved inadequate measures to contain the trend towards overfishing. There was until the end of the 1960s more doubt whether pelagic species with their generally greater abundance and share in the fish biomass were also threatened. In the 1970s it became clear that the new methods of trawling with powerful boats and of purse netting were now making big inroads to the herring and mackerel stocks: and that these densely shoaling species were now in more and not less danger of being over-fished. By the start of the 1970s it was becoming agreed that in addition to such technical measures as minimum mesh and minimum landed sizes, it was necessary also to set catch limits to prevent overfishing: and this brought in the principal of TACs (total allowable catches) which had to be set on the basis of scientific advice on the state of the stocks. This scientific advice was produced by international teams of scientists, and was based on sample monitoring of landed fish to assess stock age structures along with marine surveys of larvae and juvenile fish to estimate stock recruitment.

While the management measures detailed are generally accepted as essential to conserve the resource base in the face of the greatly expanded modern catching power, it has in practice been found difficult to bring in sufficient restraints in a timely fashion, due partly to the inevitable time-lag during the negotiation procedure, but also due to problems of organising and financing adequate enforcement of regulations. As a result fish stocks have tended to be reduced to, and to persist in, an unhealthy state. In the case of the formerly main fishery for the herring, the stock became so depleted by the later 1970s that the fishery had to be closed for a period of six years, although good subsequent recruitment allowed the fishery to be resumed on a controlled basis.

In the much changed modern situation there have been various adjustments in organisation and management of the fisheries within Britain and Scotland. An important change was the eventual repeal of the prohibition of trawling within three miles of the coast and in the firths. By the 1960s the curbs on the use of drag nets in inshore waters had been consideably relaxed to allow for the development of seine netting and prawn trawling, while there were now very few line fishermen who needed protection. Eventually in the 1980s the trawling ban was to be replaced by a system of protected static gear reserves in prescribed places, mainly to protect lobster and crab fisheries.

EXTENSIONS OF FISHERIES LIMITS AND THE EUROPEAN COMMUNITY 'POND'

It was not only in the waters around Scotland that application of the technology of the scientific age caused new pressures on fish stocks. Countries with good fishing grounds off their coasts also became increasingly concerned as there was increased development of distant-water fleets of big boats which literally had the range to fish anywhere. These distant-water fleets were built especially in the former USSR and Eastern Europe and formed an important part of a policy designed to improve the provision of the animal protein part of the diet. This however contributed to increasing fishing pressure around the whole North Atlantic, and similar pressures were exerted around most of the world's coasts. Ultimately this led to an increased questioning of the International Law of the Sea, which allowed unrestricted freedom of fishing on the high seas. The eventual reaction to this was the extension of national fishing limits to 200 miles, which meant that most of the continental shelves and main fish stocks became subject to the jurisdiction of coastal states.

Had the International Law of the Sea developed in isolation from other political developments, Scotland and Britain would have been in a fortunate position for gaining exclusive fishing rights over an extensive area of the continental shelf of North-West Europe. Under the internationally agreed rules for the drawing of median lines, Britain would have gained a full 60% of the 'pond' of the enlarged EEC in the 1970s (i.e. the EEC including Britain, Denmark and Ireland as well as the original six states). However with Britain's accession to the EEC in 1972, the consequence was the formulating of a fisheries policy for the EEC as a whole. This was a major challenge in the new world of 200-mile limits in which fish stocks were now in general inadequate for the fleets dependent on them. The decisive factor was that in the EEC there were a majority of continental countries which would have had much to lose with a system of median lines to divide up the offshore waters, and what was eventually agreed as the main basis for the new Common Fisheries Policy (CFP) was a principal of open access to all member fleets in the EEC 200-mile zone. Essentially linked to the policy of open access was a system of national quotas for main commercial species: this was and is the basis for dividing up the TACs for individual species, and the national allocations are based on track records of catches in the period 1973–78 with some adjustments for special factors: these include mainly degree of dependence on fishing at the regional or community level, and also compensation for losses in distant waters from which EEC fleets had to withdraw under the new 200-mile regime of the International Law of the Sea. The fact that under this new regime a considerable part of the North Sea is now in the Norwegian fisheries zone inevitably complicates the administrative arrangements, and involves in the first place a division of fish quotas between the EEC and Norway.

The division of a nation's quotas among its own boats is still its own affair, and the system developed in Britain and Scotland deputes the big part of this to the fishermen's own producers' organisations (or POs), each of which gets a 'sector' quota, although individual vessels still have the option of getting quotas direct from the SOAFD. The majority of the POs in Britain are in fact Scottish, and the system in this country of devolving quota management has often been quoted in the international context as a progressive example of 'co-management' in which the organs of national government act in concert with other interests in the industry.

This brings into focus one of the prominent developments of the modern period, which has been the growth of more co-ordinated action by fishermen. This has been stimulated by the necessity to organise politically in order to have their interests represented at a series of political levels – regional, national and European. The formation of these associations has also been encouraged by the

European Commission under the CFP. This is the situation that has produced the modern producers' organisations, and for Scotland as well as the Scottish Fishermen's Organisation (SFO), there are also a number of separate regional organisations and also the Scottish Pelagic Fishermen's Association, which represents boats engaged in the herring and mackerel fisheries.

Among the leading recent changes is the run-down of the former company-owned trawler sector to very small proportions. This has been due to a combination of causes. It is in part due to the extensions of fishing limits and the severe curtailing of working at the middle and distant-water ranges off foreign shores: this bore hardest on the biggest trawlers. The rise of Aberdeen as the 'Houston of the North Sea' has not directly caused a contraction of the trawling based on the city by taking over harbour space or land and buildings in the harbour area; but the oil industry has given more profitable outlets for venture capital, and there are cases of companies that were earlier involved in trawling that have found a bigger and more profitable function in oil-related work. There has also been the rise of greater efficiency in the fishermen-owned sector of the fleet, which has less of the poorer labour relations that were frequent in trawling. In the developments in fishing in Britain in the last quarter-century, even more severe than the contraction of trawling in Scotland has been the contraction in England, in which the leading distant-water trawling sector was virtually eliminated. The overall result is that in fishing Scotland has become the dominant partner in the United Kingdom, and now lands about 70% of the national catch by weight.

However the Common Fisheries Policy has been dogged all along by an over-capacity in catching, as the fleets built up during the earlier period of freedom of fishing after World War II have been estimated at 30% or more in excess of what is required to take catches at the maximum sustainable yield. Control of catching power was obviously essential, and Britain along with other EEC countries has had a programme of fleet restructuring, which in effect means fleet reduction, and poses big political as well as economic problems. In the first place a system of vessel licensing was brought in; but although this did put an effective ceiling on vessel numbers, it was still essential to allow fleet modernisation and replacement, and the effects of licensing were lessened by the transfer of licenses from smaller older to bigger new boats. However such transfers were progressively curbed by regulation.

Enforced cut-backs are particularly problematical in fishing communities in remoter areas where other employment is scarce or
absent; and along with the programme of fleet reduction there has
actually been some new construction with the help of European
regional funds. Even so, it was accepted that there should be
phased reduction in fleet tonnage and horsepower, and this is now
taking place.

FISH FARMING – HOW BIG IN THE FUTURE?

Fishing has been criticised as a primitive method of utilising
resources, in that it is a late survival of the age-old hunting method,
and exploits resources in the wild. While fish farming has to an
extent been practised for thousands of years, it has mainly been
within the more limited and easily controlled confines of fresh water.
Nevertheless, mariculture too has been in some degree practised
for many centuries, although it has been confined to the edges of
the sea in such locations as coastal lagoons and estuaries. Inevitably
fish farming generates great extra costs compared to conventional
fishing, and with intensification of production there are enhanced
problems from the incidence of disease; on the other hand it allows
a degree of control never possible in conventional fishing. In the
developed world, fish farming is in general restricted to high value
species which can justify the extra production costs. Scotland has
shared in the modern upsurge of fish farming, especially in salmon
production, and the rate of increase in the 1970s and 1980s was
spectacular. By 1991 the value of farmed production had reached
about £150 million, which was equivalent to 55% of the value of
fish landings. By tonnage, however farmed production is still only
equivalent to under 10% of landings; and it is clear that the industry
has now passed from boom to consolidation, and indications that
production has stabilised. While there is already significant farmed
production of trout and mussels, and there are prospects of extending production to other high value species like turbot and halibut,
all these species can be better farmed in warmer conditions than
those of Scotland. It is at present difficult to foresee any major
expansion of farmed production in Scotland.

CONTINUITY AND CHANGE

For at least the short and medium term it appears that conventional fishing will continue to be the main source of fish supply in

Scotland and Britain, and indeed the world. Fishing too retains its essential age-old characteristics. It still involves working on the heaving deck; it also entails irregular hours, discomfort and (at times) danger.

Fishing in Scotland has as long a history as any human activity. It embodies a rich collection of folk tradition, commercial effort and national aspirations over the centuries; these were forged to a significant extent in circumstances of international competition, and indeed they continue to give character to the fisheries today in the context of the international situation in the European Union and beyond. For the great part of history, the great issue was to make fuller use of the available resource base. The decisive change in our age is that this has been replaced by a continuing need for conservation to prevent the resource base from being overtaxed.

Select Bibliography of Main Sources

The sources used in compiling this book have been referred to in the references appended to each chapter. The main sources are repeated here along with important material not mentioned in the references.

BOOKS

Dunlop, J. (1978) *The British Fisheries Society 1786–1893*, John Donald, Edinburgh.

Goodlad, C. A. (1971) *Shetland Fishing Saga*, Shetland Times, Lerwick.

Gray, M. (1978) *The Fishing Industries of Scotland 1790-1914*, Oxford U.P., Oxford.

Hallewell, R. (1991) *Scotland's Sailing Fishermen*, Hallewell Publications, Port-an-Eilean, Perthshire.

Martin, A. (1981) *The Ring-Net Fishermen*, John Donald, Edinburgh.

Smith, H. D. (1984) *Shetland Life and Trade 1550–1914*, John Donald, Edinburgh.

Summers, D. (1988) *Fishing off the Knuckle - the Fishing Villages of Buchan*, Centre for Scottish Studies, University of Aberdeen.

Carré, F. (1988) *Pêches Maritimes et Pêcheries de L'Ecosse* (5 vols.) Université de Bretagne Occidentale, Brest, France.

JOURNAL ARTICLES AND BOOK CHAPTERS

Coull, J. R. (1988) 'The Boom in the Herring Fishery in the Shetland Islands', *Northern Scotland* 8, 25–38.

Coull, J. R. (1992) 'The Development of the Fishery Districts of Scotland', *Northern Scotland* 12, 117–131.

Coull, J. R. (1991) 'The Development of the Herring Fishery in the Peterhead District of Scotland before World War I' in *Sjöfartshistorisk Årbok 1990* (Norwegian Yearbook of Maritime History) 1990, Bergen Sjöfartsmuseum, Bergen, 119-142.

Coull, J.R. (1987)'The Engagement System during the Shetland Herring Boom, 1880-1914', *Scottish Economic and Social History*, 7, 55-65.

Coull, J.R. (1969) 'Fishing in the North-East of Scotland before 1800', *Scottish Studies* 13, 17–32.

Coull, J.R. (1992) 'Seasonal Fisheries Migration: the Case of the Migration from Scotland to the East Anglian Autumn Herring Fishery', in Fischer, L.R., Hamre, H., Holm, P., and Bruijn, J.R. *The North Sea. Twelve Essays on Social History of Maritime Labour*, Stavanger Maritime Museum, Stavanger, 127-144.

Coull, J. R. (1986) 'The Scottish Herring Fishery 1800–1914': Development and Intensification of a Pattern of Resource Use', *Scot. Geog. Mag.* 102, 4–17.

Gray, M. (1967) 'Organisation and Growth in the East Coast Herring Fishing 1800–1885', in Payne, P. L. (ed.) *Studies in Scottish Business History,* Cass, London, 187–216.

Gray, M. (1972) 'Crofting and fishing in the north-west Highlands 1890–1914', *Northern Scotland* 1, 89–114.

Smith, H.D. (1973) 'The Development of Shetland Fisheries and Fishing Communities', in Fricke, P.H. (ed.) *Seafarer and Community,* Croom Helm, London, 8–29.

Smith, B. (1992) 'Adam Smith's Rents from the Sea: Maritime Share-cropping in Shetland' in Smout, T. C. (ed.) *Scotland and the Sea,* John Donald, Edinburgh, 94–113.

OFFICIAL REPORTS

Annual Reports of the Fishery Board, 1809–1881.

Annual Reports of the Fishery Board for Scotland, 1882–1938.

Buckland, F., Walpole, S., and Young, A. (1878) *Report on the Herring Fisheries of Scotland*

Committee appointed to inquire into the State of British Fisheries, 1785 (1803) *Reports from the House of Commons*, X.

Committee appointed to inquire into the State of British Herring Fisheries, 1798, (1803) *Reports form the House of Commons,* X.

Committee of the Economic Advisory Council (1932) *Report on the Fishing Industry.*

Report of Bonamy Price and Frederick St. John to the Commissioners of the Treasury on the Subject of the Fishery Board (1856).

Report of the Royal Commission on Beam Trawling (1885)

Report of the Scottish Departmental Committee on the Scottish Sea Fishing Industry (1914).

Sea-Fish Commission for the United Kingdom (1934) *First Report on the Herring Industry.*

Index

Abbey of Holyrood 55, 56
Aberdeen 35, 132, 135, 137, 140, 142,
    143, 144, 146, 148, 152, 153, 164,
    175, 181, 189, 193, 194, 219, 221,
    242, 246, 247, 252, 267, 268, 286,
    289, 295
    fish market 143, 144
abirdenes 84
Acts of Scots Parliament 6, 65
anchored nets 56, 212
Annan 233
Anstruther 40, 46, 86, 110, 137, 138, 152,
    185, 212, 213, 232, 260, 261, 266
Arbroath 44, 45, 82, 153, 182, 212, 251,
    287
archaeological evidence 2, 6, 223–224
Ardersier 39
Armistice Gale 160, 163, 171
Auchmithie 36, 37, 45
auctions 124–125, 143, 209
Austro-Hungarian Empire 122
Avoch 39, 47, 62, 72
Ayala, Pedro de 7, 84
Ayr 35, 55, 144, 175, 185, 186, 189, 281,
    287
    Newtown of 39

bait, baiting 49, 50, 82–83, 97–101, 178,
    190
baldies 240
Ballachullish 110
Ballantrae 259, 264, 286, 287
Ballantrae Bank 18
Balta Sound 135, 208, 209, 211, 286
Baltic 75
Banff 35, 153, 212, 261, 284, 287
Banffshire 76, 153
bank herring 17–18
barometers 108
Barra 83, 86, 137, 192, 280, 281, 286, 287
barrel bounties 71, 106
barrel staves 58, 66
barrels, regulations for 58, 107
Bayble 205

Berwick 35, 233
Billingsgate 148
Bishop Leslie 4, 7, 84, 224
Blaeu, Willem 34
blue whiting 198–199
Board for Manufactures and Fisheries
    70, 275
boat fishery 71–73, 105, 106, 277–278,
    289
Boatlea 36
boats 235–250
Boddam 34, 117
Boece, Hector 7
Bohuslan herring fishery 76
booths 90
bounties 105, 106, 109, 124, 275
    barrel 106, 276, 277, 278
    cod and ling 93–95, 279–280
    tonnage 70–71, 106, 276, 277
branding 107, 109, 170
branding fees 109, 282
bream 28
Bremen 86
Bressay 197, 209
Bristol 279
British Fisheries Society 38, 73, 116, 202,
    257, 268
Broadhaven 203
Broadsea 35, 47, 84
Bronze Age 28, 224
Broonies Taing 273
Brussels 290
Buchanhaven 42
Buckie 10, 37, 42, 45, 46, 47, 62, 63, 72,
    110, 152, 154, 181, 202, 214, 218,
    232, 238, 259, 266, 268, 272
Buckhaven 258, 261, 266
Buckland, Walpole and Young Report 267,
    268
Buckquoy 214
Buncrana 218
Burghead 38, 258
burghs 35, 39–40, 251, 252, 257
Burnmouth 264

Burntisland 261, 263, 281
Burra 38, 192
Burra Haaf 88
bush rope 112, 130, 242
busses 64–71, 115, 116, 200, 236, 276,
    277
    annual round of from the Forth 67–68

Cairnbulg 37, 203, 273
Caithness 62, 72, 108, 115, 116, 118, 120,
    130, 201, 202, 203, 228, 262, 264,
    265, 276, 278, 288
Callicot 264 (see also Port of Ness)
Campbell, John 182
Campbell, William 238
Campbeltown 33, 38, 72, 86, 137, 185,
    233, 286, 289
canning 158, 169
capstans 112, 131, 242, 243
Caribbean (see also plantations, West
    Indies) 281
Carnish 264
Carradale 38
Carrick 82
Carsaig 258
carvel built boats 238
Castlebay 138, 205
Cellardyke 10, 253, 257, 264, 266
Celtic Sea 19
centralisation of markets 143–144
Civil War, the 74
clams 98 (see also scallops)
clinker built boats 235
Cluny Harbour, Buckie 259
Clyde, Firth of 3, 54, 55, 56, 59, 61, 65,
    75, 85, 86, 96, 105, 108, 115, 122–123,
    128, 134, 141, 149, 154, 179, 180,
    185, 213, 227, 236, 239, 243, 262,
    276, 278, 279, 285, 286, 289
Clyth 260
coalfish 84
coast fishery 106, 272
Cockenzie 225
cockles 23, 223
cod 9, 15, 16, 27, 29, 30, 67, 79, 82, 84,
    88, 90, 93, 94, 97, 101, 142, 175, 177,
    183, 188, 200, 206, 255, 279, 280,
    281, 282
codling 82
Coldingham 34, 257 (see also St. Abbs)
Collieston 82, 85, 98
commercial fishing 2, 54, 84, 85
Common Fisheries Policy (C.F.P.) 290,
    294, 295
commuting 221
companies 51, 142, 143
complements 109, 118
conger eel 28
conservation 291–297

continental shelf 12, 15
controlled prices 185
Convention of Royal Burghs 7, 65, 70, 275
cooks 127
coopers 123, 158, 205
cotton nets 113
Cove (Kincardineshire) 36
Cove (of Dunglass, Berwickshire) 264
Cowie 46, 47
cowper boats 67
crabs 14, 20, 224, 233–234, 293
Crail 36, 55, 251, 263
crans 109, 113–114, 118, 120, 124, 131,
    216
Crawton 35
crayfish 224
crears 66, 236
creels 50, 228–229, 246
crofter-fishermen 38, 130, 154, 188, 219
Crofters Commission 129
Cromarty 39, 62, 152, 263, 284, 287
Cromarty Firth 99, 253
Cromore 205
Crovie 37, 284
Crown, the 2, 7, 56
crown brand 107
crustaceans 14; (see also shell fish)
Cullen 38, 46, 85, 273
curers 50, 108, 127
curing aboard 67, 106, 116
curing yards 117, 126, 209
customs 56, 74

danger 9–11
Danish seining 179
Dark Age 29–30
Decca system 249
decked boats 110, 237–239
deckhands 142
demersal fish 15–17 (see also white fish)
Denmark 162, 179, 197, 198, 246, 294
Development Fund 269
dialect 39, 79
diesel boats 161, 173, 191, 192, 245–248
dip nets 80
diversification 198–199
Dock Labour Scheme 193
Dogger Bank 180
Dornoch Firth 80, 98, 99
Dover 281
Downings Bay 218
drag–netting 180, 293
dredging 273
drift net 56, 111–113, 122, 161, 181, 188,
    195, 241, 245, 246, 285
Dumbarton 55
Dumfries 282
Dunbar 251, 261
Dunbar Harbour 255, 258, 259, 260, 261

Dunbeath 115
Dundee 140, 143, 175
Dunmore 218
Dunrossness 90
Dunure 37, 47, 258
Dunvegan 280
Durness 228
Dutch 64, 69, 70, 75, 76, 83, 104
    herring fishery of 3, 54, 64–65, 206,
    254, 277

East Anglia 127, 128, 132, 160, 166, 167,
    171, 201, 209
    Scottish movements to 213–218, 220
East Neuk 42, 65, 129, 134
Economic Advisory Council 171
echosounders 249
Edinburgh 43, 225, 226, 278, 288
emigration 124, 154
engagements 63, 72, 108, 124
English trawlers 140, 144, 151
equipment 248–249
Eriskay 192
estates 35–41, 43–49, 251
Estonia 121, 170
European Economic Community (E.E.C.)
    294, 295
European Union (E.U.) 297
Exchequer Rolls 6, 34, 55, 79, 84, 85
export bounties 69, 87
exports (of herring) 73–76, 169
extension of fisheries limits 293–294
Eyemouth 40, 45, 46, 96, 98, 128, 137,
    152, 185, 189, 212, 237, 243, 252,
    261, 272, 279, 280
    disaster 10–11, 253

Fair Isle 83
family business 49–51
Faroe 95, 145, 175, 176, 219, 220, 236,
    246
Ferryden 38, 44, 110
Fethaland 88
Fetlar Firth 88
feuars, feuing 36, 43
Fife 76, 96, 117, 118, 134, 153, 192, 202,
    203, 214, 218, 220, 236, 245, 261, 262
fifie 182, 236, 237, 238, 242
fillets, filleting 183
Findochty 38, 85, 273
Findhorn 38, 261, 284, 287
Findhorn estuary 99
Findon 96
finnan haddock 9
firemen 127
first class boats 112–113, 132–133
first settlers 22–28
fish farming 234, 296
fish freezing 189, 220

fish friers 183
fish meal 198
fish merchants 47, 48, 85, 143, 189
fish migration 17–19
fish mongers 143
fish salesmen 162, 166, 168
fish scales 14
fish selling companies 143
fish shops 183
fish spears 30
fish stocks 14–20
fisheries districts 275–280
fishermen and farming 35–38
fishermen, numbers of 187–188
fishertouns 35–41
Fisherrow 35, 225, 261
Fishery Board 8–9, 93, 97, 99, 104, 106,
    107, 108, 110, 112, 121, 122, 201,
    237, 252, 253–267, 276, 277, 284
Fishery Board for Scotland 100, 127, 129,
    135, 149, 151, 158, 165, 169, 170,
    176, 218, 224, 227, 267–273, 285
fishery cruisers 108, 151
fishery inspectors 170, 276, 277
fishery officers 94, 105, 107, 109, 151,
    170, 260, 276, 277, 282
fishing:–
    a demanding and dangerous calling 9–11
    evidence for and records of 6–9
    in the context of Scottish national
    objectives 4–6
fishing tenures 87
fishing villages 35–41
fishwives 50, 97
Fittie 35, 41, 43, 44, 84, 142
flat fish 97, 146, 180
flat freight rate 185, 192
Fleetwood 186
flounder 84, 239
fly dragging 244
food chains 13
food pyramids 13
Footdee see Fittie
Fordun, John of 7
Fort William 72, 257, 264, 281, 286, 287
Forth, Firth of 3, 4, 56, 60, 61, 65, 82, 85,
    99, 108, 115, 128, 180, 211, 212, 223,
    225, 226, 228, 251, 261, 262, 265,
    266, 276, 278, 281
fourerns 88, 236
France 75, 225
Fraserburgh 38, 40, 47, 84, 85, 111, 115,
    116, 128, 129, 137, 138, 140, 143,
    148, 153, 185, 197, 203, 252, 255,
    262, 265, 268, 271, 272, 284, 286
free trade 93, 282
freshers, freshing 128, 158, 168–169, 211
Freswick Links 30
friendly societies 43

gadoids 15
Gairloch 85
gales 10
Galloway 80
Gardenstown 273 (also known as Gamrie)
Gardner engines 182, 243, 245
Gareloch Head 227
garvies 19 (see also sprats)
gear maintenance 50–51
general strike 161, 177
German merchants 87
German trawlers 144, 176, 183
Germany 127, 128, 155, 156, 157, 167,
   170, 213
Girvan 186
Glasgow 35, 75, 97, 122, 128, 148, 154,
   161, 226, 287
Glencoe 100
Gloup 88
Golspie 35
Gourdon 49, 258, 262
government guarantee schemes for herring
   160, 170–171
grant and loan schemes 191, 245, 292
Granton 140, 144, 146, 153, 175, 186,
   219, 247, 287
Gravesend 281
great lines 82
Greenland 95, 146, 175, 219
Greenock 62, 75, 96, 115, 287
Grimsay 192
Grimsby 144, 216, 246
gutters 123, 158, 164, 202

haaf fishery 86–92, 95, 101, 200, 206, 236
haaf (or half) net 30
haberdynes 84
haddock 14, 15, 80, 82, 84, 96, 97, 145,
   146, 176, 244
hake 84, 93
half catch 143
halibut 15, 95, 142, 175
Hamburg 86, 128
Hamnavoe 38
hand lines 81
hand net 30
harbours 106, 111, 143, 251–274, 276,
   289
harbours of refuge 256, 257
Hay, James 90
Hebrides (see also Western Isles) 175, 204,
   205, 234, 261, 280, 286
Helmsdale 115, 152, 182, 262, 288
herring 3, 9, 14, 84, 101, 185, 186, 189,
   191, 192, 195, 255, 292, 295
   boats, development of 110–113
   curing 58–59, 64, 107, 164–167, 201,
      202, 203, 208, 209, 211, 214, 216,
      260, 276, 277, 278

fishing 55–78, 104–138, 231, 265, 275,
   276, 289
   build–up to crisis in 123–125
   fluctuations in yield of 118–120
   seasonal rhythm of 117–118
gutting 58, 107
inshore fishing 59–63
life cycles of 17–19
nets, development of 111–112
sorting of 107
trawling 164, 196
Herring Industry Board (H.I.B.) 156, 162,
   163, 166, 169, 172, 197, 235
Highland Board 192
Hilton 44, 264
hired men 127, 159
Holland 121, 155, 157, 167, 225, 246 (see
   also Dutch)
hooks 81
Hopeman 39
houses, housing 41–43
Howth 218
Hull 142, 144, 220, 246
Humber 142, 146, 220, 246
hydraulic equipment 248, 249

ice (use of) 143, 146
Iceland 86, 95, 144, 146, 151, 155, 167,
   176, 183, 197, 214, 220, 236
inshore white fisheries 79–86
inshore herring fisheries 59–63
International Law of the Sea 198, 293, 294
inter–war depression 170, 188
Inverallochy 34, 37, 203
Inveraray 38, 264, 287
Inveravon shell mound 23
Inverbervie 233
Inverness 35, 55, 85
Ireland 96, 176, 294
   herring fisheries at 218, 220
   market for herring in 61, 75,121, 124,
      203, 281
Iron Age 28–29, 223, 224
Irvine 61
Islay 96, 281
Isle of Man 108, 209, 218, 282, 285

Jarlshof 29
jerseys 10
Johnshaven 38, 45, 85

Keiss 258
Kentra 258
Kilkeel 218
Kinlochbervie 193, 221
Kinneff 35
Kinsale 218
Kintyre 232
kippers, kippering 128, 154, 158, 168, 206

klondyking 128, 158, 198, 206, 212, 213, 221
Knap of Howar 28, 224
Knox, John 37, 72–73
Kyle 287

Labrador 86
Laing, Samuel 116
Lammas drave 59
land marks 83
lasts 73–74
Latvia 121, 170
legal rights to fishing 2–4
leisters (see also fish spears) 80
Leith 148, 189, 213, 215, 225, 232, 279, 287
lemon sole 16, 177
Lerwick 33, 128, 135, 209, 211, 219, 252, 269, 271, 278, 279, 285
Lewis 85, 92, 96, 97, 117, 204, 208, 227
licensing of vessels 295
limpets 20, 23, 26, 98, 223, 226
ling 9, 15, 16, 29, 30, 79, 90, 93, 95, 200, 206, 279, 281, 282
lithic scatter sites 24
Lithuania 121, 170
Liverpool 279
lobsters 4, 14, 20, 152, 224, 228–230, 231, 293
Loch Broom 66, 284
Loch Carron 284, 287
Loch Fyne 122, 239
   herring 61
   skiff 236
Loch Gilphead 281, 284
Loch Hourn 100
Loch Ryan 227
Loch Shildag 284, 286
Loch Snizort 100
Lochinver 193, 221
Lochs (Lewis) 86
London 4, 85, 139, 143, 146, 147, 226, 228, 229, 279
Long Island 230, 287 (see also Hebrides, Western Isles)
long lines 83
Lossiemouth 137, 154, 179, 181, 182, 232, 252, 261, 272, 287
lug sails 237, 238, 239
lugworm 98, 99
Lussa Wood 24
Lybster 115, 152, 182, 262, 265, 287

Macduff 181, 182, 287
MacFarlane's Geographical Collections 8
mackerel 14, 19, 101, 196, 197, 221, 292, 295
Major, John 7, 224
Mallaig 148, 185, 232

Manchester 148
marine scientists 12, 292
marine superintendence 106–107
mates 142
mattie herring 107
maximum sustainable yield 295
meads 83
Medieval 1, 4, 6, 35, 54, 55, 84, 86, 251
Mediterranean 86, 92, 183, 214
Mercantilist period 3
Merchant Marine 5, 185
merchants 50
Mesolithic 2, 22–28, 55, 200, 223
Methuen, James 108
MFVs 245
microliths 24
Minch 60, 64, 129, 132, 138, 205–207, 232, 278
minimium mesh sizes 106–107, 292
minimum landed sizes 292
Mitchell, Joseph 263–264
mobility 200–222
modernisation 291–292
Montrose 35, 44, 79, 85, 132, 152, 242, 282, 286, 287
   Basin 99, 100
Moray Firth 16, 61, 62, 85, 101, 118, 134, 138, 144, 149, 162, 180, 181, 182, 186, 202, 203, 214, 216, 221, 227, 229, 231, 232, 236, 244, 261, 271, 272, 279, 284
Mortensen v Peters case 151
Morton 24, 27
motor boats 151, 152, 153, 159, 161, 181–182, 213, 219, 242–245
Muchalls 35
mullones 79
multiple tenancies 44–46
mussel scalps 98, 99
mussels 15, 20, 82, 98–101, 223, 234

nabbie 236
Nairn 35, 43, 62, 259, 271
Napoleonic Wars 5, 104
Navy 5, 185
Neolithic 28, 223
nephrops 190, 198, 221, 232–233 (see also prawns)
Ness yole 236
net mending 50–52, 188
New Statistical Account 228
Newcastle 60, 228
Newfoundland 80, 86, 88
Newhaven 35, 43, 44, 84, 225, 226
   Society of Free Fishermen of 43, 226
Newtarbat estate 45
Norse 29–30, 236 (see also Vikings)
North Berwick 261
North–East 38, 45, 82, 85, 96, 100, 117,

132, 134, 137, 189, 192, 221, 242, 245, 247, 286
North East Atlantic Fisheries Convention 292
North European Plain 75, 121
North Germany 86, 92
North Isles (or Northern Isles) 20, 59, 95, 189, 229, 230, 235, 236, 263, 280, 282, 287
North Norway 15, 80, 151, 220
North Queensferry 262
North Sea 12, 13, 15, 17, 138, 144, 145, 175, 176, 179, 180, 195, 219, 220, 251, 284, 294
North Shields 142
North Sunderland 280
Northesk, Earl of 45
Northton 223
Norway 86, 88, 121, 128, 155, 167, 170, 195, 197, 211, 228, 229, 235, 236, 294
Norway pout 17, 198
nylon nets 195

Oban 144, 175, 193, 221, 232, 271
Occumster 260
oilskins 10
Old Castle 82
Old Statistical Account 63, 85
Old Whinnyfold 36
open access 294
open boats 71–73
operating subsidies 191
Orkney 67, 85, 96, 115, 116, 138, 152, 153–154, 175, 186, 192, 227, 228, 230, 236, 279, 289
   neolithic in 28
Oronsay 23, 24
otoliths 14
otter boards, otter trawl 240
overfishing 296
oysters 4, 15, 223, 224, 225–228, 230

packers (of herring) 123, 158, 164
paddle–tugs 130
Panmure estate 45, 47
Peel 218
pelagic fish 17–19, 195–198, 221, 249
pelagic fleet 197
Pennan 46
perquisites 109
Perth 35
Peterhead 36, 38, 40, 42, 43, 44, 85, 86, 97, 111, 113, 114, 115, 116, 123, 124, 128, 129, 131, 137, 138, 140, 144, 148, 153, 175, 185, 193, 194, 203, 221, 255, 257, 261, 262, 265, 268, 279, 281
Petty, Fisherton of 34, 35, 36, 45, 46
photosynthesis 13

phytoplankton 13, 14
piers: see harbours
Piers and Quays Fund 271
pilots 44
piltocks 101
Pittenweem 261
plaice 16, 79, 84, 142, 179
plankton 13–14
planned villages 37–39, 73
plantations (see also Caribbean, West Indies) 121, 203
Plymouth 196, 218, 281
poke net 30
Poland 121, 167, 170
Polmonthill shell mound 23, 223
Pont, Timothy 34
poor fishermen, help for 106, 256
Port Ellen 258
Port Erroll 117
Port Glasgow 99
Port of Ness 255, 264
Port Wemyss 38
Portessie 38
Portgordon 258, 282, 284
Portknockie 37, 273
Portlethen 37
Portmahomack 251
Portnahaven 38
Portsmouth 281
Portsoy 261, 271
pouch net 218
position of Scotland 12
prawns (see also nephrops) 239
press gang 5
primary production 12–14
Privy Council 34
producers' organisations (P.O.s) 294
Public Loan Commissioners 261, 267
Public Works Loan Fund 271, 272
Pulteneytown 42, 262
purse seine (or purse net) 195, 196, 246, 292

queen scallops 233
quotas 294

radio carbon dates 23, 24, 27
radio receivers 249
radio transmitters 249
rail transport 95, 97, 139, 146, 148, 183, 216, 286
Ramsgoe 115
Rathven 63
Rattray 36
razor shells 223
red herring 58
reduction (to meal and oil) 198
Register of the Great Seal 6, 34, 35
registration of fishing boats 108, 285

rents 45–47
 in kind 46–47
requisitioning of boats for naval service
 154–155, 171, 185
resource base 12–21
ring–net 122–123, 137, 154, 195, 239,
 244, 246, 285
road transport 183, 193
Robertson, Robert 244
Rockall 176
Roman times 223
Rosehearty 45, 117
Rothesay 115, 137
Royal Commission on Beam Trawling 149
Royal Fishery Companies 69, 74
Russia 121, 122, 167, 170

sail boats 127, 130, 131, 133, 134, 159,
 161, 241, 269
sailing trawlers 140
saisines 80
saithe 15, 23, 26–27, 79, 80, 90, 94
Salen 264
salmon 3, 7–8, 56, 84, 296
salt 58–59
solar 59, 70
salt duties 87
sand eels 80, 198
Sandend 37
Sandwick (Unst) 29
Sarclet 115, 203, 255, 262, 264
scaff, scaffie 236, 237, 238
Scallisaig 258, 264
scallops 198, 230, 233
Scalloway 128
Scalpay 192
Scandinavia 55
Scarborough 216
scientific research 9
Scots Parliament 6, 65
Scottish Fishermen's Organisation (S.F.O.)
 295
Scottish Office Agriculture and Fisheries
 Department (S.O.A.F.D.) 294
Scottish Pelagic Fishermen's Association
 (S.P.F.A.) 295
Sea Fish Commission 160, 172
seals 23–24
seasonal migration 200–218
seine net 154, 162, 174, 179–183, 190,
 192–194, 221, 246, 247, 248, 293
sessile species 22, 23, 233
settlements 33–53
share ownership and allocation 60, 61–2,
 126–127, 134, 143, 161
shell fish 1, 3–4, 19–20, 189, 198,
 223–234, 246
shell middens 23–24
Shetland 8, 54, 64, 80, 83, 86–92, 94, 95,
 96, 97, 117, 129, 130, 135, 137, 138,
 152, 154, 160, 166, 175, 180, 185,
 189, 192, 197, 200, 203, 204, 206–211,
 221, 227, 280, 286, 289
 haaf fishery in 86–92
 herring boom in 123
 method 48–49, 87–88
shrimps 239
Sibbald, Sir Robert 59
sinker stones 81
sixareen, sixern 90, 91, 236, 237
Skara Brae 28
skate 15, 79, 95, 175, 177
Skipness 264
skippers 142, 160
Skye 227, 280
smacks 86, 94–95, 280
small lines 82
Smith, Adam 71
smoked haddock 96–97, 146, 183
Society for the Free British Fishery 70
sole 14, 79, 82, 146, 179
Solway 30, 227, 239
sonar 249
Sound, the 74
South Queensferry 262
Spain 96
Spanish market 96
spawning 14, 15, 18, 19
sprats (see also garvies) 19, 197, 198
squatters 8, 34
squid 101
St. Abbs 34, 37 (see also Coldingham)
St. Andrews 100, 262
St. Combs 42, 47
St. Ives 94
St. Margaret's Hope 115, 280
St. Monance 258, 261, 263
static gear reserves 293
Staxigoe 115, 203
steam drifters 113, 131–138, 155, 157,
 161–163, 172, 178, 181, 182, 190, 212,
 216, 219, 220, 245, 269
steam liners 133, 142, 175, 177, 191
steel boats 241–248
Stenness 88
Stettin 121
Stevenson, Thomas 260, 263
Stonehaven 153, 251, 261, 270, 272, 287
Stornoway 33, 72, 86, 128, 129, 137, 138,
 189, 192, 204, 205, 212, 233, 269, 271
Stranraer 272
Stromness 205, 206
Stronsay 116, 138, 154, 280, 282
subsistence fishing 2
subsidies 191
surnames 49
Sub–Atlantic period 27–29
Sutherland 115, 228

sweep nets 56–57, 80
Sweden 76
Swedes 75
synthetic fibres 195, 248

tacksmen 47
Tain 98, 99
Tarbert (loch Fyne) 38, 287, 289
Tay, Firth of 99
tee–names 49
teind fish 46, 79–80
telegraph 267
Telford, Thomas 42
tenancy, tenure 43–49
Thames 246
three–mile limit 149, 151, 285–286, 293
Thurso 282
tidal traps 27, 30 (see also yares)
Tobermory 38, 257, 281
Tongue 281, 282
Torry 36, 42, 142, 143
total allowable catches (T.A.C.s) 292, 294
trawlers, trawling 93, 96, 101, 130, 133,
    135, 139–152, 174–176, 179, 193, 194,
    249, 285–286, 295
  beam 149, 240
  controversy 140, 149–151
  herring 164, 195
  light 198
  mid–water 195, 198, 292
  nephrops (or prawn) 192, 232–233,
    247, 293
  pair 195, 247
  steam 96, 139–152, 218–219, 240–241,
    247, 286, 289
  stern 246
  white fish 139–152
turbot 84
turnpike roads 93
tusk 94
types of fish stocks 14–15

Ullapool 38, 196, 221, 257
Union Harbour, Anstruther 261, 266–267
Unst 208, 280, 282

Usan 44, 100
USSR 198, 293

Vikings (see also Norse) 235, 237

wadmel 86
Walls 208, 282
Washington Report 260
Waterford 218
weekly commuting 221
West Highland lochs 68, 75
West Indies (see also Caribbean,
    plantations) 70, 76, 121
West Loch Tarbert 227
Western Isles (see also Hebrides) 65, 95,
    128, 130, 154, 160, 166, 189, 192, 229
Westness (Rousay) 30
Whaligoe 37, 115
whales, whaling 24, 116, 135
Whalsay 192
whelks 20, 23, 224
Whinnyfold 37
Whitby 94, 281
white fish, white fisheries 3, 79–103,
    139–152, 174–184, 189, 190, 191, 221,
    249
White Fish Authority (W.F.A.) 245
white fish bounties 93–95
White Sea 219
Whitehaven 218
Whitehills 36, 182, 261
whiting 15, 16, 183
Wick 40, 42, 63, 108, 110, 111, 114, 115,
    116, 118, 120, 128, 129, 137, 153,
    182, 202, 257, 260, 261, 264, 265,
    268, 272, 278, 284
Widewall Bay 116
winches 179, 190, 240, 243, 248
winkles 23

yares 30
Yarmouth (also Great Yarmouth) 213, 276,
    278, 281
Ythan estuary 98

zooplankton 13, 14
zulu 238, 244